*The investigation of
the physical world*

T0296331

The investigation of the physical world

G. TORALDO DI FRANCIA

Istituto di Fisica Superiore
Università di Firenze

CAMBRIDGE UNIVERSITY PRESS

Cambridge
London New York New Rochelle
Melbourne Sydney

Published by the Press'Syndicate of the University of Cambridge
The Pitt Building, Trumpington Street, Cambridge CB2 1RP
32 East 57th Street, New York, NY 10022, USA
296 Beaconsfield Parade, Middle Park, Melbourne 3206, Australia

Originally published in Italian as *L'Indagine del Mondo Fisico*
by Giulio Einaudi editore, Torino, 1976
and © Giulio Einaudi editore, 1976
Now first published in English by the Cambridge University Press,
1981 as *The Investigation of the Physical World*
English translation © Cambridge University Press, 1981

Printed in the United States of America
Typeset by Maryland Composition Co., Inc., Glen Burnie, Maryland
Printed by Publishers Production International, Bronx, New York
Bound by Arnolds Book Bindery, Reading, Pennsylvania

Library of Congress Cataloging in Publication Data

Toraldo di Francia, Giuliano, 1916–

   The investigation of the physical world.

   Translation of L'indagine del mondo fisico.

   Bibliography: p.

   Includes index.

   1. Physics – Philosophy.   2. Irreversible
processes.   I. Title.
QC6.T6513   530'.01   80–12791
ISBN 0 521 23338 0 hard covers
ISBN 0 521 29925 X paperback

Contents

Contents

Preface to the English edition

Five years have elapsed since the first Italian publication of this book – a fairly long time when considering the rapid pace of the progress of contemporary physics.

Although the main purpose of *The investigation of the physical world* is not to supply the latest technical information but to discuss the place of physical science in the context of modern culture and to consider its epistemological implications, some revision and updating was necessary for this present English edition. In fact, discussing today's epistemology on the basis of yesterday's science would be rather awkward.

Of course, the most important discoveries and advances that occurred during these intervening years concern astrophysics and particle physics, and accordingly, some changes have been made and new material has been added, especially in Chapters 4 and 5.

The author at this time has also taken the opportunity to clarify or to correct several points throughout, rendering this book a better edition.

G. TORALDO DI FRANCIA

Istituto di Fisica Superiore
Università di Firenze

Preface to the Italian edition

This book has its origin in the seminars on the fundamentals of physics, which I have given for several years to philosophy students at the University of Florence. This should indicate the purpose of the book, as well as the class of reader initially proposed to reach. However, the result has been a much broader scope in text and consequently a wider and more varied group of possible readers.

First of all, I hope to reach the cultured reader who is aware that by accepting the absurd dichotomy between the two cultures, one simply arrives at a nonculture – the fiction of two bodies of doctrine, neither of which can rightly reflect the articulate, but indivisible, intellectual endeavor of our society.

It is often said that science is difficult and requires an amount of technical knowledge, especially in mathematics, which the average educated person does not possess. It is usually believed that an attempt to remedy this situation can be made by means of *popularization*. But there are different kinds of popularization, and I must state at the beginning that I am generally not enthusiastic about those books that present science as *science for children*.

In this book the reader will not find physics for children, with the inevitable and tedious references to *Alice in Wonderland*, but will find instead physics for adults. Those who are interested in making a real intellectual effort to understand modern philosophy, music, and poetry should find it reasonable to devote an equally serious effort to acquiring scientific culture. But the nonspecialist is met halfway; only a few simple concepts of high school algebra are required, whereas an attempt has been made to explain the more advanced material as clearly as possible. Perhaps being "simple" may result in being "simplistic," but it is a risk, which, I believe, is worthwhile.

For high school teachers of physics or general science this book may be useful in helping to familiarize them with a philosophical approach to science, which, although not aimed at *superseding* that of standard college or university textbooks, will favor a new and deeper insight into the subject. Thus the gap between what the philosophy instructor teaches and what the physics instructor teaches can be narrowed and at the same time can provide the means to satisfy the many intellectual curiosities and cultural needs of intelligent students.

In addition, my intent is to reach science students, too, particularly students of physics. This book aspires to represent for them an introduction to the critical analysis of the foundations of their subject, as well as to help them dissolve the dangerous illusion that many current doubts and perplexities concerning the value of science have only arisen today, instead of having had a long history.

There are many people whom I wish to thank for their assistance, friendship and expertise, and contribution, in one way or another, to the making of this book.

I am greatly indebted to Milla Baldo Ceolin, Carlo Ceolin, Claudio Chiuderi, Maria Luisa Dalla Chiara, Aldo De Luca, Bruno de Finetti, Corrado Mangione, Mario Polsinelli, Willard V. Quine, Guglielmo Righini, Giorgio Salvini, Matthew Sands, who have read all or part of the manuscript and have suggested essential improvements. Among the many people who, through conversations, discussions, seminars, or correspondence, have helped to clarify some important points, I would like to mention E. Agazzi, E. Bellone, A. Borsellino, L. Bulferetti, M. Bunge, E. Casari, C. Cellucci, M. Cini, R. S. Cohen, G. Cortini, D. Costantini, L. Geymonat, M. Jammer, P. Rossi Monti, W. Shea, B. Touschek, R. Wójcicki. Above all, I am extremely grateful to the many students who, by continuous, intelligent, and stimulating discussion, have enabled me to examine, verify, or improve my "investigation of the physical world."

G. TORALDO DI FRANCIA

Istituto di Fisica Superiore
Università di Firenze

1 *The method of physics*

1.1. Introduction

In this book I shall describe the principles of physics. Physics is a science that has developed enormously during the last four centuries, apparently at an increasingly rapid rate. In recent years physicists have at times felt that they had reached a dead end. However, this may be a result of having too close a perspective – "He who is at the foot of a mountain sometimes cannot see the summit." Perhaps in some areas we are following a dried stream, whereas at a short distance there may be a new and rich stream of knowledge, which we are about to discover. Or maybe it is simply a matter of having to stop occasionally and to *digest* the enormous quantity of knowledge we have acquired before proceeding to obtain more. In any case, one cannot deny that new and important discoveries are being made even today, so that the prophecies of doom that have sometimes been heard appear to be at least imprudent. Only history will decide to what extent our period has been fruitful or barren.

What we can say for certain is that the quantitative effort expended by each generation so far in the field of physics research has greatly exceeded that of the previous generation. Apart from some fluctuations, the number of physicists, as well as the number of institutions devoted to physics, and the number of countries concerned with physics, have steadily increased. Most of all, the number of printed pages giving information on physics research has grown amazingly, year after year.

If all this could be based solely on the growth of and spread of interest in scientific knowledge, then there would be cause for gratification. Unfortunately, this is not so. The main stimuli that have progressed scientific research in the last century have been the desire for military power and the needs of industrial competition.

Wars have become extremely scientific. The military have realized for a long time now that the most powerful armed force is the one having the greatest amount of scientific knowledge and the capacity to use it. Scientists willingly or unwillingly, directly or indirectly, have been enlisted en masse and incorporated in the war machines of the great powers.

1

At the same time, and not independently, industry has also become scientific. In order to withstand competition and to capture more of the industrial market, industry must keep products up-to-date by making them more functional, or simply more attractive. By applying scientific and technical knowledge, a company can hope to secure the market over its competitors that do not yet have the know-how. Another value of scientific knowledge is that it enables the worker's productivity to be increased to the maximum.

This commercialization of science, along with its reduction to a mere tool, has had an important effect on the behavior of the research worker. The *competitive* aspect of what once was disinterested and independent intellectual activity has lately undergone considerable development. This does not only apply to the military forces or industrial laboratories of the developed countries. Even institutions such as universities or scientific organizations of the less industrialized countries have been unable to resist the influence of the general climate in which science is developing.

No doubt, there are still researchers who work (or *believe* they are working) for the pure love of knowledge. However, they have much less of a decisive role in establishing custom and can avoid the competitive system to even a smaller extent.

The young person beginning a scientific career today, particularly in physics, believes he is taking part in a race. His goal must be to get there before the others, at whatever cost – to learn his job and to accumulate a certain number of papers as soon as possible. But does the "runner" in the race notice the scenery around him? No; by being so intent on winning he sees and thinks of nothing. Sometimes the young scientist even forgets the motives for entering the race. Unfortunately, this is the sad fate awaiting many researchers: lack of culture and alienation.

Many, of course, are able to react and somehow preserve their humanity. But it is a hard-earned victory against the environment, one that cannot be achieved without character and determination.

The situation just described has been reflected in the way that physics is taught and presented in textbooks. Students must acquire a large amount of precise technical information as quickly as possible. There is no time to reflect and examine critically what is being done. A standard and aseptic method of teaching has been set up that, in the hands of the best authors, can yield excellent results in terms of the purpose for which it is intended. But this method fails to satisfy some deep cultural needs of students and surely does not stimulate them to examine what they are doing more closely. In a sense, then, they are encouraged to choose the easiest way. For as C. Weizsäcker (1971, p. 110) rightly observes: "Science is easier to practice than to understand. It is easier to be a physicist and acquire correct knowledge of physics than to explain what exactly one does when doing physics."

There are many good books on the philosophy of science, epistemology, and the foundations of physics.[1] To some readers these might seem to refer to an activity separate and different from that of the physicist, to a kind of reflection from the outside on the science of the past rather than on science *in the making*. Sometimes scientists seeing physics from the inside cannot easily recognize in these writings the features of that science with which they are most familiar. They remain particularly perplexed when the validity of current methodologies, which they and their colleagues hold to be indispensable, are criticized or "refuted" without the author offering a valid alternative to be applied in the laboratory or on the theorist's desk. Common sense, then, prevents scientists from reading further when they come across an extreme case in which the philosopher implies that physicists, in spite of their successes, have "racked up" a lot of mistakes!

On the other hand, there are a number of historical and scientific works of great interest, in which some of the most eminent modern physicists have discussed the procedure, value, and results of science. But here, apart from a few worthy exceptions,[2] one gets the opposite impression. The philosophical thought often appears to be out of focus, bound to rather simple and outdated schemes, expressed in vague terminology. In extreme cases one may witness the efforts made by a great mind in order to discover and describe (badly) what Kant discovered and described (well) almost two hundred years before. This is not to mention those final chapters where the author without the least embarrassment draws conclusions on morals or on free will!

In my opinion, the epistemology of a given science is inseparable from that science; conceptually, even if not always chronologically, the birth of epistemology is simultaneous with the birth of science. Every advance in science is an advance in its epistemology.[3] Starting with this conviction, I have applied a somewhat unusual formula. I have tried to introduce the principles of physics, keeping in mind at all times their epistemological aspects and the critique of their foundations.

Discussions are never carried out in the abstract, but are always made in light of concrete subjects that suggest the analysis. Apart from Chapter 1 which covers general and preliminary concepts on the method of physics, the epistemological discussions are usually found interspersed with scientific expositions, and the two are developed together. As specific knowledge broadens and allows more profitable analysis, fundamental problems are raised and examined; some are taken up several times but at different levels.

Philosophy of science and history of science are closely related. One cannot develop a philosophy of science without reference to history nor can one write a history of science without consciously or unconsciously considering a philosophical view.[4] For this reason, historical references, wherever relevant, will be provided. But I wish to emphasize that this is not an historical work and that I am a physicist and not an historian of physics.[5] The most important historical concern of this book is to present physics *as it is*. The philosopher of science (or even the physicist) cannot always resist the temptation to criticize a physics that never existed or which, in any case, does not exist today. Having performed this operation, one is also tempted to go on to give his opinion on how physics *should be*. This method is not very profitable; it is much better to accept the nature of physics as an historically defined object and to proceed from there to a critical assessment. Of course, one can object and say that I cannot portray physics *as it is* but only *as it seems to me*. This is true, but at least I will not attempt to present physics as I *want* it to be.

The awareness of the important relationship between scientific thought and the social and political context in which it develops is very much alive for some scholars today.[6] This is a very positive fact, which regrettably many scientists have not yet learned. Although this theme is relevant, it will not be dealt with at length in this book, because it might extend the field of coverage too far. On the other hand, I think that even with the object of an in-depth study of the relationship between science and society, one should be aware of what science really is.

Given the object of this book, one should not be surprised that it does not contain *all* of physics. There are enormous gaps, even at the elementary level, that have been left intentionally. In support of this, I would like to say that I consider my readers sufficiently intelligent not to have to drink all the water in the sea in order to realize that it is salty! *Completeness* of information on the contents of physics can be found elsewhere, in excellent treatises written at all levels. Of more significance, however, is the fact that although physics is an empirical or, rather, an *experimental* science, the reader will find almost no description of apparatus or experiments here. One might think that this is a fine limitation for someone who wants to present physics *as it is*! So far I have failed to build up an *epistemology of the laboratory*. To explain what a laboratory is would be like trying to explain what music is to a person who has never listened to music. One learns in a laboratory; one learns how to make experiments only by experimenting; and one learns how to work with his hands only by using them. The first and fundamental form of experimentation in physics is to teach young people to work with their hands. Then they should be taken into the laboratory and taught to work with measuring instruments – each student carrying out real experiments in physics. This form of teaching is indispensable and cannot be read in a book. For this reason, I shall assume that the reader is familiar with some experimental physics learned in high school. But on the other hand, do not be misled into believing that physics as described here is a purely theoretical science developed a priori!

The methodologies of the various sciences have many common features; but they also present some specific differences that cannot be ignored. Very often people have used means of criticizing the methodological peculiarities of physics by taking examples from sciences completely different from physics. As such a procedure can confuse the issues, the reader should know that we will be dealing *only* with the methodology of physics in this book. However little I believe in the truth of rhetorical expressions such as "the unity of science,"[7]: I do believe that the unity of science should not be turned into the "confusion of science."

For the same reasons, one should not disregard the fact that the methodology of physics has evolved over the centuries. Some writers seem to believe that the twentiety-century physicist behaves exactly like his colleagues of the nineteenth or even of the eighteenth century. This is absolutely wrong! The modern physicist also has naïve beliefs and prejudices, but they are different. Physics today is *modern physics* and to discuss its procedures by continually calling up venerable ghosts such as epicycles, phlogiston, caloric, and so on is not very illuminating, except possibly from a purely historical point of view.

1.2. What is physics?

As the object of our study is physics, it seems appropriate to *try* to define first what is meant by physics. However, physics, like all other sciences, is not easily definable. Perhaps we should not pay much attention to definitions because of the risk of accepting deep meanings and concepts where there is a convention. The convention being made only to be sure that we are all discussing the same thing. But, as already emphasized, because we shall be dealing solely with physics, its methods and contents, a definition is needed.

I believe there are two possible ways to define physics. The first is to consider it as simply made up of those topics with which physics

has been concerned during various historical periods.[8] Thus we have
Aristotle's physics, Galileo's physics, Newton's physics, Maxwell's
physics, and so on; each being different in part from the others. Without
doubt, this purely historical method is acceptable, but it can lead to
some difficulties, for not all these authors called their science physics
(in the course of history different names such as world system, natural
philosophy, etc. have been used). It may be more interesting to ask
why physics has shown certain peculiarities, tackled certain problems,
and had certain contents during different historical periods, and also,
why the stress has shifted elsewhere today.

The term *physics* comes from Greek and means "everything that is
concerned with nature." Therefore physics should cover all the problems of nature, but, in fact, it does not. Although it might have been
true in the time of the Greek or medieval civilizations, a break called
the *scientific revolution* took place between the sixteenth and seventeenth centuries. Since that time, in a rather approximate way, physics
has dealt only with those parts of the natural world where biological
processes are not considered. However, one can be much more precise.

Even if it were not clearly stated at first, people started to term
physics as *all that could be profitably studied by using a certain
method*. This method we attribute today mainly to Galileo. Naturally,
this has a certain degree of convention, so we say, once and for all,
that Galileo's contribution was neither the only one nor the first.[9] However, no one before him had formed and expressed such clear and
precise ideas on science. Galileo discovered a method that when applied to a given class of problems, gave excellent results. Since then,
all that could be studied with this method has constituted the unified
and well-defined science that we currently call physics.[10]

This therefore is the second way in which to define physics: It is
defined according to the *method* rather than to the *contents*.

As a result, physics did not include, especially at the beginning, some
important parts of the study of nature. In particular, a large number
of problems concerning living organisms were excluded. Also, other
branches of research such as chemistry, mineralogy, and geology did
not seem to fit quite well within the limits of application of the method.

One should notice an important point. Although there has been a
period in which some sciences have gradually separated from physics,
today we are witnessing a sort of reflux. There is a tendency to bring
some branches of science, which previously were far removed, back
again into the sphere of physics. The main reason is that whereas in
the past the method could be applied fairly easily to only a certain class
of problems, today it is possible to extend its application to an ever
increasing number of disciplines. Because physics has interpreted all

the fundamental laws of chemistry, chemistry, in a sense, has become a chapter of physics. Solid-state physics, a relatively recent science, has brought the specific methods of physics into mineralogy, even if descriptive mineralogy still exists as an historical inheritance.

Traditional geology exists, too, but *geophysics* is becoming more and more important.

Concerning biology, there are already a considerable number of problems such as those dealt with by *molecular biology*, which can be approached with methods very close to those of physics. An interdisciplinary science has been born called *biophysics*. We are still very far from referring to biology as "physics," but it is not inconceivable that one day, perhaps not too far in the future, all of biology will become a chapter of physics.[11]

Apart from the historical and methodological characterizations of physics, there is a fairly important question regarding its *contents*, the answer to which can shed some light on why physics is approaching other sciences that were completely separated in the past. Is physics concerned with single *objects* and contingent *facts* or, rather, with general *laws*?

Galileo was definitely interested in both single objects and facts (e.g., the physical nature of the moon) and general laws (e.g., the fall of material bodies). But after Galileo there has been an increasing tendency to think that physics should be concerned only with general laws; to ascertain or to describe factual situations was thought to belong in the province of other sciences, such as astronomy, geology, and natural history. However, in modern times it has become very difficult to stick precisely to this distinction (see §5.1). Increasingly, one has the impression that the general laws of the universe are inseparable from its apparently contingent structure. As we shall see, strong doubts in this sense can arise even from the second law of thermodynamics (see §3.16), not to speak, of course, of cosmology.

We shall deal first with that aspect of physics that is concerned with general laws, calling this the *nomological* aspect. Later, we shall show that the *factual* aspect is sometimes closely connected with or even inseparable from it.

1.3. A first approach to the method

Having explained what physics means, and having shifted the accent from content to method, let us stop now to examine this method.

A scientific method cannot adequately be discussed if it is divided from the science to which it applies. To understand fully the significance of the brief survey of the method of physics that we present here, one should see it at work on concrete examples, which will be given in later chapters.

To put the following discussion in the proper perspective, we reiterate that this is not a *history* of physics. The historian dealing with Galileo's method must refer to that author's writings[12] and determine from them what Galileo himself intended his method to be. We, instead shall take the philosophical viewpoint of one who comes three and a half centuries later,[13] by analyzing what Galileo's methodological lesson means to us. As a result, we shall feel authorized to say, with some caution, even things that Galileo never explicitly said nor stated in that form.[14]

Because our interest for the time being is purely methodological, we shall consider only one aspect of Galileo's work. But do not forget that there is much more than method in Galileo. The battle and the victory against the Aristotelian and theological conception of science implies much larger problems than those that can be faced by using the method of physics. However, having discovered this method, and having shown that it works well, Galileo had a powerful weapon for following up the victory on an even larger scale.

Galileo's method is an *experimental* method. What does this mean? At an elementary level, the explanation is very easy, but it becomes much more complicated when a deeper insight is sought.

In a first and simplified version, one can say that the method consists in relying on facts or, rather, in taking experience as a guide and not in proceeding with abstract and a priori theories that are not based on experimental evidence. By saying only this, however, we do not clarify Galileo's great discovery, and do him an injustice.

To approach the center of the problem, one must first notice the distinction between *observation* and *experiment*. We can use a generalization that, although not always true, is useful as a reference. Before Galileo's time the scholar observing a phenomenon had, as it were, the role of a spectator, or of a witness; after Galileo, the scholar not only listens to what nature spontaneously says, but interrogates it. This changes observation into experiment and provides the main key that opens the door to the modern conception of physics: Nature is interrogated and humans formulate the questions.

What is the importance of all this? Today, a posteriori, we can describe Galileo's approach by saying that he worded his questions in a biased way. This is not to say that he required prefabricated answers from nature; but that his questions fitted at least partly with the answer, which he managed to guess. Galileo's questions were certainly not haphazard, which is why he referred to "sensible experiments" rather than to just "experiments."[15] What distinguishes a sensible experiment from just an experiment?

A physical phenomenon generally depends on a number of different factors or *parameters*, as we call them today. Some of them are essential, whereas others are secondary, and may disturb the phenomenon one wishes to study. When formulating a question, Galileo first simplified the problem by stripping the phenomenon of all those secondary parameters that would otherwise complicate the answer, rendering it incomprehensible. The problem was then formed in such a way as to make it depend on only a few parameters; therefore one could study nature's behavior as a function of only these parameters.

At this point, we should ask, "How did Galileo know that by carrying out the experiment in a certain way the phenomenon was stripped of those secondary parameters that he called 'external and accidental impediments'?" This question leads us to a consideration that probably did occur to Galileo, although he did not make it explicit.[16]

The following example may clarify this point. How can we be sure that the fall of a strangely shaped body, part stone and part wood, half red and half blue, part cold and part hot, through a dense atmosphere, crossed by irregular air currents, has less right to represent the natural world than the fall of a simple ball in empty space without any of the previous complications?

In other words, it would appear that nature should not "like" the latter phenomenon better than the former and that both should have exactly the same native rights.

Where is the difference? The difference lies in the intervention of the researcher, who does not simply ask for an answer from nature, but asks for an *intelligible* answer.

One need not necessarily be ready to accept the whole Kantian approach to appreciate, at least in general terms, the truth stated here from the preface to the second edition of *Critique of Pure Reason*:

When Galileo caused balls, the weights of which he had himself previously determined, to roll down an inclined plane; when Torricelli made the air carry a weight which he had calculated beforehand to be equal to that of a definite volume of water; or in more recent times, when Stahl changed metals into oxides, and oxides back into metal, by withdrawing something and then restoring it, a light broke upon all students of nature. They learned that reason has insight only into that which it produces after a plan of its own, and that it must not allow itself to be kept, as it were, in nature's leading-strings, but must itself show the way with principles of judgment based upon fixed laws, constraining nature to give answer to questions of reason's own determining. Accidental observations, made in obedience to no previously thought-out plan, can never be made to yield a necessary law, which alone reason is concerned to discover. Reason, holding in one hand its principles, according to which alone concordant appearances can be admitted as equivalent to law, and in the other hand the experiment which it has devised in conformity with these principles, must approach nature in order to be taught by it. It must not, however, do so in the character of a pupil who listens to everything that the teacher chooses to say, but of an appointed judge who compels the witnesses to answer questions which he has himself formulated. Even physics, therefore, owes the beneficent revolution in its point of view entirely to the happy thought, that while reason must seek in nature, not fictitiously ascribe to it, whatever as not being knowable through reason's own resources has to be learnt, if learnt at all, only from nature, it must adopt as its guide, in so seeking, that which it has itself put into nature. It is thus that the study of nature has entered on the secure path of a science, after having for so many centuries been nothing but a process of merely random groping (translated by N. K. Smith).

Some students reject with horror all that allegedly leads to idealistic attitudes and the negation of objective reality. However, asking an interlocutor to answer our question and to speak our own language (instead of an unknown one), is not the same as believing that the interlocutor does not exist and that we are concocting an answer of our own.

Thus we become acquainted with a procedure that modern physics has worked out in a very stimulating way. An experiment represents a question that is not independent of the nature of the questioner. It is formulated by that observer and leads to an answer suited to the observer. One must think of an encounter between subject and object, the scientist on one side and nature on the other. The researcher can manage to insert some subjective elements to describe phenomena that are seen to take place objectively in the outside world, but as far as nature is concerned, it would be nonsense to introduce preferential elements. Out of this complicated babel of words, the observer begins to select those nearest to his own language. In order to be understood, nature must speak the language of the observer.

To be honest, Galileo has expressed himself in apparently antithetic terms. There is a famous passage in the *Assayer* in which he says:

> Philosophy is written in that great book which ever lies before our eyes, I mean the universe, but we cannot understand it if we do not first learn the language and grasp the symbols in which it is written. This book is written in the mathematical language, and the symbols are triangles, circles and other geometrical figures, without whose help it is humanly impossible to comprehend a single word of it, and without which one wanders in vain through a dark labyrinth.

Literally, therefore, it is nature that forces its own language on the observer, and not the other way around. Often when speaking of this, Galileo's Platonism is mentioned (see e.g., A. Crombie, 1950; W. Shea,

1972, p. 155; C. Weizsäcker, 1971, p. 113). But we should realize that it was difficult for Galileo in the seventeenth century to express himself in any other way, even had he not been convinced that nature really spoke its own language, independent of the observer. I agree with Geymonat that:

> His appeal to mathematics rests upon a methodological rule; the philosoph-ical defense of this rule does not interest him, but is left in the distant depths of the debate. The desire to read more than this into the *Assayer*, to see there a retrogression to mathematical Platonism, may hide from us the living heart of the work (Geymonat, 1965, p. 110).

Whatever Galileo's metaphysics, I believe the best justification we can give today for the methodological canon described is the one already mentioned: Out of the profusion of words uttered by nature, the sci-entist selects those that can be understood.

Thus we reach another essential aspect of the method of physics, directly derived from Galileo's teaching; that is, the formulation of questions in a *quantitative* form. The physicist must manage to attach numbers to phenomena and to natural quantities. Here the concept of *measurement* is introduced.

I would like to state at this point that although, as a result of a long training, we find it extremely natural today that questions are put in quantitative form, this was an enormous step forward in the seven-teenth century. As A. Koyré rightly says:

> It is very strange: two thousand years before, Pythagoras had proclaimed that number is the actual essence of things, and the Bible had taught that God had based the world upon number, weight and measurement. Everybody re-peated this, nobody believed it. At least, up to Galileo, nobody took it seri-ously. Nobody ever tried to determine those numbers, those weights and meas-urements. Nobody tried to overcome the practical use of numbers, weights, and measurements in the imprecision of everyday life – counting the months and animals, measuring distances and the fields, weighing gold and corn – in order to transform this into an element of precise knowledge (Koyré 1961; see also Crombie, 1961).

In this connection Galileo's work may be compared with that of Francis Bacon. Probably the main factor that prevented Bacon from becoming the founder of modern empirical science was his total in-comprehension of the role mathematics should play in it.

One could also notice that the prescription to make *numbers* cor-respond to natural quantities does not faithfully represent Galileo's thought. He talks about "triangles, circles, and other geometric fig-ures," instead of numbers. But it can be noted that Descartes put ge-ometry into a quantitative form when (just about that time) he discov-ered analytical geometry. Had Galileo used the method of analytical geometry, he would have been able to talk about *algebraic equations* instead of circles, lines, and triangles. Hence putting things into geometric form meant taking, if only implicitly, a quantitative attitude.

During the course of the following centuries, and especially in our own, the quantitative aspect in physics has become a little less exclusive.

But here one must be careful not to fall into the misunderstanding of those who affirm that Galileo's conception of the universe has been abandoned, or should be abandoned. In my opinion the situation is this: Physics has departed from a strictly quantitative approach, exactly to the same extent as mathematics itself has ceased to be only the study of quantity. But mathematics has never ceased to be the indispensable base of physics.[17] Furthermore, if it is true (particularly today) that there is much more in physics than just numbers and the relations between them (e.g., symmetries), then it is also true that in order to be able to conceive or discover this "much more," one must always start by making measurements and handling numbers. The central point of this second feature of the method is that to each natural phenomenon and its physical quantities there corresponds a set of numbers. This correspondence is brought about in practice by means of *measurement*.[18]

How does one carry out a measurement? Let us take the simplest example: the measurement of a length. First, we must set up the *unit*; to do this, we choose a rod and agree that it represents the *unit of length*. Second, by applying a well-known and elementary procedure, we compare the rod with the object we want to measure. We note the integral or fractional number of times the rod is included in the length under consideration. Granting that this concept is clear and unequivocal (which, as we shall see later, is not completely true), the number thus obtained represents the *measure* or the *value* of the length.

The important fact that characterizes physics is that the relations occurring between natural properties can be brought back to relations between the numbers that represent their measures. It is not only a question of characterizing objects, as when we say: a certain object weighs so much, has such a length, lasts such a time, and so on. The great discovery consists in having found that between these numbers there hold some mathematical relations, which are always valid, in the sense explained in §§1.4 and 4.21.

If in the example of a freely falling body we record both the distance covered (measured with a given length unit) and the time it takes (measured with a given time unit), we can note that the numbers expressing distance values are proportional[19] to the squares of the numbers representing the corresponding time values. When we say for short that the distance is proportional to the square of the time we mean that the numbers of a certain set (values of distance) are proportional to the squares of the numbers of another set (corresponding values of time).[20]

The two most salient features of Galileo's method have thus been brought out: Questions to be put (1) in a simple form with only essential parameters and (2) in a quantitative form. The latter allows us to obtain sets of numbers between which there holds a mathematical relation. As will be seen in §1.10, there is something beyond this in the method of modern physics; something at which perhaps Galileo had guessed without fully realizing its importance.

1.4. The value of the method

We shall now discuss an essential characteristic of the method just described – the excellent way it works.

Many objections may be raised (and have actually been raised) to Galileo's approach. Some people may claim that it does not have a genuine cognitive value, because it does not bring us nearer the intimate *reason of things*. Many physicists may be ready to accept this, within reason; as long as one does not attempt to derive an absurd conclusion such as that the method is void of *any* value whatever.

The amazing point is that for the first time since the discovery of mathematics, a method has been introduced, the results of which have an *intersubjective* value! One can argue considerably about the doubtful role of *consensus* in scientific theories. Nevertheless, the fact remains that after Galileo no sensible person who has taken an unbiased look at the experiments will affirm that (within the limits we shall discuss later) a freely falling body does not cover distance proportional to the square of time. Intersubjectivity has often failed to be acknowledged because it was believed that physics could, or wanted to, solve a number of problems that fall outside its province, and therefore cannot be solved.[21]

The problems that can be solved by the methods of physics are in a well-defined class, whose boundaries will be better determined in the following chapters. One may think that the results obtained are very modest. But, on the other hand, it must be recognized that they are certain. This is an interesting fact, especially when thinking of other subjects that deal with apparently deeper and more fundamental problems, which are far from reaching intersubjectivity.

Physics can predict with certainty that some phenomena characterized by some numbers are connected with some other phenomena characterized by some other numbers. But what is this *certainty*? It is not *apodictic* certainty in the Aristotelian–Kantian sense, characterized by logical necessity, or by necessity a priori. There is no apodictic assertion in physics nor could there ever be one.

Here, of course, we have the venerable problem of the value of induction. The doubt that has been expressed frequently (particularly by Hume, very clearly) cannot be ignored. The fact that the sun has risen in the morning, a hundred, a thousand times, does not give the apodictic certainty that it will also rise tomorrow. There is no objection to this. Yet no one gets up early in the morning mainly to see if the sun will rise. Why?

Hume had already noticed that the inductive process can be justified by using the postulate of the *uniformity of nature*. But the concept is rather vague and deserves a deeper analysis. Furthermore, physics itself has taught us that too often statements whose truth was allegedly required by this uniformity of nature have turned out to be incorrect.

In a later section (§4.21) we shall discuss the general problem of inductive inference at greater length. For the present, we shall only set forth a limited form of the postulate of the uniformity of nature, on which all physics is based. This postulate is the *space–time invariance* which, in simple terms, can be stated as: *Physical phenomena take place in a given place at a given time, exactly as, under the same conditions, they would take place in any other place and at any other time.*

The importance of this postulate goes much further than physics in the strict sense of the word. It is a necessary condition so that we can survive. Indeed, when we walk, talk, or eat, we must trust that physical nature (including our bodies) behaves today in this place as it has behaved in the past in other places. In everyday life, of course, it is only necessary that this invariance be respected within a limited space–time region around us. The extension of the postulate to the whole universe is a scientific generalization for which Galileo is mainly credited. Incidentally, Galileo had difficulty in convincing some scholars of his time that physical laws were the same in the heavens as on the earth.

What is important to note is that the postulate of space–time invariance in conjunction with the Galilean precept that phenomena should depend on only a few easily controlled and essential parameters, has led physicists to discard the classical procedure of induction. It is sufficient to study how a *single* stone falls, to derive how *all* stones fall, and will fall.[22] This is a feature of physics of which many people are still unaware. Its experiments are essentially *unique*. When a particular experiment is repeated, it is merely to improve precision, or to gather statistical information, and not to check the uniformity of nature which is taken for granted.

We have discussed a postulate. Now what value can we ascribe to this postulate from an epistemological point of view? One might say that the validity of the postulate is proved by the fact that the physical phenomena observed have always respected it. But this would be like justifying induction by using an inductive argument. Hume's threatening ghost would appear, urging that we should know better!

What we seem to be confronted with here is a structure of our mind. But we should not think of the Kantian a priori with its necessity and immutability. Instead it is one of those structures that, according to the theory of cognitive development or genetic epistemology (J. Piaget, 1970), arise and develop in the child on contact with the outer world. In this connection I think we can agree with K. Popper's remark:

> When Kant said that our intellect imposes its laws upon nature, he was right – except that he did not notice how often our intellect fails in the attempt: the regularities we try to impose are *psychologically a priori*, but there is not the slightest reason to assume that they are *a priori valid*, as Kant thought (Popper, 1972, p. 24).

The *subjective* interpretation of the space–time invariance is supported by the fact that each time someone has seen, or imagined to have seen, a corpse come to life, he has never thought of a nonuniformity of nature, but of a miracle – of a supernatural event! On the other hand, the *objective* interpretation is also possible, because miracles do not happen at all times and no miracle has ever been documented in such a way as to convince all sensible people. In other words, it is legitimate to suppose that our *subjective* postulate is based on the real *objective* behavior of nature, but we shall never be able to prove it.

In any case, I cannot help expressing the opinion that the laws of logic and mathematics are not in the least privileged, compared to those of physics. In order to say $2 + 2 = 4$, I must ask myself a question and perform an inner experiment, although rapidly.[23] Until now, each time I have made this experiment, I get the same reply. And I know that the same has happened to all humans. But how can we be sure that tomorrow, on making the same inner experiment, we shall not find a different reply from today's? Who can assure us that humans on landing on Mars will not begin to think $2 + 2 = 5$? No one can! It is a question of faith, and of an ensuing postulate.

In other words, the assertion that mathematical propositions are *analytical* or, as Leibniz put it, that they are true "in all possible worlds," does not help, because, strictly speaking, among all possible worlds there also should be those worlds in which humans think differently.[24]

Consequently, saying that the laws of physics are not sure, has the same value as saying that humans can be sure of nothing. Such a statement has, of course, a great philosophical importance. Scepticism cannot very easily be put aside. Fortunately, however, it has never prevented humankind from building up science. And it is the structure of physical science that we want to investigate here, not the most general reply to scepticism.

1.5. The operational definition

Physics is primarily concerned with *physical quantities* and with the relations existing between them. Examples of physical quantities are length, volume, time, force, mass, electric charge, and so on. But in defining these entities, one realizes that it is an extremely difficult task. Try, for instance, to define length or time. You will soon be convinced that it is a tremendous problem. You may even conclude that physicists deal with entities for which they use terms of completely unknown meaning.

Let us suppose, however, that we have given a reasonable definition of a certain physical quantity. As soon as we start to measure it, another problem arises: How can we be sure that what we are measuring is really the quantity we have defined?

From §1.3 we know that the quantitative attitude is essential to physics. Things are investigated only in as far as it is possible to measure them, and not with the impossible goal of discovering their intimate essence. The aim, at least at first, is a much more modest one, and consists in comparing some measurements with others in order to discover the constant mathematical relations existing between them. Thus starting from the known results of some measurements, we may be able to predict the results of some future measurements. At this point, one may ask whether it is really necessary to define what we are measuring in some way, other than by exactly prescribing how to measure it.

Thus we arrive at the concept of *operational definition: A physical quantity is defined by prescribing the operations that are carried out in order to measure it.*

From Galileo on, I believe, this concept has been much more intimately and necessarily tied to the development of physics than is generally admitted. First, the concept was, at least in part, applied by many great physicists in the past, although it has not come to full maturation until the twentieth century, thanks to H. Dingler, A. S. Eddington and, especially, to P. W. Bridgman.[25] Second, the operational point of view is not to be conceived of as a philosophical position. It merely represents a *methodology*; an extremely fruitful methodology, however, which is able to produce new knowledge, as clearly shown by A. Einstein's first article on relativity. By analyzing the way in which time is measured (and the simultaneity of two events established) in two different inertial frames, he found some fundamental results. No brooding on a purely aprioristic time concept could have led to so much. Analogously, results of great importance were achieved by W. Heisenberg, by analyzing the simultaneous measurement of two conjugate quantities (§4.13).

Operationalism has often caused lively discussion and argument in the last few decades. This is due (1) to the fact that some people have sought dogmatically to extend its range of application much further than necessary[26] and (2) to the emotional aversion that several workers have shown for it. Their criticism is, more often than not, based on a series of misunderstandings. In order to deal with the subject adequately, a long discussion would be necessary; here, however, I shall limit myself only to clarifying a few points.

First of all, let us emphasize that we are not at all pretending to eliminate all intuitive concepts of length, time and so on; we only remark that length, time, and so on do not become physical quantities until we know how to measure them. To be sure, I have a concept of beauty and can discuss it; but beauty cannot represent a physical quantity because I do not know how to measure it.

Next, it is important to keep in mind that difficulties and misunderstandings can arise when the operational definition is not strictly limited to physical quantities but is extended to all the concepts of physics, or of science in general.

What is meant by a physical concept? Sometimes the term is used to denote a general type of object of physics, such as a solid or liquid body. But this can, no doubt, be defined, where necessary, by requiring the values of certain quantities measured on the body (such as elasticity, viscosity, or compressibility) to fall within certain limits. But, in fact, these definitions, unlike those of physical quantities, have very little importance. No one is concerned by the fact that glass can be classified among either solids or liquids, according to the prescribed definition. Obviously, the discussion becomes more difficult when the object is an atom, a nucleus, or a photon. This subject will be dealt with more fully in §4.1.

Sometimes reference is made to certain properties that *prima facie* could seem to be qualitative, as when one says: "This thing is green." But such properties could very well be specified by measurements. Instead of saying that an object is green, one could say that its reflection coefficient shows a certain behavior as a function of the wavelength of light.[27] If one does not usually do this, it is only because it is not worthwhile.[28]

A hopeless confusion arises when the term *physical concept* is applied to some mathematical entities employed in theoretical physics, such as the *hamiltonian* or the *state vector* (which will be discussed in Chapter 4). The state vector is by no means a physical quantity and cannot be measured. It represents either a vector in Hilbert space or a complex function (depending on the definition), that is, a mathematical entity used in calculations which arise in quantum mechanics. Numbers are used in physics to make calculations. But who would think that numbers are physical concepts? The claim that all concepts of science should be operationally defined overshoots the mark, and can readily be criticized. How can one operationally define *homo sapiens* or an economic depression? But let us return to physical quantities.

An objection often raised against the operational definition of physical quantities concerns those quantities that are ordinarily evaluated by first measuring a number of different quantities and then by performing some mathematical operations on the values thus obtained. Are such pencil and paper operations to be included in the class of those operations that give the name to, and define the meaning of, the operational method? Some people argue that this conclusion is unacceptable. It is customary to minimize the problem by distinguishing *primary* from *secondary* quantities. The former quantities are those based on a direct operational definition, whereas the latter quantities are those whose values are obtained by mathematics from the values of other quantities. This position is usually accepted by convention, in spite of its inelegance and the essential arbitrariness with which each quantity is assigned to either category. It seems to me, however, that this dichotomy is void of any fundamental justification, as will be shown in the next section. In principle, all physical quantities are to be conceived of as primary quantities, even though it may be *convenient* to derive the values of some of them from the values of other quantities. We shall see this in the case of velocity (§2.2), work (§2.5), momentum (§2.6), and so on.

A further objection that appears to be serious is that the operational approach is *circular*. Popper argues: "As to the circularity of the operational definition of length, this may be seen from the following facts: (a) the operational definition of *length* involves *temperature* corrections, and (b) the (usual) operational definition of *temperature* involves measurements of *length*" (Popper, 1959, p. 440).

It is easy to answer this criticism by observing that every dictionary is circular, that dictionaries are, nonetheless, useful to those who want to employ words correctly, and that there is no reason that the dictionary of physics need have the special characteristic of being noncircular.[29] But there is another factor.

Strange as it may seem, this kind of circularity embodies an essential feature of physics. This concept may be understood better after dealing with the precision of measurement (§1.9) and the domain of validity of a physical law (§1.10). We shall now describe a procedure that is natural in physics. We first measure length with a rod, whose temperature is unknown. We then use this rod to measure the thermal expansion of fluids and to build the scale of a thermometer. At this point we use the thermometer to check the temperature of the rod in order to keep it constant. The rod is then applied to build a better scale for the thermometer, and so on. Usually, at every step, we obtain a better accuracy, and each step has its own value and can be used in *physics*, without having to wait for the next step. If this were not true, physics could never be applied, for, as is evident, we could never really take the final step!

Some scholars do not like the operational definition because of the fact that it does not measure things in order to reveal their intimate essence, but with the more modest aim to compare the results of those measurements in order to discover the mathematical relations existing between them. Many people consider this aim too modest. We shall often have to deal with similar objections.

Let us emphasize that our intent is not to ridicule in any way the scholar who seeks to pursue more important or deeper aims. Many physicists do, indeed, pursue such aims. But virtually all physicists firmly believe that even when this can be done the only hope of success lies in securing, as a first step, the modest, but sure and intersubjective, knowledge that physics can supply. To this end, we must make sure (1) that we are all talking about the same things and (2) that we can apply the method of physics to these things. This can only be obtained by starting from the operational definitions of physical quantities.

1.6. The language of physics

The operational definition of physical quantities gives us the opportunity to mention the language of science.[30] Although this subject cannot be treated thoroughly here, we shall illustrate a few essential concepts.

Let us start with the remark that the terms of ordinary language often present two kinds of shortcomings:

1. They denote several different, even if well defined, concepts (e.g., the word "post" in English).

2. They denote hazy and not well-defined concepts or, alternatively, concepts for which it is difficult to envisage a definition encompassing

all of the possible meanings (think, e.g., of the words "time," "fear," "law").

With such a language it is obviously impossible to build up a science aiming toward a high degree of objectivity and precision. In order to overcome this difficulty, one generally sets up a precise *scientific language*, suitable to the intended branch of science.

The most exacting demands are met by a formalized language. This consists of a set of symbols, along with some rules of *formation*, which allow the symbols to be put in a sequence, forming a *sentence*. In addition, there should be some *correspondence* rules, which make symbols (or strings of symbols) correspond to observable objects or properties. Finally, there are rules of *deduction*, which enable us to pass from one sentence to another, carrying out the process of reasoning.

Such an abstract tool is usually not required in physics. In general, ordinary language is used, in which we seek to assign an unequivocal meaning to every term. To this end, the concepts under (2) can be subjected to a process called *explication*.[31] From *explicanda*, they become *explicata*. This pedantic Latin terminology is used in order to clarify that it is not a question of the ordinary explanation of terms and concepts.[32] This kind of explanation should make explicit everything that is implicit in an intuitive concept. But this is not the task of the explication in science.

The *explicatum* is not a *synonym* of the *explicandum* and, consequently, cannot be substituted for it in every context. It embodies a more precise but *poorer* concept than the *explicandum* and can be substituted for it only in a well-determined context. It does not even necessarily represent one of the normal meanings of the *explicandum*, because the application to that context is not considered in the ordinary language. For instance, it is not always agreed that the everyday concept of "number" should include the notion of "class of classes." The latter, however, can be advantageously used in arithmetic in order to derive some rigorous consequences that are not entailed by the intuitive concept.

In particular, the purpose of the operational definition of physical quantities is by no means to explain what such quantities are. The purpose is to define a number of concepts in such a way that students may know what they are discussing and in turn can discuss it in the same way as other students.

It is obvious that the ordinary concept of "time" is richer than that defined operationally. It is richer, but more vague! The fact that an explication is valid only relative to a given context should explain that – even within the compass of physics – explications and, consequently, concepts, can change as theories change (see Pearce and Maynard,

1973). No wonder, then, that I. Newton's "time" is not identical to A. Einstein's "time."

An important distinction that has not yet been clearly grasped by all scientists is the one between the *intension* and the *extension* of a term.[33] We shall illustrate this distinction with an example. The intension of the term "planet" embodies the *set of properties* that characterizes planets, that is, material objects of a certain size and shape, revolving around the sun in nearly elliptic orbits. The extension is instead represented by the *set of individuals* that we call planets, namely: Mercury, Venus, Earth, Mars, Jupiter, Saturn, Uranus, Neptune, Pluto.

The extension of a term may even consist of a single individual; for instance, this is the case for the intension "the stars closer to the earth than a light year," whose extension consists solely of the sun. Moreover, a perfectly admissible intension may have *empty* extension; for instance, this is true today of the intension "incandescent planet."

We can construct as many intensions as we wish, provided only that they are consistent with both logical laws and the known laws of physics. But often it is very difficult to ascertain the corresponding extensions, partly because in order to determine an extension, it is nearly always necessary to assume a good deal of theory. For instance, Frege talked of the intensions "morning star," and "evening star," which have one and the same extension (i.e., the planet Venus). It is clear that in order to carry out this identification, one must know the theory of planetary motion (or something equivalent).

An important point to emphasize here is that in order to establish a certain extension, we must assume that the individuals under consideration are *distinguishable* and *permanent*. This is all right in mathematics, but not in modern physics! We shall see this when dealing with elementary particles. Now we mention a result that will be clarified in §4.7. Although it is true, say, that "morning star" and "evening star" have the same extension, the sentence: – the intensions "incident particle" and "scattered particle" have the same extension – may be *undecidable* (neither its truth nor its falsity can be demonstrated).

1.7. Observables or theoretical constructs?

At this point it is important to describe a subject held close by some logical positivists and which has been the topic of considerable writing. Many scientists maintain that besides the terms of *logic* and *mathematics*, two kinds of terms are used in science: *observational* and *theoretical*. The former refers to objects or properties that are immediately observable, whereas the latter refers to theoretical constructs that are not directly observable.

The distinction has never been too clear, and different interpretations have been given by different authors. One has the impression that even some of its original proponents have, in time, weakened its significance and importance (Carnap, 1966, p. 258). Others, it seems to me, have concluded that it is a question analogous to the one about the sex of angels (see, e.g., Putman, 1962, p. 240).

Again, we shall be concerned only with physics and its method, and therefore shall abandon our examination of what purpose the distinction between observational and theoretical terms can have in other sciences. Some scholars, familiar with the many ramifications to this problem, would perhaps appreciate a longer discussion of this subject, but this would be digressing too far.

In physics the question can essentially be put under two subheadings; that is, whether the quality of being *observational*, or *theoretical*, applies:

1. To physical quantities (all or some);
2. To physical objects, such as electrons, protons, and so forth.

The second subheading will be examined when we describe sub-atomic particles. Here we shall consider some misunderstandings that also affect the first.

Trouble arises mainly because the expression "directly observable" is not sufficiently analyzed. For example, the assertion that whereas the weight w and the volume v of a body are directly observable, its specific weight $s = w/v$ is not directly observable.[34] The problem to some extent is related to that of primary and derived quantities.

What does it mean to *observe* an object? Today we know fairly well the chain of phenomena that occurs when we look at an object. We shall discuss this in detail later (§2.16); here, however, we shall merely summarize some simple ideas. Visible radiation, say, from the sun, falls on an object (Fig. 1.1) and thereby is scattered in all directions, with greater or smaller intensity. This scattered light carries information about the object.

One way of deriving what information is contained in the scattered light can be to detect it in the various directions. For example, we could measure with a number of directive photocells, placed a great distance apart from the object, how much light, and of what color, is scattered in direction A, how much in direction B, and so on. The fact that in direction D (shade) the intensity is zero, is also a piece of information.

According to traditional optics, the *rays* coming from the object are thus examined one by one. Knowing the laws of optics, we can piece together all the results of these measurements and the visual properties of the object can be derived by a long, yet feasible, computation. All

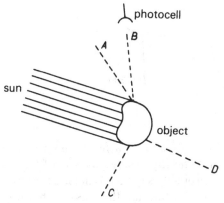

Figure 1.1

the rays coming from a given point of the object would have to be singled out and their intensities added in the respective colors. Today this operation would be carried out by an electronic computer. Those who believe in the distinction between observables and theoretical constructs would find great difficulty in saying that we have directly observed the object in this way.

Computers or data processors can be broadly divided into two categories: *digital* and *analog*. A digital computer uses numbers, more or less in the same way as we do when calculating. An analog computer, on the other hand, avails itself of an appropriate physical phenomenon that simulates the one to be investigated, so as to pass directly from the input data to the result without using numbers.

In the previous example, there is a simple and extremely efficient type of analog computer which is called a *lens* (Fig. 1.2). For each point P of the object, the lens collects all the rays coming from it and sends them to a point P'. The image of the object can be observed on a screen placed at P'.

At this point the screen can be replaced by a sensitive plate and thus the camera has been discovered. But nature has long ago revealed this kind of data processing; indeed, the eye is made exactly like a camera. When the instrument used is the eye, we usually say that *the object is directly observed*.

But what difference of principle distinguishes the eye from another instrument or computer? Perhaps the fact that it is made of organic matter? For a moment, let us assume this absurdity, and claim that by using our eyes we see the object directly. Next, let us reflect that usually the light from the object does not travel through empty space, but through the air. Suppose we do not place much importance on this

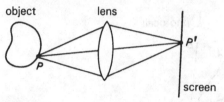

Figure 1.2

fact; then imagine that we are looking at a fish swimming in a pool. Do we see it directly, even if the rays carry out part of their journey through the water? We shall probably say yes. What if it is an aquarium, so that part of the journey occurs through the glass? What if one side of the aquarium is slightly convex, so that we see the fish slightly magnified? What if this side is replaced by a lens? What if instead of one lens we use two lenses forming a microscope? We have thus passed with continuity from direct observation to observation through an instrument.

Therefore it is absurd to think that the distinction between direct observation and observation by means of an instrument can have a precise and essential meaning. We have also shown that an instrument can (and usually does) include an analog computer. So why not a digital computer as well? What conceptual difference is there between a lens and a set of electronic circuits?

It should be clear that in any observation a data-processing stage naturally exists.[35] This is why I agree with those who say that observational terms do not exist; there are only theoretical terms.[36] Why, then, continue with this distinction?

In particular, returning to the previous example, we can construct an instrument that directly measures the specific weight of a body, in many different ways. But what substantial difference can there be between the case where the calculation is made by the instrument and the one where we do it ourselves?

These considerations should show that there is no difference, in principle, between primary and derived quantities. They can all be operationally defined. The operational definition may or may not include the help of an instrument or the performance of a calculation; to believe that there is a sharp difference between the two cases is a misconception.

At this point some people might think it necessary to ban the two expressions *primary quantities* and *derived quantities* from the physics vocabulary. This is not so. This distinction, even though conventional (in that it is arbitrary to choose which are the primary or derived quantities), can be preserved for the sake of convenience. Practical examples will be given later. But we affirm this does not mean that some quantities can be operationally defined and others cannot.

1.8. How many physical quantities are there?

One difficulty often encountered is that one and the same physical quantity can have several equivalent operational definitions. For example, a length can be defined by measuring it with a rod or by using optical triangulation, or radar.

Strictly speaking, one can maintain with P. Bridgman that every different set of operations defines a different physical quantity. If this is the case, the equality of the measures obtained by different methods should be interpreted as a physical law, linking these quantities. This view is not very popular among physicists because it is not very practical. If we were to accept it, we would have a very large number of physical quantities, perhaps with different names, and memorizing them would be quite a task.

Another point of view that is fairly widely accepted, owing to its apparent usefulness, is that a physical quantity is defined by the *class* of all its possible operational definitions (Agazzi, 1969, p. 128). To be quite clear, in this case we shall refer to a *generalized operational definition*. It is a point of view adhered to by virtually all physicists, usually without saying so explicitly. All this seems quite sensible, but one must be careful. Things are not so simple!

When I say that the width of this table is 2 m, or that the distance from the earth to the sun is 149 million km, or that the distance between adjacent atoms in a given crystal lattice is 10^{-8} cm, am I discussing the same physical quantity? Notice that neither the distance from the earth to the sun, nor the space between the atoms can be measured with a rod.

We must face the following. The size of the table can be measured in many ways, for example, with a rod or by optical triangulation. In the class of the possible ways of measuring there is at least one (optical triangulation) that is also valid for measuring the distance earth–sun. On the other hand, the direct comparison method (rod) can be used down to small distances, and by using a microscope, to extremely small distances. At this point one can start to measure the sizes of objects by using the diffraction of electromagnetic waves or particles; this method is valid down to the interatomic distances (diffraction of X rays, scattering of neutrons). Thus we arrive at the situation illustrated diagrammatically in Figure 1.3, where several classes of procedures for measuring lengths are represented with Venn diagrams.[37]

Each class contains all the methods of measurement that, for lengths

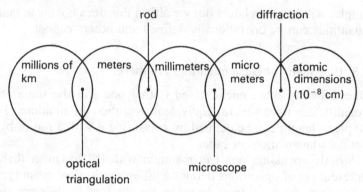

Figure 1.3

belonging to a certain interval, yield identical results. One class contains the methods valid for *millions of km*, another class, the methods valid for sizes of the order of a *meter*, and so on. It is essential that two adjacent classes have a *nonempty intersection*, that is, a part in common. In this case it seems possible, by proceeding from one class to the next, to match the units properly, and to define the generalized physical quantity by the class that results from the *union* of all the classes represented. This is the class including all those procedures of measurement that belong to at least one class of Figure 1.3.

This procedure seems very simple. However, it turns out that the necessary matching across the intersections is not always possible. There are conspicuous examples of this in the transition from macrophysics to microphysics. There are several procedures yielding one and the same result when applied to measuring the diameter of a billiard ball, which give different results when applied instead to the measurement of the diameter of an atom. As a result, in the case of the atom, one must be prepared to accept the statement that, in general, different procedures that are equivalent in classical physics, actually define different quantities. Hence the problem of the operational definition of physical quantities is very complicated.[38]

We shall return to this subject when we have enlarged on some other concepts.

1.9. The precision of measurements

By using a common meter stick, divided into centimeters and meters, one is not able to measure distances with the precision of a thousandth of a millimeter (which is called *micrometer* or sometimes *micron*, as denoted by μm). One can resort to a standard meter and a good microscope, but even in this way, for example, a precision of the order of 10^{-8} cm cannot be obtained.

The common judgment concerning these facts is expressed in the following way. There exists a real number a (with an infinity of decimal places) that represents the *true* or *exact* measure; but every time we try to reach this truth, that is, every time we carry out a measurement, the grossness of our means introduces an *error* unknown to us. We find a certain number b, which we call the *value* of the quantity, and we say that the real measure a lies between $b - \epsilon$ and $b + \epsilon$, that is,

$$b - \epsilon \leq a \leq b + \epsilon \qquad (1.1)$$

where ϵ represents a small positive number.

Now it should be clear to those who understood the reason for the operational definition of a quantity that the expression "the true value a," where a represents a real number does not have any meaning in physics, and therefore is avoided.[39]

A more adequate wording follows.[40] A measuring device does not supply *one* real number a, but a continuous *interval* of numbers, of which $b - \epsilon$ and $b + \epsilon$ are the upper and lower bounds.[41] Only conventionally can one say that the true measure a can be anywhere within this interval and that ϵ stands for the error of the measurement. Actually this only means that if the measurement is repeated with an instrument that has a much smaller ϵ, the new interval can fall anywhere inside the old interval.[42]

If this does not happen, we simply say that the new instrument defines a different physical quantity. Here and in what follows, we shall omit the *systematic errors*. Strictly speaking, these represent merely errors in the application of the established operational rules (generally due to mistaken construction, inaccurate setting or improper use of instruments).

Using more appropriate language, we should state that ϵ does not stand for the error, but for the *precision* of the method of measurement or of the apparatus. However, once the concept has been understood, one can use the term error, as in fact physicists often do. Sometimes, conventionally speaking, we say that the value of the quantity measured is a, within precision ϵ (or 2ϵ). Then we usually write:

$$measure = a \pm \epsilon \qquad (1.2)$$

We notice that the specification of ϵ should *always* either *explicitly* or *implicitly* accompany the result of a measurement. We can dispense with the explicit specification when (as in most cases) the implicit one is perfectly clear. For instance, when we say that the distance between two places is 342 km, we generally mean that it is 342.5 ± 0.5 km, or 342 ± 0.5 (according to conventions).

It is usual to stipulate that when measurement gives the result $a \pm \epsilon$, we can also write $a \pm \epsilon'$ with $\epsilon' > \epsilon$.[43] The result $a \pm \epsilon$ gives more information than $a \pm \epsilon'$, because the former entails the latter, and not vice versa. Hence it is convenient to take the smallest value for ϵ among those allowed by the available instruments. Therefore between two measuring apparatuses, we shall, as a rule, prefer the one with the smaller ϵ.

To every measurement procedure, when the apparatus used to carry it out is specified, there corresponds a value of ϵ – sometimes predictable, sometimes experimentally determined. Therefore the classes that are schematically represented in Figure 1.3 also depend on ϵ. Each one contains all the procedures yielding identical results, within a prefixed ϵ. If ϵ changes, then the content of the class can usually change, too. We shall discuss this kind of consideration again in §1.11.

A case that seems to be an exception to what has been said so far is the one when the measurement consists of *counting*. Suppose, for example, that we are interested in knowing how many particles are contained in a given volume of space. The result is a whole number, apparently *exact*; for an integer is also a *real* number. When talking about elementary particles, we shall see (§4.22) that even in this case the situation is not so simple as it appears. However, in classical physics one can assume that the result of a count is a whole number with $\epsilon = 0$.

Let us now consider a very common kind of measurement, which has become of paramount importance, especially in modern physics. This is when one can give (or wants to specify) only *the order of magnitude* of a physical quantity. A few examples may help to explain this concept. The numbers 3,120,000; 4,750; 0.08 can be written as 3.12×10^6, 4.75×10^3, 8×10^{-2}, respectively; their orders of magnitude are 10^6, 10^3, 10^{-2}. In general, a number is represented by a number between 1 and 10 multiplied by a power of 10. The important thing, and what measures the order of magnitude, is the *exponent* that is given to 10. It is therefore, by *convention*, an integer. One can say, then, that *a physical quantity has the order of magnitude of the number that measures it* (with a convenient unit).

Of course, the measurement would be more precise if one gave the entire logarithm to the base 10 of the number, instead of only the integral part of it. The fact is that (1) sometimes one does not know how to measure more than the order of magnitude and (2) sometimes one does not want to specify more than the order of magnitude.

The latter case is very interesting and deserves a little reflection. How is it that although we have a certain amount of information, we choose to give less? This is largely because nature itself invites us to do so; the more we know about the structure of the physical world, the more we recognize the importance of the order of magnitude. We shall see this in detail in subsequent chapters. The universe appears to be structured on a number of *quantitative levels*, which turn out to be also *qualitatively* different from one another (e.g., universe, systems of galaxies, galaxies, solar systems, stars and planets, molecules, atoms, nuclei, elementary particles). What distinguishes one level from another is the order of magnitude. Perhaps we can say that this fact justifies Hegel's intuition, taken up by Engels, according to which *quantity* in its gradual variation transforms (at a *nodal* point) into *quality*.

Nevertheless, the choice of 10 and its powers is contingent and largely depends on the fact that we have ten fingers. If we had twelve, we would probably have a duodecimal numeration and the base would be 12.[44]

Finally, consider the case of a variable base that tends to infinity (or to zero). If a function $f(n)$ tends to infinity when n tends to infinity and (from a certain value of n onward) $n^{k-\epsilon} < |f(n)| < n^{k+\epsilon}$, k being a positive constant and ϵ a positive number, however small, one says that $f(n)$ is of the order of n^k, and writes $f(n) = O(n^k)$. Likewise, if $f(n)$ tends to zero for $n \to \infty$ so as to have $1/n^{k-\epsilon} > |f(n)| > 1/n^{k+\epsilon}$, one says that $f(n)$ is of the order of $1/n^k$ and writes $f(n) = O(1/n^k)$.

1.10. The limits of the validity of a physical law

We are now able to introduce a concept that is an essential part of physics. Although most likely understood by Galileo, it is only during recent times that its central and necessary role has become clear. We have already commented on the quantitative aspect of the method. A typical law of physics consists of a mathematical relation, constantly found to exist between the measures of the various quantities that take part in a phenomenon. As an example, let us examine the simplest kind of law encountered in physics. Let A and B represent two quantities involved in a phenomenon, and a and b their respective measures. If one finds by experiment that the numbers a, b are constantly tied by the relation

$$a = b \tag{1.3}$$

one can say that this equation embodies a *physical law*.

Note, incidentally, that such an example is neither as unusual nor as trivial as it would seem. For it includes, as particular cases, all the fundamental *conservation* laws of physics, which will be dealt with later. In a conservation law, A and B represent one and the same quantity, before and after a certain phenomenon has occurred. For instance, one can think of mass, before and after a chemical reaction has taken place. At any rate, the following discussion, possibly with simple modifications, holds for all the laws of physics, even those of a very different form.

We shall now show that the traditional equation (1.3) is fairly naïve and misleading. For it behooves us to ask with what procedure can it be established that equation (1.3) is valid.

Evidently, we must begin by carrying out the measuring operations of A and B. We know, however, that far from yielding two real numbers a and b, they always supply us with two *intervals* $a' \pm \epsilon_a, b' \pm \epsilon_b$, where ϵ_a, ϵ_b represent the precisions of the instruments used. As for the values a' and b', they will virtually never be found exactly equal. Accordingly, the physicist does not seek to verify whether $a' = b'$, but examines the relative positions of the intervals $a' \pm \epsilon_a$ and $b' \pm \epsilon_b$.

Let us consider an interval $\epsilon = |b' - a'| + \epsilon_a + \epsilon_b$ (Fig. 1.4) representing, as is evident, the minimum interval that includes both original intervals. The result of the measurement is consistent with the equation

$$a = b \pm \epsilon \tag{1.4}$$

Hence we can say that a is equal to b, within the precision ϵ.[45]

Now let us clarify some doubt that could arise in observing Figure 1.4, where the two intervals have been drawn so as not to overlap. To be sure, from the results of measurement one can draw a more precise law than equation (1.4). For example, in the case of Figure 1.4, one can put $k = b' - a'$, $\epsilon = \epsilon_a + \epsilon_b$, and write the new law as $a = b - k \pm \epsilon$. This is more precise than equation (1.4) because its ϵ is smaller. But this does not invalidate the correctness of equation (1.4).[46] As a rule, the physicist selects the most precise law allowed by his measurements, so that the two intervals of Figure 1.4 are overlapping. But it is not necessarily so in all cases.

It will be stressed, once and for all, that a relation of the type as equation (1.4) is the legitimate form that can be taken by a physical law, whereas equation (1.3), which expresses the equality of two real numbers, the results of two measurements, cannot have any meaning in physics.[47]

The specification of the *limits of validity* of the law, or of the precision ϵ with which it has been verified represents an integral and essential part of the physical method. Strictly speaking, in no serious paper on

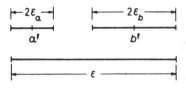

Figure 1.4

physics should a law such as equation (1.3) be stated without an indication of the limits of its validity. In practice, such an indication can be omitted when (1) the law is well known and everyone is acquainted with the limits within which it has been verified or (2) the law is new but those who are familiar with the instruments used can readily figure out what degree of precision has been reached by the experimenter.

The assimilation of these concepts can effectuate a profound change in attitude toward physics for many people. One must overcome the traditional misconception whereby physics is closely associated with mathematics in its *method*, instead of being simply associated with it as a *user*. In many cases, faulty teaching in high school is responsible for such misconceptions.

As a result of this misguided attitude, one could say that the laws of physics are never *exact* or that they are never *sure*. Referring to equation (1.3), which is not part of physics is a mistake. Of course, a law of that kind can never be exactly verified. But the physical equation (1.4) represents, instead, an exact and exactly verifiable statement. It is not in the least an unreliable relation. If deduced by a correct physical procedure, it expresses a truth. The *proposition* synthesized by equation (1.4) states precisely that *every time measurements are carried out with a given apparatus, two numbers a, b are found which differ by less than* ϵ. It is a *true* sentence, in the most strict and up-to-date sense of the word, as clarified by A. Tarski (see, e.g., Tarski, 1935, 1944, or Dalla Chiara, 1974, p. 65). It is difficult to see what is unsatisfactory, in an epistemological sense, in this statement.[48]

Also, it should be understood that when a physicist states the most exact law that his measurements will allow him to make, he is implicitly declaring that he absolutely does not want to affirm what the outcome would be if the measurements were made with better precision, that is, with $\epsilon' < \epsilon$. Therefore he claims certainty within the limits stated, but does not have to commit himself outside these limits.

Of course it can happen (and often has happened) that an improvement in accuracy, made possible by the progress of experimental equipment, may suggest abandoning an already established and universally recognized law in favor of a new one. Let us take as an example our very simple equation (1.4). Let us suppose that it represents the most accurate law warranted by the means at our disposal. Perhaps when the experimental equipment becomes more refined and allows us to reach a better precision $\epsilon' < \epsilon$, we shall find that $a = b \pm \epsilon'$ is not correct. We are thus forced to seek a new relation between a and b which is verified within ϵ'. For example, suppose we find

$$a = f(b) \pm \epsilon \qquad (1.5)$$

where $f(b)$ is a suitable function of b.[49]

In the past when we met this situation, we said that the law $a = b$ was false or had been disproved, whereas the law $a = f(b)$ was probably true. It is unfortunate that even today many people take this attitude, which does not make much sense. The law $a = b$ could never have been true, because as we have already argued, it is void of physical meaning and the same, of course, can be said about the law $a = f(b)$. The law $a = b \pm \epsilon$ continues also to be true after the experimental equipment has been refined. No progress in physics could disprove that a equals b within the precision ϵ, as it could never disprove that a turns out equal to $f(b)$ within ϵ'.

Often the name *revolution* is given to the most significant progress in physics, meaning that, with this, one renounces a mistaken past, so as to embrace a new and at least possible truth. It is essentially a misconstruction, grounded on the naïve assumption that laws "without ϵ," such as equation (1.3), can have any meaning whatsoever in physics and therefore can represent a truth.

Incidentally, it is curious to notice that even those who take this attitude, are often compelled to go back on their word. For example, let us take the case of an engineer who is convinced that relativity and quantum mechanics have exposed the nonvalidity of Newton's mechanics. When designing the mechanical parts of a car, the engineer will nonetheless use Newtonian mechanics, because he is really convinced that for the ϵ's occurring in his case, Newtonian mechanics is perfectly valid.

An argument of this kind is likely to cause the physicist to be accused of having an *instrumental* view of science. But a little reflection on what we have just mentioned is sufficient to realize how superficial this accusation is. A statement such as that contained in (1.4) has *cognitive* value![50]

It is not only a question of agreeing with a formal criterion of truth, like the one already mentioned; it is something more conceptual. Take, for example, the law discovered by A. Lavoisier, according to which the total mass of reagents, before and after a chemical reaction, is equal. Today relativity teaches us that this is not quite true. If we take measurements with extreme accuracy, then we should find that in an exothermic reaction a small fraction of the mass is lost. Must we therefore affirm that Lavoisier's law is false or that it has lost its conceptual and fundamental importance? If we decide on this, then we are throwing away very important information on nature's behavior! It would be ridiculous to teach physics and chemistry without using Lavoisier's law. The fact that physics can only supply statements of this type (i.e., with ϵ), may dissatisfy those who, following the useless fetish of real numbers in physics, are prepared to renounce the only results of intersubjective value. However, it does not displease the physicist, who finds these results, for he has given up such a fetish (without regret).

To complete the treatment of the limits of validity of the laws of physics, we must now add an essential element that has so far been omitted for the sake of simplicity. We have said that equation (1.4) refers to a given phenomenon in which the quantities A and B are involved. To be really useful, however, it must apply to a whole class P of phenomena. The specification of this class is clearly necessary for the assignment of the limits of validity.

Usually, P is specified by both describing qualitatively the type of phenomena and assigning the upper and lower bounds for the values a and b, as well as for all parameters involved. For instance, one will say that in order for nonrelativistic formulas to be valid for given values of ϵ, the speeds involved must not exceed a certain limit (see §2.19 and following sections).

Regarding class P, one could repeat word for word what has been said about ϵ. Sometimes one is able to perform experiments in a novel situation, not included in the original class P. If the law is still found to be valid, then one simply says that class P turns out to be larger than was first thought and also includes the new experiment. Otherwise the new experiment is not part of class P; it exceeds the limits of validity of the law so that a new law must be sought for it. Once the new law has been found, it may happen that all the phenomena in class P satisfy the law. In this case we say that the new law has a class of validity $P' \supset P$, larger than that of the old law. Thus it should be clear that it is unwise to assert that the new experiment has *refuted* or *falsified* the old law. It should be said, instead, that the new experiment falls outside the domain of validity of the old law, which still remains perfectly valid for class P.

Sometimes, lacking better information, one provisionally assumes that a law, verified inside class P, also is valid outside class P. This is called an *extrapolation*. Extrapolation often has considerable heuristic usefulness. But every physicist knows that it is not a rigorous procedure. Physicists are sometimes accused of extrapolating their laws to fields in which these laws are not applicable. This charge is based on a lack of information. Extrapolation may sometimes be used carelessly by writers, philosophers, historians, politicians, and so on; not by physicists (at least not by good ones!).

1.11. The procedure of classical physics

We can now outline a general scheme of the procedure adopted in classical (nonquantum) physics for the formulation of the laws.

Suppose that we are investigating a class P of phenomena, involving the quantities A_1, A_2, . . ., A_n, which, by convention, will simply be denoted by A_n. Let their measures have precisions ϵ_1, ϵ_2, . . ., ϵ_n, or just ϵ_n for short.

It may be the case that within those precisions, the values a_n of the A_n are found to satisfy the relation

$$f_1(a_n) = 0 \qquad\qquad (1.6)$$

where $f_1(a_n)$ denotes a function of all variables a_n.[51] The expression "within those precisions" means that it is possible to find for each A_n a real number a_n', included between $a_n - \epsilon_n$ and $a_n + \epsilon_n$, such that $f_1(a_n')$ turns out to be *exactly* equal to zero. Hence strictly speaking, equation (1.6) is true only in a conventional sense, and not in the usual mathematical sense. The physicist will say that the *law* governing the phenomena considered is represented by equation (1.6), within the precisions ϵ_n.

In order to visualize the procedure, let us take the case of two quantities A_1, A_2. In a rectangular diagram, we plot the values a_1 of A_1 as abscissas (i.e., on the x axis) and the corresponding values a_2 of A_2 as ordinates (i.e., on the y axis) (Fig. 1.5). From each point thus obtained we draw the horizontal lines $\pm\epsilon_1$ and the vertical lines $\pm\epsilon_2$ so as to obtain a cross. If the experimental points are sufficiently closely spaced, all the crosses are contained within a strip. The points whose coordinates a_1', a_2' satisfy the equation $f_1(a_1', a_2') = 0$ fall within this strip; in other words, the curve represented by $f_1(a_1, a_2) = 0$ must be contained within the strip.[52]

Obviously, f_1 is not necessarily the only function satisfying the desired conditions. Generally, many other functions can do as well.[53] There exists an entire set G of functions, each one suitable to the purpose. This set will evidently depend on both the class P of the phenomena considered and the values ϵ_n of the precisions with which the A_n are measured. The set will therefore be denoted by $G(P, \epsilon_n)$. The

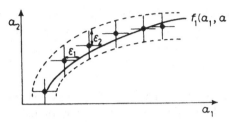

Figure 1.5

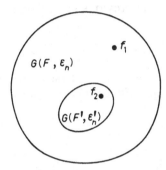

Figure 1.6

function f_1 is an element of G, which symbolically is expressed by

$$f_1(a_n) \in G(P, \epsilon_n) \tag{1.7}$$

In Figure 1.6 the extension of set G is represented by a Venn diagram.

The selection of f_1 out of all the functions contained in G, has a practical value and is not of theoretical interest. One usually seeks to choose the simplest,[54] most convenient, or even most elegant form. Such criteria are perfectly justified; what is unjustified is the relation that a number of physicists assume exist between the criteria and the truth content of the formula. Sometimes physicists say, with humor, that the simpler a formula is, the nearer it comes to the truth. But they should know very well that this is not so.

Physical truth is represented by the complete equation (1.7), that is, by the fact that f_1 belongs to the set of functions contained within the strip.[55]

Suppose now that we refine our measuring equipment so as to obtain the new precisions ϵ_n', such that

$$\epsilon_n' \leq \epsilon_n \tag{1.8}$$

Further, we either leave unchanged the class of phenomena considered or widen it by adding new phenomena. In symbols we consider a new class P', such that

$$P \subseteq P' \qquad (1.9)$$

We say that P is a *subclass* of P'. The set of the possible functions becomes $G(P', \epsilon_n')$. It is clear that this new set cannot be *larger* or contain more elements than the old one. All that can happen is that a number of functions that were all right before, now cease to be acceptable. Hence we have a *subset* of the previous set. In symbols we write

$$G(P', \epsilon_n') \subseteq G(P, \epsilon_n) \qquad (1.10)$$

In agreement with the criteria just discussed, the physicist will select a function $f_2(a_n)$ out of the new set and will say that the physical law is represented by

$$f_2(a_n) = 0 \qquad (1.11)$$

within the precisions ϵ_n'. We also write

$$f_2(a_n) \in G(P', \epsilon_n') \qquad (1.12)$$

It might be the case that $f_1 = f_2$ but, generally, one is forced to choose a new function that is more complicated than the previous one.

It is now evident that if we continue on to more precision, ϵ_n'', and to a larger class P'', the set of the possible functions will shrink further, and we then have

$$G(P'', \epsilon_n'') \subseteq G(P', \epsilon_n') \qquad (1.13)$$

From this new set the physicist will select a new function f_3, such that

$$f_3(a_n) \in G(P'', \epsilon_n'') \qquad (1.14)$$

and so on.

Many students still entertain the idea that once class P is assigned and maintained constant:

1. The limiting set $G_0(P)$ does necessarily exist such that

$$\lim_{\epsilon_n \to 0} G(P, \epsilon_n) = G_0(P) \qquad (1.15)$$

2. The limiting set G_0 contains one single element, that is, only a well-determined function $f(a_n)$.

All this is expressed by saying that if the measures were exact, the only true law would result.

$$f(a_n) = 0 \tag{1.16}$$

However, considering $\epsilon_n \to 0$ is meaningless, as will become clear when we study quantum mechanics (Chap. 4). At any rate, it should be evident by now that we do not have to wait for centuries to pass, when $\epsilon_n = 0$ will be attained, before asserting a number of true propositions. The statement expressed by equation (1.7) is true, the statement expressed by equation (1.12) is true, and so on. However, because each one of these statements represents a truth, we should not use the term *approximate law*. If this expression were only meant to remind us that there is an ϵ, its intent would be trivial; ϵ is an essential part of any physical statement.

What may happen instead (and has often happened), is that as one goes on with the succession $\epsilon_n \geq \epsilon_n' \geq \epsilon_n'' \cdots$, one arrives at some $\bar{\epsilon}_n \neq 0$ for which it is impossible to find any function $f(a_n)$ in accord with experience. One must recognize that the set $G(P, \epsilon_n)$ is *empty* or, in symbols

$$G(P, \bar{\epsilon}_n) = \emptyset \tag{1.17}$$

There are then two possible ways out:

1. One should realize that for $\epsilon_n \leq \bar{\epsilon}_n$ the process also depends on some other quantities B_n, upon which it did not depend for $\epsilon_n > \bar{\epsilon}_n$.[56] In this case, all one has to do is to take the B_n into account, and to look for a set of suitable functions $f(a_n, b_n)$ that also depend on them. This procedure is successful in a large number of cases.

2. One should assume that there are no functions $f(a_n, b_n)$, depending on both the old and the new quantities A_n, B_n, and not clashing with experimental data. In this case one usually says that for $\epsilon_n < \bar{\epsilon}_n$, the process no longer occurs in a *deterministic* way; hence one has to look for a new kind of law (see quantum mechanics, Chapter 4).

The generalized operational definitions of physical quantities and the formulations of physical laws are strictly interconnected, and give rise to a circularity, to which we now turn.

It is clear that the formation of the classes of equivalent operations representing a given quantity (Fig. 1.3), is based on a number of physical laws, already known. Saying that operations A and B give one and the same result within ϵ, amounts to asserting a law such as equation (1.4). Because on refining the measurements and decreasing ϵ a situation may arise where equation (1.4) is no longer valid, we have to accept the possibility that any two adjacent classes of Figure 1.3 may shrink and cease to have a nonvanishing intersection. What was one and the same quantity, may therefore split into two different quantities. For example, today we can measure a mass, either by means of a spring balance or by evaluating the acceleration conferred to it by a given force; in both cases we obtain the same result. Should we one day find different results, we would be forced to recognize that *gravitational* mass and *inertial* mass are different quantities.

Alternatively, the classes of equivalent procedures of measurement may become wider when we start to consider some group of operations that had not been investigated before; thus two quantities that were previously different, may become identical! For this to happen we have but to discover a novel law of the type such as equation (1.4).

We can give two important examples of this occurrence. Einstein's law $E_0 = mc^2$ (§2.24) substantially states that the energy of a system can be measured by a balance or, alternatively, that a mass can be measured by the energy yield when it disappears. Should not, as a result, energy and mass represent one and the same physical quantity? Let us take Planck's law $E = h\nu$ (§4.5), stating that frequency can be evaluated by measuring an energy, and vice versa. Why do not energy and frequency represent one and the same quantity? To be sure, there is no compelling reason, and in both cases we usually prefer to maintain the distinction between the two quantities; we then discuss the physical laws that bind them together.[57] It is a legitimate convention, but no more than a convention.

In conclusion, we can state the following. From a rigorous point of view, each group of operations of measurement defines its own physical quantity, different from all others. For practical reasons, however, it is expedient to make use of a number of laws such as equation (1.4), in order to group many different quantities under the same name. But we should not forget that this can be done only in a well-determined historical period when certain precisions ϵ have been attained and certain laws are known. In any case, the choice of which quantities are identical and which are different remains largely arbitrary and conventional, but scientists tend to be conservative.

However, the reader should not conclude that the whole of physics is conventional. It is only a part of the *language* of physics that assumes an arbitrary choice. Any language is, by nature, conventional, but the content expressed by it may not in the least be conventional. Conventionalism will be discussed later in §2.29.

1.12. The mathematical functions used in physics

We saw that out of the set of the possible functions describing a physical law, the physicist selects the one that seems to fit the purpose best.

We now want to illustrate the main criteria guiding this selection in the form generally taken by the laws of physics.

Let us recall that a function $f(a_1, a_2, \ldots, a_n)$ of the variables a_1, $a_2, \ldots, a_n$ is assigned when one and only one value of f is made to correspond to any n-ple of values of the variables (in a given domain). Note that this general definition has nothing to do with a possible rule enabling us to compute the value of f, starting from the values of a_1, $a_2, \ldots, a_n$. What we can derive from experience is a correspondence between the values a_n and the value of f. When f depends on only one variable, such correspondence can be visualized by plotting it on a graph (within the accuracy allowed by this device).

However, since Galileo, we have been accustomed to think that a function should be represented by a rule of computation, that is, by a set of operations to be carried out on the a_n, in order to get the value of f. We want to stress once more that this is necessary only because the human mind does not have better means to master the function; but we believe that such computations are not inherent in nature.[58]

The simplest operations that we can carry out are addition and multiplication, whereby we arrive at the concept of *polynomial*. Let us discuss this further with an example.

In the case of two variables a_1, a_2, the expression

$$p = k_1 a_1 + k_2 a_1^2 a_2 + k_3 a_1 a_2^3 \tag{1.18}$$

represents a polynomial. Each addendum, or *term*, is the product of a constant coefficient $(k_1, k_2, \ldots,)$ times some powers of the variables. By adding the exponents of the variables appearing in a term, one obtains the *degree* of that term. Thus in equation (1.18) the first term is of the first degree (or *linear*), the second term is of the third degree, and the last term is of the fourth degree. The highest degree appearing in a polynomial is called the *degree of the polynomial*.

The role of polynomials is important in physics. There is a theorem, due to C. Weierstrass, which is valid for a very wide class of functions, virtually for all functions encountered in physics. In simple words, the theorem states that *in a given domain every function can be approximated, as closely as desired, by a polynomial of sufficiently high degree*. This shows that once the precisions ϵ_n are assigned, any physical law, taken in a given interval, can be represented by a polynomial. Indeed, polynomials are used for a great number of elementary laws. Precisely, one resorts to a polynomial every time the number of its terms can be limited (say, 1 to 3). Should the polynomial turn out, instead, to be too long and cumbersome, one would seek a more synthetic formula by using trigonometric functions, exponentials, logarithms, and so on.[59]

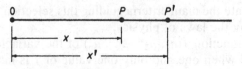

Figure 1.7

The reader acquainted with infinitesimal calculus may be surprised by our implied assumption that all the laws of physics are represented by functions, as in equation (1.6).

We could discuss this subject at length but it would take us far afield. We prefer to show by a concrete example that even when the symbols of differential calculus are used, the *elementary law* can still be expressed in the form of equation (1.6). We shall start by illustrating what we mean by an elementary law.

Often, two quantities of the set A_n are really one and the same quantity A, measured at two different places (or times). Let the two places be represented, say, by points P, P' (Fig. 1.7), specified by their distances x, x' (*abscissas*) to a given point O (*origin*) taken on the same line. Let us measure A both at P and P', with the results a and a', respectively. Experience suggests quite often that when P and P' are not very far apart, the difference $a' - a$ should be proportional to $x' - x$; the proportionality constant depends on both the location and the other relevant quantities. We therefore have[60]

$$a' - a = f(\bar{x}, a_1, a_2, \ldots, a_h) (x' - x) \qquad (1.19)$$

where $\bar{x}$ characterizes the location and $a_1, a_2, \ldots, a_h$ stand for the other quantities involved (to be again denoted by a_n for short).

For the sake of simplicity, let us assume that the quantities A_n, whose values appear in $f(\bar{x}\, a_n)$, do not depend on x, that is, on the place where they are measured. There remains, however, this problem: How should we choose the value of $\bar{x}$ to be inserted in f? Does it represent the abscissa at P, at P', or at an intermediate point?

It is usual to solve this by taking the distance PP', or the difference $\Delta x = x' - x$ so small, that $f(\bar{x}, a_n)$ varies only unappreciably in passing from $\bar{x} = x$ to $\bar{x} = x'$; hence f can be considered as constant within the whole interval Δx.[61] For reference, let us assume that $\bar{x}$ coincides with the abscissa x at point P. Indicating by Δa the difference $a' - a$, we can rewrite equation (1.19) as

$$\Delta a = f(x, a_n)\Delta x \qquad (1.20)$$

This is what a physicist would call the elementary law of the phenomenon.[62]

We can now ask ourselves how small Δa must be in order to be considered sufficiently small. Obviously, this can change from time to time, depending on the precisions of all the measurements involved. The lack of a unique interpretation for equation (1.20) is inconvenient. Fortunately, it can be circumvented by applying the concepts and the notations of differential calculus. First, let us divide both sides of equation (1.20) by Δx, and write

$$\frac{\Delta a}{\Delta x} = f(x, a_n) \tag{1.21}$$

Second, let us forget for a moment the physical interpretation and follow the reasoning of the mathematician. A mathematician, when confronted with an expression $\Delta a/\Delta x$, where the increment Δa of a function of x is divided by the corresponding increment Δx, habitually inquires whether this ratio tends to a well-determined *limit* when Δx (and consequently Δa) tends to vanish. If this is the case, the limit is denoted by da/dx and is called the *derivative* of a with respect to x.[63] Thus the definition of derivative can read

$$\frac{da}{dx} = \lim_{\Delta x \to 0} \frac{\Delta a}{\Delta x} \tag{1.22}$$

where the notation should be self-explaining.

By replacing the left side of equation (1.21) with the derivative, one conventionally writes

$$\frac{da}{dx} = f(x, a_n) \tag{1.23}$$

With a nonrigorous but suggestive phrase, one usually says that the *small* increments Δa, Δx have turned into the *infinitesimal* increments da, dx.

What meaning does this mathematical notation have in physics? How can one talk about $\Delta x \to 0$, whereas the accuracy of a measurement is always finite?

Equation (1.23) merely represents a symbolic way of saying that because the smallest Δa and Δx that can be measured renders equation (1.20) valid, we conventionally assume that it is also valid for any smaller values of the increments. According to the present state of the art, this assumption is not verifiable; but at any rate, it is not contradicted by experience. On the other hand, it is very convenient, for the assumption enables us to utilize a powerful mathematical tool (i.e., differential calculus). However, one should be careful not to forget that this is merely the result of a convention.

Suppose now that we want to evaluate the difference $a' - a$, in a case when P and P' are not very close to each other, or when $x' - x$ is not very small. To this end, let us divide the interval PP' into a number of very small intervals, by means of the intermediate points $P^1, P^2, \ldots, P^k$, having the abscissas $x^1, x^2, \ldots, x^k$.[64] It is assumed that $x < x^1 < x^2 < \cdots < x^k < x'$ and that all the intervals $\Delta x^i = x^{i+1} - x^i$ are sufficiently small for the elementary equation (1.20) to exist.

Omitting a_n for simplicity, and putting $\Delta x_0 = x^1 - x$, we can write successively,

$$a_1 - a = f(x)\Delta x^0$$
$$a^2 - a^1 = f(x^1)\Delta x^1$$
$$a^3 - a^2 = f(x^2)\Delta x^2 \tag{1.24}$$
$$a' - a^k = f(x^k)\Delta x^k$$

By adding all left sides and all right sides, we get

$$a' - a = \sum_{i=0}^{k} f(x^i)\Delta x^i \tag{1.25}$$

where the symbol on the right side indicates the sum of all the terms that are obtained by putting $i = 0, i = 1, \ldots, i = k$.

This is as far as the physicist could go should he bother to specify the values Δx^i which, to him, are sufficiently small. However, he can, also in this case, make use of a convenient mathematical notation

$$a' - a = \int_{x}^{x'} f(x)\, dx \tag{1.26}$$

which is termed the *integral* of $f(x)$ from x to x'. The integral represents the limiting value of the sum equation (1.25), when all Δx^i become smaller and smaller and tend to vanish (consequently, their number tends to infinity).

Applying equation (1.26) has again a conventional meaning for the physicist. But it is very useful because mathematicians have elaborated on a number of methods for evaluating integrals and for using them.

Of special interest to the physicist are *differential equations*, which bind together an unknown function and its derivatives. For example, if $f(x)$ is known, equation (1.23) represents a simple differential equation for a, when a is conceived as an unknown function of x.

There are several methods for solving a differential equation, that is, for finding the unknown function for assigned *initial* and *boundary* conditions. When one does not succeed by a direct approach, one can resort to numerical methods or a computer.

The infinitesimal calculus is an indispensable and familiar tool to the physicist, since the time that I. Newton and G. Leibnitz gave birth to it with a view to solving the mathematical problems of physics.

However, one should be careful not to jump at the wrong conclusions. Calculus serves the purposes of physics so well, that one can easily be led to believe that the operations of *limit*, equation (1.22) or (1.26), may have a physical, instead of a purely mathematical, meaning.[65] Once the integral is evaluated or the differential equation is solved, the physicist must remember that he has started from equation (1.20) (or an analogous equation), which is only verified within a certain precision ϵ; an easy method, which he must apply, enables him to assess how much the original precision will influence the ϵ of the result. Woe to him, who is persuaded by mathematics to believe that one can take more from calculus than what one puts into it!

Bishop Berkeley, who started a vigorous argument against infinitesimals, was probably the first to notice the trap hidden behind the excellent performance of the new calculus. In mathematics, things were satisfactorily[66] straightened out in the last century as well as in ours.[67] However, the situation is different in physics. Quite often the suggestion has been made that some of the great difficulties of contemporary physics may be attributed to the unwarranted assumption that there is a physical continuum. A form of *graininess* or *discreteness* has been discovered for mass, energy, angular momentum, and other quantities, to be dealt with in microphysics. There has also been some discussion on elementary interval of length or of time, but no firm result has been reached as yet. At any rate, nothing authorizes us to ascribe a meaning to the physical continuum.

1.13. The units of measurement

In the previous sections the quantitative approach was introduced as an essential feature of physics. One cannot adequately deal with physical quantities without being able to attach numbers to them by a process of measurement.

Measurements can be of different types and consist of many diverse sets of operations, as we shall show in the following examples. Adhering, however, to our commitment of illustrating only essential concepts and of putting aside technical details, let us consider, by way of example, the measurement of a length. According to a well-known procedure, the distance to be measured is compared with a rod by counting how many (integral or fractional) times the latter is contained in the former.

If one examines this simple procedure critically, one soon realizes that it can give rise to many difficulties and to delicate questions.[68]

However, we shall not pursue this discussion here. What we want to stress is that no one doubts that it is possible to affirm with certainty that, say, this table is 181 ± 1 cm wide.

What is the *rod* we are talking about? Clearly, in this as in other similar cases,[69] one takes a quantity of the same kind as that to be measured, embodied, as it were, by a material object. In the case of length the rod is the material object, and its longitudinal size represents the *unit* of length. With a convenient, if somewhat inaccurate, expression, it is customary to say that the rod *is* the unit of length.

Choosing the units of measurement represents a delicate and rather involved problem. Many international committees have worked on it for more than a century. We shall confine ourselves to a few remarks.

First of all, one must emphasize the complete freedom and arbitrariness with which the units can be established. Until one gains a deeper insight, especially in microphysics, one gets the impression that there do not exist in nature any privileged objects, noncontingent and not subject to change, that are particularly suitable as standard units. This freedom has resulted in virtually all civilized peoples selecting *human-size* units. Think, for instance, of units of length such as *foot*, *inch*, *step*, and so forth. With this choice, the properties of the everyday objects that we see and touch turn out to be measured by not very long numbers, that is, by numbers with not too many decimal figures. These are the orders of magnitude of things at human level.

However, there is at this stage a troublesome shortcoming. *Anthropometric* units are not precise, uniquely determined, and reproducible; as a result, they have given rise at different times and places to an incredible variety of diverse systems.

The beginning of unification is to be credited to the French Revolution. A panel formed by J. C. Borda, A. Condorcet, G. L. Lagrange, P. S. Laplace, and G. Monge was entrusted with the study in 1790. Their work resulted in the *metric system*, which after several ups and downs, was adopted by a considerable number of countries.

In the English-speaking countries, where the revolution did not reach and where industrialization was already in progress, measuring systems remained anchored to less convenient and less rational units. Only recently has the door been opened, and these countries are now gradually switching to the metric system.

The differences in the systems of units concern only everyday life and practical applications, but has long ceased to plague physics. In physics, the metric system has been universally adopted. The latest codification of this system has given rise to the *international system* (Système Internationale, or SI) adopted in 1960 (see, e.g., Chiswell and Grigg, 1971).

The fundamental units of mechanics are the *meter* (m), the *kilogram* (kg) and the *second* (s), giving rise to the MKS system. Quite often, especially in microphysics, one takes as fundamental units a submultiple of the meter and a submultiple of the kilogram, adopting the system *centimeter* (cm), *gram* (g), *second* (s), or CGS.

In the rest of the book I shall mainly refer to the CGS system. Some people might frown at this choice and will argue that the MKS system is the one that is being universally adopted today. In a sense they are quite right. Nonetheless, there are very good reasons for my choice. First of all, if one wants to depict physics as it is really being developed now, one must recognize the fact that most physicists working on fundamental problems (microphysics, relativity) still use CGS and are very reluctant to switch over to MKS. Is this based merely on intellectual laziness? I do not think so. Physicists who live up to a high professional standard are usually willing to adopt what is rational and useful. The question is a little more involved.

As long as one remains confined to classical mechanics, an argument between supporters of the two systems is futile. Multiplying by a power of 10 is not a frightening task. But the MKS system has been selected mainly because it fits very well with the former practical units of electricity. To this end, it is completed by a fourth unit, the *ampere* and thus becomes the MKSA system. For practical applications it is the only reasonable system to be adopted today. It would be nonsense, say, for an electrical engineer to use the CGS system. However, I believe that many physicists are displeased about the privilege granted electricity. They may argue that the electromagnetic interaction is just one of the four fundamental interactions known today (see §4.26). Presumably, they feel that neither the MKSA nor the CGS system are entirely satisfactory; accordingly, they are waiting for another "natural" system to be found, one more suitable for fundamental physics.

As previously noted, the selection of units and their standards has been largely arbitrary.[70] As a result of the progress of science, physicists have come to realize that it is possible in a number of ways to establish some *natural* standards, which prove to be stable and more reproducible than the previous human-size standards. Thus, for example, wavelengths and the frequencies of some spectral lines emitted by atoms are now known with amazing accuracy.

Today the standard of length is represented, in a sense, by the wavelength of a particular line emitted by the noble gas krypton (precisely by the Kr 80 isotope); for it has been stipulated that 1 m should equal 1,650,763.73 times that wavelength. Further, it has been established that 1 s should equal 9,192,631,770 times the period of vibration of the radiation emitted by the alkali metal cesium (precisely by the Cs 133 isotope) in a particular transition.

As to the mass, it might seem expedient to take as standard the mass of a given elementary particle, such as the *proton* or the *electron*. However, the accuracy with which these masses can be measured is still too poor to render such an operation advisable. Hence the kilogram remains anchored to the iridium–platinum standard, preserved by the Bureau Internationale des Poids et Mesures at Sèvres (France).

As a by-product of this search for natural units, we can derive a conclusion of some interest: human beings are definitely contingent entities, very far from anything that is fixed and constant in nature. We are either too large (e.g., our size and mass as compared to those of elementary particles) or too small (e.g., the speeds we can attain as compared with the speed of light). Is this contingency, perhaps, an essential feature of our singular nature?

At first, one might believe that there should be as many units as there are physical quantities and that all these units should be independent of one another. But this would entail considerable complications, for the number of possible physical quantities is extremely large (virtually infinite). Further, as we shall see in a moment, we would be forced to learn one particular number (possibly with many decimal places) for each law of physics. Consequently, it would be virtually impossible to remember the exact expression of many fundamental laws. Fortunately, most fundamental laws of physics are given a peculiar form which permits great simplification.

We said that in physics the main and simplest form a function can have (or can be given) is represented by a polynomial with only a very small number of terms. A polynomial with one single term (in the case of, say, two quantities A, B) would give us the law $k_1 ab = 0$; this could hardly be useful, for it simply entails that either a or b equal zero. Much more significant is a polynomial with two terms, giving rise to the relation

$$k_1 ab + k_2 cd = 0 \qquad (1.27)$$

where four quantities A, B, C, D are assumed to be involved. Two of them could be identical. If, for instance A is identical with B, one has to replace ab with a^2. We can divide both sides of equation (1.27) by k_1; thus by putting $-k_2/k_1 = k$, we get

$$a = k \frac{cd}{b} \qquad (1.28)$$

Apart from the number of quantities involved, which can vary, this is the actual form taken by virtually all the elementary laws of physics.

The constant factor or *coefficient* k depends, obviously, on the units adopted. To see how this occurs, let us assume the validity of equation (1.28) where a given value of k has been established. If we want, for example, to halve the unit for A, the value of a is doubled. Consequently, in order for the law to remain valid, the value of k must be doubled, too.

Hence all we can derive from the experimental investigation is that a is proportional to cd/b. The proportionality constant k depends on how we select the units of measurement; its value is in no way forced on us by nature. The reader is asked to clearly realize that the elementary laws of physics express proportionalities, not equalities.[71]

Let us suppose that when we proceed to establish the experimental equation (1.28), the units for the quantities A, B, C have already been selected, so we are only left the choice for D. We may make use of this opportunity and fix the unit for D in such a way as to render $k = 1$; thus the law, equation (1.28), takes the simple form

$$a = \frac{cd}{b} \tag{1.29}$$

The practical usefulness of this device is only too evident. Remembering equation (1.29) is easy, whereas in order to remember equation (1.28), one should have to memorize the numerical value of k (say, $k = 0.2461$). It would be a very troublesome task to perform for all the elementary laws of physics, quite aside from the fact that every time technical advances enable us to make more accurate measurements, we should have to revise the value of k or add more digits to its expression.

It is clear that in order to apply the preceding procedure, one must begin by establishing, in a more or less arbitrary way, the units for a number of quantities. This must be done before we can make use of the laws of physics, in order to help us choose suitable units for the other quantities and to eliminate the coefficients k in the laws such as equation (1.28).[72]

As already mentioned, in mechanics one chooses length, mass, and duration as starting quantities, with the (CGS) units, *centimeters*, *grams*, *seconds*. These units are called *fundamental*,[73] whereas those fixed in the same way as the unit for D in equation (1.29), so as to have $k = 1$, are called *derived*. A number of examples will be given as we proceed.

Unfortunately, such a simple and convenient procedure cannot be applied in every case, and one can immediately understand why. Once the three fundamental units have been fixed, a physical law connecting length, mass, and duration with a fourth quantity can be used to fix the unit for it. Two such laws can serve to fix the units for a fourth and a fifth quantity, and so on. In general, N physical laws can serve to fix the units for $N + 3$ quantities. What happens if we discover at this point that these same $N + 3$ quantities are connected by a further physical law, not included among the N laws already considered? There is nothing we can do: The last law will be written with a coefficient k whose value is forced on us by experience.

In this case physicists might say that k is a *universal constant*. This bombastic name is rather unfortunate. It may disguise the fact that *equalities* such as equation (1.29) are man-made and not inherent in the physical world. Experience can only furnish proportionalities; hence *every physical law has its own universal constant*! Using this name only for the k's of the laws discovered last can be misleading.

1.14. The dimensions of physical quantities

We shall now illustrate by a particularly simple example the tendency that we have to forget the conventional character of the approach and to consider as inherent in nature the man-made forms of some relations. Let us take the elementary case of the relation between length and surface.

In this case, of course, the relation could be discovered by pure mathematical speculation, rather than by experiment. However, this fact will be ignored because it is of little significance here. Proceeding in the manner of the physicist, we may perhaps cut out a number of similar plane figures (such as equilateral triangles or circles) from a piece of cardboard. By weighing these objects, we discover that (within the precision of the measurement) the surfaces are proportional to the second power of the values of corresponding lengths of the figures (such as sides of the triangles, radii of the circles). We then must write

$$s = kl^2 \tag{1.30}$$

where s stands for surface and l for length. The constant factor k depends on the units adopted, as well as on the shape of the figure and on the particular distance measured on it. We can eliminate k (i.e., make $k = 1$), but only for one particular shape.

It has been convenient to take as unit surface the *square* whose side has unit length. Hence for the surface of a square of side l, we obtain

$$s = l^2 \tag{1.31}$$

This choice, to be sure, is perfectly sensible and is suggested for many good reasons, but it is by no means necessary. Nothing would prevent us from using as unit surface the surface of a circle of unit radius, so as to make equation (1.31) valid for a circle of radius l.

At any rate if one chooses the square, the factor k disappears only when dealing with squares. For the circle, we must introduce $k = \pi$, for the equilateral triangle $k = \sqrt{3/4}$, and so on. Should we prefer the circle of unit radius, π would disappear for circles, but would reappear for squares, in that we should have $s = l^2/\pi$.

The concept that the usual form, equation (1.31), should be necessary arises, as a rule, during the first school years, owing to a misguided method of teaching. Unfortunately, the universally accepted terminology does not help in making critical reflections. Indeed, one is even led to confuse the square and the second power; *to square* is equivalent to raising to the second power. It is imperative to inform schoolchildren that had the unit surface been fixed equal to the circle of unit radius, one would have to discuss *circling* rather than *squaring*. Ironically, one can think that had Galileo lived before Euclid, and had he put his free-fall coefficient $\frac{1}{2}g$ equal to 1, the square of a number today might be called the *fall* of a number!

There is a further and very useful convention, which calls for a critical consideration and which may be illustrated by means of the elementary example, equation (1.31). Suppose we want equation (1.31) to remain valid, however we choose the unit of length. If we multiply this unit by 2 or by 3, the unit of surface must obviously be multiplied by 4 or by 9, respectively. Usually, we say that the dimension of surface equals the second power of length. An analogous convention will be made for all physical quantities that we shall have to encounter.

Some quantities turn out to be independent of the units and are called *adimensional*. Their measures are expressed by *pure numbers*. For example, the measure of an angle in *radians* (ratio of the arc to the radius) does not depend on the unit of length or on any other unit, as is evident. As a result, angles are adimensional quantities.

This convention about the dimensions is very useful when one wants to check the validity of an equation. Both sides of an equation representing a physical law must have the same dimensions. Otherwise the law instead of having general validity, would be verified only for a particular choice of the units, which is absurd. In this way one can often check by inspection, whether a given law is admissible or definitely incorrect. As an example, consider the assertion: "by multiplying the diagonal of a face of a cube by the cube diagonal, one obtains the volume of the cube"; this is certainly wrong, because it requires the second power of a length to equal the third power of a length.

Be careful! It is also easy in this case to be held back by words and to forget the conventionality of the language. Sometimes one says hastily: "Surface *is* the square of a length" or "Volume *is* the cube of a length." One should not interpret such expressions literally, thus becoming convinced that they imply something essentially inherent in nature.[74] Such a belief is false, as is evidenced by the fact that by suitably juggling conventions, one can confer to a given quantity any dimension one wants.

The word *convention* has often occurred in our discussion. Let us once more make it clear that our conventions concern language only. We stipulate conventions only where the physical world grants us this freedom. For instance, in the case of the adimensional constants, we do not enjoy this freedom. Actually, many scholars believe that the pure numbers that appear in microphysics should have profound significance, independent of conventions, which some day we shall be able to understand. Some theoretical physicists are even steadily working on this problem.[75]

1.15. Theories, hypotheses, models

The word "theory" is frequently used in physics, although its meaning is not unique and well established. In the following chapters the reader will find many illustrations of what is generally meant by a physical theory. But let us anticipate a few useful distinctions and clarify some misunderstandings.

The most exacting form of theory is an *axiomatic* theory, a concept taken from mathematical logic (see, e.g., Henkin, Suppes, Tarski, 1959; Sneed, 1971). Strictly speaking, one must set up a formal language, a number of primitive sentences, named *axioms*, from where one can derive other valid assertions of the theory by means of the *rules of deduction*. The terms of the language must be interpreted through some *rule of correspondence* with the objects and the physical phenomena; at least some of the derived assertions should be verified by experience.

However, physical theories are usually subject to much looser conditions.[76] Ordinary language is largely used, with the exception of a few technical terms that are rigorously specified. Axioms can take different forms. Some of them are given as physical laws of the general type, equation (1.6). The rules of deduction (unless stated to the contrary) are those of classical logic and of mathematics. Deduced propositions can also take the form of equation (1.6).

Experience sets the *domain of validity* of the theory, as will be seen more precisely later on. The theory is accepted by the physical community only after having ascertained that its validity domain is sufficiently wide, at least so as to include all known experiments. This last concept is perhaps not very precise; however, it seldom gives serious trouble. Until this verification has been made, it is much better to discuss a *hypothesis* rather than a theory. Nevertheless, as already remarked, current use is not always so precise.

In this connection we want to stress a point. Sometimes, besides a number of propositions that are actually used to derive experimental results, one finds among the axioms some independent and gratuitous assertions whose verification or refutation is deferred to a future and more advanced state of science. A typical example is represented by the *elastic theory of light*, very popular in the first decades of the last century, according to which light consisted of mechanical waves, propagated within a special elastic medium. Experience at that time could only verify that one was confronted with a wave propagation, obeying certain equations. The mechanical nature of the waves was to be verified in the future, perhaps by finding a way to observe the material medium in which the waves were propagated or a way to measure its displacements from equilibrium. In this case, it would also be much better to discuss a *hypothesis* rather than a *theory*. It is important to keep in mind that when the elastic theory of light is mentioned, one should understand something essentially different from the electromagnetic theory of light. The latter has been a hypothesis in J. Maxwell's days, but was later confirmed as a *theory* by H. Hertz's experiments, which at the same time, refuted the elastic hypothesis.

Special attention should be paid to the use of the word *model* in physics.[77] It is often used by science philosophers, and even more often by physicists, in very different contexts. Moreover, its meaning has undergone a rapid evolution, of which not everybody seems to be aware. Hence very different meanings can coexist, giving rise to much confusion.

Originally, when talking about a model one probably had in mind a *material reproduction* of a physical system of interest, on a reduced or expanded scale, so as to render it more accessible to observation or study. Thus, for instance, the globe is a model of the earth; further, everybody may have seen a model of a crystal where the atoms are represented by little balls of different colors, arranged in such a way as to reproduce the structure of the crystal. Models of this kind are largely used for technical purposes (by engineers, architects, and so on), and there is considerable theory about them.

The physicist, however, when talking about a model, rarely only has in mind a *material* reproduction; he refers rather to an *ideal* or conceptual reproduction, or to an inner visualization (see, e.g., Bunge, 1973).

The most celebrated historical example of this interpretation is embodied by the planetary model of the atom. When such a model was first imagined, and also for some time after, one could think of a *faithful*, even though ideal, reproduction of reality. The atom was invisible, only because it was too small; but if it were magnified, it should transform into a system of visible "balls," having well-defined positions and shapes, as is the case for the objects we see and touch every day. When people began to understand that the situation was not so simple, problems arose.

Microscopic (or subatomic) particles do *not* have the same properties as the *macroscopic* objects with which we are familiar; forming an *intuitive* or *visual* idea of them means falsifying reality. At this point the concept of model underwent a transformation. Physicists began to think that it did not necessarily indicate a faithful copy of reality, but was a *fictitious* system that can help us to represent a reality that cannot be visualized.

But once this inexactness was realized, it was unavoidable for the concept of *model* to experience further evolution and to become purely abstract. Indeed, in contemporary particle physics, a model is nearly always a *mathematical model*, having very little to do with visualization. It does not represent a full-fledged theory, because we are generally far from proving that it agrees with all the experimental evidence. Neither does it represent a *hypothesis*, for it at most describes a particular side of reality, although remaining separate from, or even in contradiction with, some previously well-established theories.[78] Frequently, it merely represents a heuristic tool.

As already remarked, this evolution of the concept of model is not always considered. Sometimes authors stress the great importance of the idea without bothering to acknowledge how vague and confused it is. Many emphasize that the physicist can only deal with mere models. However, the interpretations given to this assertion can be utterly different in different contexts. They range from the Kantian view, according to which we reason only about phenomena, whereas the thing in itself remains necessarily unknown to the position of dialectic materialists who maintain that our knowledge mirrors reality, however only in part, because reality is inexhaustible.

I believe that in this situation the term "model," which is often quite useless, should be used as little as possible.

2   *The physics of the reversible*

2.1. The divisions of classical physics

It is customary in traditional textbooks on physics to make a distinction between classical and modern physics. The former includes mechanics, thermology, acoustics, optics, electricity, and magnetism, whereas the latter consists of relativity and quantum mechanics, as well as atomic, nuclear, and particle physics. This old-fashioned subdivision has mainly an historical significance and its conceptual value is very doubtful. We know today that there are no watertight compartments in physics and that everything is linked to everything.

Of course, when writing a book on any branch of knowledge, one tends to divide the subject matter into chapters and sections. Essentially this is due to the fact that the natural process of human thinking is *diachronic* rather than *synchronic*; that is, it is characterized by successive, rather than simultaneous, acquisitions. This time spread can reasonably be used in two ways. One can either set forth the subject matter, sticking largely to historical chronology or work out a subdivision based on such conceptual affinities as may aid comprehension. Obviously, this latter arrangement rests much more on our way of reasoning than on the intrinsic nature of the subject. But how could it be otherwise?

However, even if we follow this procedure, we are left with considerable freedom in selecting the proper ordering of the subject. Accordingly, it should hardly be necessary to stress that the criterion adopted here is not the only one possible.

The division into classical and modern physics will be replaced by the distinction between *macrophysics* and *microphysics*. Microphysics is concerned with the structure of microobjects such as molecules, atoms, nuclei, and fundamental particles, and with their mutual interactions. It can be adequately dealt with only by *quantum mechanics*. Macrophysics is concerned instead with objects much larger than molecules, for instance, with human-size objects. As a rule, phenomena on this scale are adequately described by the laws of classical physics, that is, by laws having the form previously discussed (§1.10). But the reader should be warned that there are some exceptions (such as *lasers*, *superconductors*, *superfluids*, and so on) where the behavior of a macroscopic system can be understood only by a quantum treatment.

The phenomena studied in macrophysics can be divided into two quite different classes, comprising the *reversible* and the *irreversible* processes, respectively.

Suppose we are watching a movie with the film running in the reversed direction. If the scene represents, say, a train in motion, at first we do not notice anything terribly exceptional. Anyone may have seen a train engine pushing instead of pulling the carriages. From this sole fact, we cannot infer that the film motion is reversed. For we are watching a reversible process, or a process that can go either way.

If the engine is a steam engine, however, we notice steam[1] being formed in the air and entering the funnel. We immediately infer that the film is run in reverse; for we have observed an irreversible process going in the *wrong* direction.

Mechanical and electromagnetic phenomena are reversible, and will be dealt with first, whereas thermodynamic phenomena are irreversible, and will be discussed in the next chapter.

Acoustics and optics can be considered as parts of mechanics and electromagnetism, respectively, and do not call for separate treatment.[2] Relativity represents nothing but the completion of classical physics, hence it is conveniently treated right after mechanics and electricity.

2.2. Velocity and acceleration

In nearly all elementary textbooks on physics, velocity is introduced as the ratio between distance (traveled by a body) and time (spent in traveling that distance). In my opinion, this is an awkward way to introduce velocity and does not help to clarify ideas. This unnatural encounter with velocity–defined as a derived, rather than a primitive, quantity–represents, I believe, one of the first intellectual traumas experienced by a young person being taught physics at school. Such traumas are likely to convince the pupil for a while, if not for ever, that physics is an abstruse subject, quite remote from reality.

According to J. Piaget:

> In the child we observe a precocious intuition of speed, independent of duration, and founded on the merely ordinal notion of overtaking (order of succession in space and in time, with no reference to distance covered nor to duration), whereas temporal intuitions, in particular simultaneity, seem always to be associated with speed ratios (Piaget, 1970, p. 98).

Presumably, this has always been true, but today it is strikingly evident. Any modern child knows what speed is much before being able to tell the time.

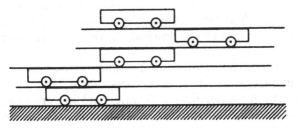

Figure 2.1

The problem is, of course, to make the intuitive concept of speed correspond to a precise measurement procedure. This is not impossible, although very inconvenient.

First, we have to establish a unit for velocity and a standard of it, consisting, say, of a little car provided with a speed governor, and moving along a track (see Fig. 2.1). Next, we define twice that velocity by means of a second and identical car whose track is fixed to the first car. Similarly, one can define velocities three times, four times unit velocity, and so on. It is not impossible to devise even fractional measures.

As is evident, this procedure is absolutely nonviable. From a conceptual point of view, however, all that matters is to recognize that it is possible to define velocity as a primitive quantity, not necessarily referred to something else.

If we now measure the distance l traveled by one of our cars, having velocity v, during time t, we discover the physical law[3]

$$l = kvt \tag{2.1}$$

is valid. Distance is *proportional* to velocity multiplied by time. As usual, the constant of proportionality k depends on the units adopted.

By applying a procedure previously discussed, (§1.13), we select the unit velocity in such a way as to have $k = 1$. As a result, we can write

$$l = vt \tag{2.2}$$

A body has unit velocity when it travels 1 cm in 1 s; such velocity is called *c*enti*me*ters *pe*r *s*econd and is conventionally denoted by cm/s or $\mathrm{cm \cdot s^{-1}}$.

At this point, one may easily be led by equation (2.2) to assert that velocity *is* length divided by time. As already discussed in §1.14, such an expression is purely conventional and does not correspond to an intrinsic property of the physical entities concerned. However, once the convention has been established, it will be more convenient to measure velocity by first measuring l and t and then by taking their ratio l/t than by resorting to the unwieldy device depicted in Figure 2.1.

We shall see several more instances of this procedure later. Now we have a physical quantity A whose direct operational definition is possible, but leads to great practical difficulties and to very inaccurate measurements. In this situation, if a physical law is discovered that connects A with other quantities $B, C, \dots$, involved (and is, of course, valid within a certain ϵ), we can assume this law as a *definition* of A, by writing down an *exact* equation (without ϵ), such as equation (2.2). Thus the equation ceases to express a physical law and becomes a mere definition (but still an operational definition, to be sure). There is nothing to be criticized in this procedure; but we should be careful not to mistake its significance.[4]

Until now, we have implied that the speed is constant, or does not vary during the time interval t. If the speed varies, the ratio l/t would clearly represent only the *average* velocity during the trip. In each small space interval Δl of the distance, the body will have a different velocity, obtained by dividing Δl by the corresponding small time interval Δt. The smaller Δl and Δt are, the more certain we are that the speed in the interval is constant within the best possible accuracy. As already explained (§1.12) for vanishing Δl and Δt, we can go directly to the limit and can write

$$v = \frac{dl}{dt} \tag{2.3}$$

We usually say that velocity is the derivative of space with respect to time.

In a *uniform motion* the velocity is constant and equation (2.2) applies. But this is only a particular case. In general, one must use equation (2.3); v is different at different times and the motion is *nonuniform*. In the nonuniform motion the rate at which v is varying is interesting to know. The corresponding quantity is termed the *acceleration*.

I believe that children today have a fairly clear and intuitive idea even about acceleration. But I shall not repeat the argument set forth about velocity. Let us just state that acceleration is *defined* in terms of velocity and time, by writing

$$a = \frac{dv}{dt} \tag{2.4}$$

Acceleration is therefore the ratio between the velocity increase and the time during which the increase takes place, or the time derivative of the velocity. It is expressed in cm/s/s or cm/s^2 or cm·s^{-2}, for short.

From equations (2.3) and (2.4) the acceleration proves to be the derivative of the derivative of space with respect to time. Thus we can say that acceleration is the *second derivative* of space with respect to time. In symbols, this is denoted by

$$a = \frac{d^2 l}{dt^2} \qquad (2.5)$$

When l is originally expressed as a function of t, mathematics shows how to evaluate its first derivative v and then its second derivative a. Obviously, in a uniform motion the acceleration vanishes ($a = 0$).

If all the motions encountered in nature were uniform, the path of physics would have been much smoother. However, the very first natural motion to catch the attention and arouse the curiosity of scholars was the motion of free fall, which is nonuniform. Precisely, Galileo discovered that the motion is accelerated with constant acceleration; it is a *uniformly accelerated* motion. This acceleration, usually called gravity acceleration, is denoted by g and has the value $g = 981$ cm/s^2.[5]

The law of gravity is very simple, but was not simple to discover. First, as free-falling motion is very fast, it is extremely difficult to measure with sufficient accuracy, without the aid of modern equipment. Galileo resorted to the clever device of using an inclined plane, along which motion is still uniformly accelerated, but with a smaller acceleration. Second, air resistance complicates the phenomenon and conceals the essential fact that the acceleration is identical for all bodies. As is generally known, a piece of paper falls more slowly than an iron ball. Galileo correctly guessed that this was due to a secondary factor, which had to be eliminated if one wanted to arrive at a fundamental law. Indeed, one must consider the fall *in a vacuum* and not in the air. The fact that Galileo could not perform the experiment in a vacuum because he lacked the means to do so,[6] is worth a little thought–he arrived at the correct law by guesswork and by a lucky extrapolation!

Let a body at rest start to fall. Since velocity increases by g every second, the body after t seconds, will have attained the velocity

$$v = gt \qquad (2.6)$$

Let us plot this law in a diagram (Fig. 2.2), where the abscissas represent t and the ordinates v. We obtain the straight line OV.

During the small time interval Δt, where the velocity is v, the distance

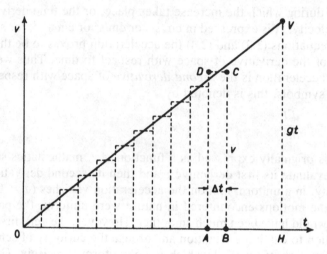

Figure 2.2

Δl covered by the body will be given by $\Delta l = v\Delta t$; this product represents the area of rectangle $ABCD$. Identical reasoning can be repeated for all intervals Δt into which the total interval OH can be divided, and on each Δt a rectangle can be constructed. The sum of all the Δl's thus obtained represents the total distance l. But the sum of all the rectangles equals the area of triangle OHV, whose base and height are t and gt, respectively. Hence we derive

$$l = \tfrac{1}{2}gt^2 \tag{2.7}$$

We have thus carried out the evaluation of an *integral* (§1.12) without resorting to any knowledge of mathematical analysis.

2.3. Curvilinear motion

The motions investigated so far were implied to be in a straight line. This is, obviously, the particular case where velocity always has the same direction. In order to deal with more general cases, it is necessary to introduce *vectors*–mathematical entities capable of describing those physical quantities that have *directions* (displacements, velocities, accelerations, forces, and so on).

A vector is graphically represented by a portion AB of a straight line (see Fig. 2.3), having specified length and direction; the latter is marked by an arrow. The length or *modulus* of the vector is proportional to the measured value of the quantity to be represented, whereas the direction coincides with that of the physical quantity.

Figure 2.3

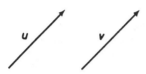

Figure 2.4

One can operate with vectors by applying *vector algebra*. For now it will be sufficient to know only two facts. First, two vectors u, v[7] are defined as being equal when both their moduli and directions are equal (see Fig. 2.4), regardless of whether or not the vectors coincide (or have the same starting point). Second the sum of two vectors u, v is, by definition, the diagonal w of the parallelogram built on the two sides u and v (see Fig. 2.5). This definition is not unjustified. It has been chosen for the very good reason that the vector quantities encountered in physics are added in that way, as proven by experience.

The simplest example is represented by displacements. Let us consider a man who is standing still at point A on the deck of a ship (see Fig. 2.6). If the ship is sailing, the motion of the ship will carry him to B in time t.

Alternatively, if the ship is standing still and the man walks on the deck, he will arrive in the same time t, at C. If we now let both motions take place simultaneously, the man will plainly end up at point D, obtained by adding the vectors AB and AC, which represent the displacements. In the same manner one can add velocities, forces, and so on.

It is known that a point in space can be specified by referring to a system of rectangular axes x, y, z (see Fig. 2.7), sometimes called a *Cartesian system of coordinates*. Such a system can also be used to specify vectors. To this end, the starting point of the vector v to be specified is placed at the origin O of the axes; the coordinates of the end point P are called the *components* of v and are denoted by v_x, v_y, v_z, respectively. Thus the components represent the orthogonal projections of v onto the axes; it is evident that they are sufficient to determine the vector.

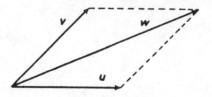

Figure 2.5

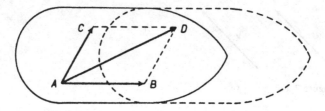

Figure 2.6

An important result of introducing vectors is that a motion cannot be considered as genuinely uniform unless its velocity v is also constant in direction. Otherwise, the motion will be nonuniform even in the case when the modulus v is constant.

For example, consider a point moving along a circle at a velocity of constant modulus (see Fig. 2.8). At A and B the velocities will be represented by v_A and v_B, respectively. These two velocities are different, because they have different directions. Let us draw from A vector AD equal to v_B; we see immediately that v_B can be obtained from v_A by adding vector CD (or AE). As we are steadily adding something to the velocity, there is an *acceleration*. This is called the *centripetal acceleration*, because it points toward the center of the circle. By using an infinitesimal argument,[8] one finds that the centripetal acceleration is given by

$$a = \frac{v^2}{r} \tag{2.8}$$

where v represents the modulus of v and r is the radius of the circle.

It is evident that a centripetal acceleration will arise in any curved (even noncircular) motion.

2.4. The laws of dynamics

Galileo's mechanics is still mostly *kinematics*. Motion is analyzed without a real understanding of the physical factors that give rise to it. The

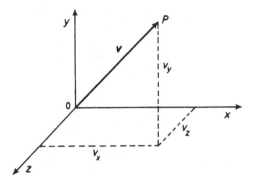

Figure 2.7

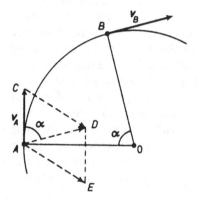

Figure 2.8

science that studies motion in relation to its *cause(s)* is called *dynamics* and is essentially due to Newton.[9]

This traditional distinction gives rise to a number of difficulties. For instance, the notion of cause is not at all straightforward in modern physics (see §4.20). As will be illustrated later, there is a tendency to replace the concept of causal evolution with that of *structure* of the physical world. For the sake of simplicity, we shall postpone this critical discussion, and consider here the traditional concept.

The fundamental physical quantity in Newtonian mechanics is *force*.[10] Today physics does not give force the same importance it had in the past. The fundamental quantity appears instead to be energy (§2.5), whereas force is now a somewhat conventional and artificial entity. Probably, the reason that force first appeared as a fundamental physical entity is anthropomorphism. Indeed, every time we want to

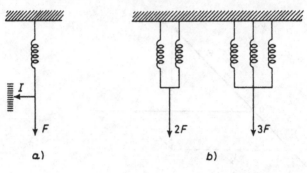

Figure 2.9

displace an object, we experience a muscular sensation, which is translated into words by saying that we apply a force to the object.

It is also possible to have an intuitive concept of energy. When
overworked and tired, we feel as though we had spent something; precisely, we have spent energy! Unfortunately, in this respect the human
body is not very favorably designed. If we press a wall with force, the
wall, of course, does not move, but we get tired and expend energy
all the same. Contrary to what happens with simple machines, the
human body can consume energy without doing work. The fundamental
property of energy, represented by its conservation, appears to be
ostensibly violated. This may probably be the reason that energy was
not first introduced on an intuitive basis. We shall return to this subject
later.

In order to change the intuitive concept of force into that of a physical
quantity, we must indicate how to measure it. We can refer to the
dynamometer in its most elementary form. A spring (Fig. 2.9(a)) has
one end fixed, and at the other end bears an Index I, capable of moving
along a scale. By applying a force at F, we can make the index move
along the scale, which can represent a measure of force. We stipulate
that as long as the spring is not extended too much, the force is proportional to the displacement of the index. This represents Hooke's
law (*ut tensio sic vis*), which holds for elasticity.

Honestly speaking, in order to establish Hooke's law, one should
already be able to measure force, so we seem to be caught in a vicious
circle. However, all we have to do is to define two or three times the
force of one spring by means of the force of two or three identical
springs, acting in parallel and set at the same tension (Fig. 2.9(b)). The
details of this operation are obvious.

When a force is applied to a free body, it sets the body into motion.
The quantitative law governing this phenomenon was put in a precise

form by Newton and is known as the *second law of dynamics*. First of all, it says that the acceleration of the body is proportional to the force, so we can write

$$F = ka \qquad (2.9)$$

Second, it is found that the coefficient k is not the same for all bodies but is proportional to the *mass* or to the amount of matter contained in the body. This concept is somewhat vague. It can be clarified perfectly only in the case of *homogeneous* bodies. For instance, two identical lead balls when put together have twice the mass of one single ball; two liters of water have twice the mass of one liter; n hydrogen atoms have n times the mass of one atom; and so forth. When, however, we deal with different materials, it is easy to start a vicious circle. This criticism has been raised several times.[11]

Let us try to find a solution by hanging the various bodies to the dynamometer (at a fixed place on the earth's surface) and by stipulating that their masses are proportional to their weights, that is, to the forces exerted on the spring. Only with the theory of relativity can we conveniently get out of this difficulty.

The unit of measure for mass is the *gram* (g), nearly identical to the mass of 1 cm^3 of distilled water.

At this point, one can simply write

$$F = ma \qquad (2.10)$$

where m stands for mass. Of course, by doing this one has fixed the unit for force as being that force that confers to the mass of 1 g the acceleration of 1 cm/s^2; it is called the *dyne*.

The weight P of a body is the force that pulls it downward. Because it gives rise to the acceleration g, we have

$$P = mg \qquad (2.11)$$

In everyday life one usually expresses weight in g or kg (or lb), instead of in dynes. Strictly speaking, of course, this is wrong; however, due to the proportionality equation (2.11), it does not really matter.

The *first law of dynamics* or law of *inertia*, is nothing but a simple result of the second law of dynamics. For if $F = 0$, one must also have $a = 0$, or a constant velocity; hence a body not subject to any force is either at rest or moving with uniform and straight motion.

It is also customary to state a *third law of dynamics*. Not too precisely, one can say that to every *action* there corresponds an equal and opposite *reaction*.

We shall be in a better position to discuss this law when we consider momentum. Here we only want to mention an important example. A body not subject to any force can only move uniformly in a straight line. To make it move along a curved path, we must exert on the body a *centripetal force* equal to the mass times the centripetal acceleration equation (2.8). By doing this, we get the impression that the body, in turn, acts on us with a *centrifugal force* that is equal and opposite to the centripetal force. It is as though we had to produce the latter, in order to counteract the former. If one of these forces is called *action*, the other must represent the *reaction*.

2.5. Work and energy

As previously noted, the use of the human body to perform work suggests a first and intuitive concept of energy. The worker feels that he expends something and consequently experiences fatigue.

Suppose that we must drag a heavy body a distance l. Everyone would be ready to admit that the cost of transportation should be proportional to l. For if we double l, labor must be doubled. Moreover, it is evident that the cost is proportional to the force F applied to the body, because twice that force will require two people instead of one.

All this explains that the total work W should be proportional to force times distance, or

$$W = kFl \tag{2.12}$$

Because measuring work by its cost or human fatigue is a very crude procedure, it is preferable to apply the method used for velocity and to *define* work by equation (2.12), which consequently ceases to represent a physical law. Precisely, one can write

$$W = Fl \tag{2.13}$$

and take as a unit of work the work done by a 1 *dyne* force, when its point of application shifts by 1 cm. This unit is called the *erg*; therefore erg = dyne × cm.

A word here about the direction in which the body is displaced. When F and l have the same direction, there is no problem and we apply equation (2.13). However, what happens in other cases? The fact that equation (2.13) cannot be valid in every case becomes evident, for example, when climbing or descending the stairs. In both cases the force applied by the legs is the same (roughly speaking, the weight of our body) and the distance covered is the same, however, the fatigue is quite different.

A precise definition of work can be conveniently given as follows. Let us consider the two vectors: displacement l and force F (Fig. 2.10), having different directions. One defines the work as being equal to the

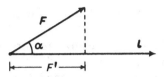

Figure 2.10

product $F'l$ of the displacement by the *projection* F' of the force onto the direction of displacement.[12] If the angle α between both directions varies, work is maximum when F and l are parallel, vanishes for $\alpha = 90°$, then becomes negative.

This definition may not seem very reasonable when considering the work as performed by the free human body. But it appears justified when we use some of the simple machines devised specifically for increasing efficiency.

As an example, consider a man carrying a suitcase on a flat road. According to definition, he is not doing any work, for the force (weight) is perpendicular to the displacement. Nevertheless, he gets tired. However, man has invented the wheel; he puts his suitcase on a cart, depending on the smoothness of the road and how greased the hubs of the wheels are, the easier the task becomes.

Negative work can be reasoned in a similar way. When descending the stairs, a person does negative work. What does this mean?—that he can use this process to *obtain* rather than to spend work. All he has to do is to cling to a rope that passes on a pulley tied at the other end to a heavy body; by descending, the man can hoist a weight virtually equal to his own, thus obtaining work.

As a conclusion, we can say that this definition of work is all right when something is done and no *waste* is involved. What form the waste usually takes is discussed in thermodynamics. In other words, we can also say that our definition corresponds to the *minimum* fatigue we must experience in order to accomplish a given mechanical task when allowed to use the most suitable (passive) machine.

The concept of work we have introduced, along with the concept of energy soon to be presented, can illustrate a viewpoint some scholars maintain with well-founded historical reasons. Work and energy were introduced and defined at the end of the eighteenth and in the early nineteenth centuries, primarily for economic reasons. At that time, the industrial exploitation of the steam engine was starting, and one wanted to measure the value of the machine's output, so as to compare it with the cost of fuel. In other words, the needs of production and of newborn capitalism forced science, from outside, to consider new concepts, new roads, new problems. A similar situation was due to arise again and again, until today.[13]

This is an important criterion followed by several authors for the interpretation of the history of science. As is true for all valid criteria, it may become false and even ridiculous when converted into a dogma, through which one should be able to explain *everything*.

Let us now pass from the concept of work to that of energy, by applying the force F to a body of mass m. If the body is free to move, it will acquire the acceleration a, given by

$$F = ma \tag{2.14}$$

If the force is constant in time, the motion is uniformly accelerated, so that, by equation (2.7), the body will travel in time t a distance l given by

$$l = \tfrac{1}{2}at^2 \tag{2.15}$$

Multiplying the lefthand sides and the righthand sides of each of the last two equations, we obtain

$$Fl = \tfrac{1}{2}ma^2t^2 \tag{2.16}$$

By equations (2.6) and (2.13) we have $at = v$ and $Fl = W$, respectively, and can also write

$$W = \tfrac{1}{2}mv^2 \tag{2.17}$$

As a conclusion, if we know the final velocity v and the mass m of the body, we can derive the work needed to set it in motion, even without knowing F and l. All that is required is to form the expression $\tfrac{1}{2}mv^2$.

Further, if we want to stop the body, we must apply a force in the opposite direction, which will cause a *deceleration* until the body comes to rest. This force performs negative work, so the body yields positive work. In other words, one can obtain work from a moving body by stopping it. It is an easy matter to show that such work is again equal to $\tfrac{1}{2}mv^2$.

In this case we say that the work W is spent to give the body *kinetic energy* E_{kin}, expressed by

$$E_{kin} = \tfrac{1}{2}mv^2 \tag{2.18}$$

This quantity, in turn, expresses the capacity of the body to do work because of its motion. The body can return all the work expended in setting it into motion.

A body can also do work because it is in a certain *position* (e.g., the water held by a dam) or in a certain *configuration* (e.g., the spring of a wound watch). The work gained in this way is called the *potential energy.* For example, the potential energy E_{pot} of a weight P, placed at a height h, is evaluated by multiplying the force P by the displacement h. Thus

$$E_{pot} = Ph \qquad (2.19)$$

Obviously, this also represents the work performed to bring the weight to the height h. Hence potential energy, like kinetic energy, *stores*, as it were, the work done on the body and is capable of returning it in an identical amount.

If the body is dropped and falls freely, we know by equation (2.17) that when it reaches the ground, it will have a velocity v that $W = Ph = \frac{1}{2}mv^2$. Hence the potential energy has vanished, but in its place we find an equal amount of kinetic energy.

One can easily prove that at any intermediate location during its fall, the body has lost as much potential energy as it has gained kinetic energy. Hence the sum of potential energy and kinetic energy is a constant and is called the *total energy* E_{tot}, so that

$$E_{tot} = E_{kin} + E_{pot} \qquad (2.20)$$

We conclude then that *the total energy remains constant* during the process.

One can show that this law is not only valid in the particular case selected in our example, but is also respected by all the phenomena of mechanics. In due course we shall see that when passing from merely mechanical processes to more general phenomena, we can define other forms of energy in such a way that the sum of all the energies involved in any physical phenomenon is a constant. This is the celebrated law of *conservation of energy*. It was simultaneously discovered by several authors about the middle of the nineteenth century (see, e.g., Kuhn, 1962).

2.6. The invariants

With the law of conservation of energy we have hit on the first case of an *invariant*. We have found a quantity that does not vary during the course of physical processes. Invariants are very important in modern physics, which has largely become a *hunt for invariants*.

No wonder, philosophers, starting from the pre-Socratic period, long realized that some invariants must exist. This represents a necessary condition for us to be able to understand something about the world. If πάντα ρεî, if everything flows and changes all the time, how can we talk about anything? As soon as we start to discuss one object, it has changed and turned into another thing.[14] Do we begin by talking about one subject and end by talking about another? What sense does all this make?

Moreover, if we affirm that everything changes, we can ask: What is changing? What is the subject of the verb *to change*? It must be something permanent,[15] which represents continuity between before and after. This something has long been called the *substance*; however, today we are no longer able to ascribe a plausible meaning to this term. Physics has encouraged us to go from the study of naïve and not well-grounded intuitions to the search for objective truth. Invariants are the best heirs of the old concept of substance(s). Energy holds the foremost place among invariants.

But energy, even though very important, is not the only invariant. In classical mechanics there are at least two more invariants of great interest, which will be described presently.

Energy represents the first example of a quantity whose intuitive notion and operational definition are not too clear at first; the prominence of energy in physics is recognized only a posteriori, mainly because of its general property of invariance. In such cases, it is definitely advisable to give up our restriction that all physical quantities be primary quantities, operationally defined. Conveniently one can forget about aesthetics and can introduce secondary quantities whose measures are obtained by mathematical operations from the measures of other quantities.

An important case of this kind is that of *momentum*. The momentum *p* of a body of mass *m* moving with velocity *v* is defined as

$$p = mv \tag{2.21}$$

Note that it is a vector quantity.

Momentum, the same as energy, obeys the conservation law, and therefore is an invariant. This is an experimental law that can be verified in collision processes. This verification may have to be done with either the collision of macroscopic bodies, such as bowls or billiard balls, or the collision of subatomic particles. If n bodies of masses $m_1, m_2, \ldots, m_n$ collide and their velocities before and after the collision are indicated with $v_1, v_2, \ldots, v_n$ and $v_1', v_2', \ldots, v_n'$, respectively, one finds that

$$\sum_{i=1}^{n} m_i v_i = \sum_{i=1}^{n} m_i v_i' \tag{2.22}$$

The (vectorial) sum of the momenta before collision equals the sum of the momenta after collision.

By *collision*, one must understand a very general idea that includes as a particular case the sudden and instantaneous shock we usually

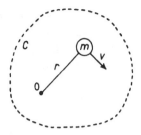

Figure 2.11

think of when we hear the word. When two bodies collide they influence the motion of each other in some way. For example, a comet coming from outer space and going back after having been deflected by the sun's attraction, has collided with the sun. Its momentum changes, but the sun's momentum undergoes an equal and opposite change. Naturally, due to the huge mass of the sun, as compared with that of the comet, the velocity of the sun will vary only in a negligible way.

It is of interest to note that the fundamental roles of energy and momentum, which in a sense are more natural than that of force, were surmised, if only slightly and implicitly, very early in the history of mechanics. Indeed, a famous argument occurred between Cartesian and Leibnizian scholars (see, e.g., Jammer, 1957) as to what was the most appropriate measure of force, the former school preferring the expression mv and the latter preferring mv^2! However, it was not yet the right time for workers really to appreciate the "hierarchical" primacy of *conserved* quantities with respect to all the others. Compared with these abstract constructions, the concept of the direct *cause* of the motion, embodied by the Newtonian force, seemed much more understandable and acceptable.

Conservation of mechanical energy is a result of the second law $F = ma$ and is virtually equivalent to it. Conservation of momentum can also have an interpretation in terms of forces. We must introduce now the third law of dynamics or the law of action and reaction, mentioned previously. When two bodies A and B collide, if A exerts force F on B, B in turn will exert force $-F$ on A. It can easily be shown that momentum conservation is a consequence of this and of the second law. Furthermore, note that the law of inertia (first law) can be considered as a consequence of momentum conservation.

Let a mass m, occupying a small volume, revolve around O (Fig. 2.11) at a constant distance r. If v is its velocity (perpendicular to r, of course) the product mvr is termed the *angular momentum* of m.

More generally, let us consider an extended body C rotating around an axis perpendicular to the drawing through O. Let us divide C into many small portions $m_1, m_2, \ldots, m_n$, located at $r_1, r_2, \ldots, r_n$ and revolving with velocities $v_1, v_2, \ldots, v_n$, respectively. The angular momentum K of C is defined as the sum of all angular momenta of the elementary portions with respect to O, namely,

$$K = \sum_{i=1}^{n} m_i v_i r_i \qquad (2.23)$$

Experience shows that this angular momentum is also conserved.

This property is fairly well known on an intuitive level. When a physical system rotates, rotation becomes faster as the masses come nearer to the axis and slower as the masses recede. Consider, for instance, a diver, when he rolls himself up or stretches, compared to the analogous movements of a skater, of the water vortex near the sink of a basin, and so forth.

More precise, angular momentum is defined as a vector, having the direction of the axis of rotation. It is exactly this vector that is conserved. Consider a spinning top, a gyroscope, artillery bullets (which are rotated in order to maintain orientation).

Classically, one can derive the conservation of angular momentum from first principles, that is, from the laws of dynamics. As a matter of fact, this is possible as long as we stick to the definition, equation (2.23). But today we know that elementary particles are endowed with a measurable angular momentum or *spin* (§4.11) that cannot be defined by equation (2.23). The actual quantity conserved is the sum of all the spins and all the angular momenta defined by equation (2.23). Therefore it is evident that angular momentum is a quantity of primary and fundamental character. Its conservation embodies one of the great laws of nature.

As a conclusion, we can perhaps say that mechanics is still based on three fundamental laws. However, they are no longer the three classical laws; instead they are the three conservation laws of energy, momentum, and angular momentum–even more general than the three classical laws.

2.7. Action at a distance

The intuitive concept of force as the immediate cause of motion seems to imply that the force should act by direct contact (in the same way as when we push a body), or, at any rate, should act through an intermediate medium (such as a rope or a stick) which gradually transmits the action from one region to the next, with continuity.[16]

Ironically, it was Newton himself who discovered that the situation can be different, when he discovered the law of *universal gravitation,* which embodies the best example of *action at a distance.*

This law states that all masses attract one another – more precisely, any two masses m_1, m_2, separated by distance r, pull each other with a force F given by

$$F = G \frac{m_1 m_2}{r^2} \qquad (2.24)$$

where G represents a constant;[17] its value in CGS is known today to be $(6.673 \pm 0.003) \times 10^{-8}$.

The universal gravitation is, first of all, responsible for the fall of free bodies. A body falls, because the earth with its mass attracts it.

If M denotes the mass of the earth and m the mass of the falling body, the latter's acceleration will be found by equation (2.10) to be $F/m = GM/r^2$. It is therefore independent of m and is the same for all bodies, at the earth's surface (i.e., at the same distance r from the earth's center), as was first set forth by Galileo. Note that this most remarkable circumstance is due to the fact that the mass appearing in equation (2.10) (*inertial mass*) and that appearing in equation (2.24) (*gravitational mass*) are equal. This property is not at all evident *a priori*, but has been confirmed so far by all experiments, up to an amazing accuracy.

The greatest success of equation (2.24) was represented by its capability to account in detail for the motion of the heavenly bodies. Roughly speaking, one can say that the moon continues its nearly circular orbit because the attraction of the earth's force balances the centrifugal force; the same holds for the motion of the earth around the sun.

From equation (2.24), however, one can derive much more than this. First, by applying mathematical analysis, one can find the complete Kepler laws for the motions of the planets around the sun; then one can evaluate, with enormous accuracy, the perturbations caused by the planets on each other.

What is the cause of the attraction of one body on another? Newton declared that he had not been able to discover the cause and did not want to frame any hypothesis (*hypotheses non fingo*).[18] Thus the equation (2.24), represents a case of action at a distance. It has resisted all previous attempts to explain it away by means of a fluid medium (*ether*) or of more complicated methods. Thus scientists began to realize that what really matters is the law's ability to account perfectly for all gravitational phenomena.[19] The matter-of-fact attitude first introduced in this way was largely adopted later by physicists and had an important bearing on the development of science.

Today, action at a distance, with a qualification to be seen in quantum mechanics, is present everywhere in physics. Material bodies are known to consist of atomic particles held together by forces acting at a distance, the most important of which is the electric force. Its law was discovered by Coulomb more than a century later than equation (2.24) was, and has virtually the same form.

As we know, a body can be electrically charged. Two charged bodies either attract or repel each other. As a result, there must be two kinds of electric charge, conventionally named *positive* and *negative*. Two charges of opposite sign attract each other, whereas two charges of the same sign repel each other. Coulomb's law states that the force acting on each one of two charges Q_1 and Q_2, a distance r apart, in a vacuum, is given by

$$F = k \frac{Q_1 Q_2}{r^2} \tag{2.25}$$

By putting $k = 1$, we can write

$$F = \frac{Q_1 Q_2}{r^2} \tag{2.26}$$

We have thus established that the unit of electric charge is the charge that, when placed 1 cm apart from an identical charge, repels it with the force of 1 dyne. It is called the CGS *electrostatic unit*.

At this point we can understand the usefulness of calling the two kinds of electricity positive and negative. To this end, we stipulate that a plus sign for F in equation (2.26) indicates repulsion, whereas a minus sign indicates attraction. All we have to do is to evaluate the product $Q_1 Q_2$ taking into account the signs of the charges and the ordinary rule for the sign of the product. Furthermore, consider a body carrying an electric charge Q_1, which exerts force F on Q_2 at a distance r. If the body is given the additional charge $-Q_1$, that charge will exert the force $-F$ on Q_2. Hence the resultant force will vanish. But we can also say that the body has received the total charge $Q_1 - Q_1 = 0$; the force vanishes because the charge is zero.

Concerning the analogy of the two great laws, equations (2.24) and (2.26), a magnetic Coulomb law has also been postulated, although based on an unwarranted assumption. The attraction and the repulsion between magnetized bodies were accounted for by means of positive and negative magnetic charges, analogous to the electric charges. Two

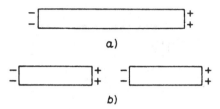

Figure 2.12

magnetic charges M_1 and M_2 should act on each other with the force

$$F = \frac{M_1 M_2}{r^2} \tag{2.27}$$

where we have implicitly defined the measuring unit. However, no one has ever been able to detect magnetic charges experimentally.

2.8. Do magnetic charges exist?

We are now faced by a most intriguing problem. Paradoxically, one can say that equation (2.27) would be perfectly correct and verifiable if only magnetic charges did exist; but unfortunately, they do not!

Let us take this opportunity to illustrate a problem that is quite frequently encountered in modern physics. This concerns the existence of certain physical objects, which show only some and not all of the properties that characterize the existence of everyday objects.

Experience shows that even granting that magnetic charges exist, the positive charges are always associated with as many negative charges, and vice versa. For example, let us consider a magnetic needle (such as a compass needle), or a long rod with positive charge at one end and negative charge at the other (Fig. 2.12(a)). If we break the rod in the middle (Fig. 2.12(b)), we find two magnetic rods, each one having positive charge at one end and negative charge at the other. Unlike the case of electricity, the result is that for magnetism it is impossible to charge positively or negatively a whole body. Presently, we shall see why this is so.

What, then, is the meaning of equation (2.27)? How can it be experimentally verified?

It is necessary to set up an experiment such as that shown in Figure 2.13. Two long and thin magnetic rods are arranged so as to have the two ends B and C very close, while the other two ends A and D are far apart. In this situation one finds that B and C ineract in accord with

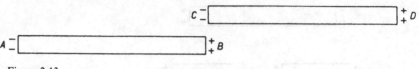

Figure 2.13

equation (2.27). The farther apart the ends A and D are (whose action is neglected), the more precisely the equation is satisfied.

But if equation (2.27) can somehow be experimentally verified, what is the meaning of the assertion that magnetic charges do not exist? This means that unlike electric charges, magnetic charges are not *separable* and cannot be isolated.

Until now it is uncertain whether magnetic *monopoles*, that is, magnetically charged particles, can be experimentally detected. Such monopoles, imagined by P. Dirac back in 1931, have long been sought. Some researchers have even announced that they have found monopoles. Most scholars, nonetheless, are still fairly sceptical. Let us therefore forget monopoles and consider instead ordinary, dipolar magnets.

Would it really be unreasonable to say that such magnetic entities that have properties analogous to those of the electric charges exist, except for the property of being separable? Actually, stating the condition that something should necessarily be separable in order to exist, might be a mere prejudice, or the result of a convention.[20]

Clearly, this problem is closely connected with the problem of theoretical *terms*,[21] previously discussed (§1.7). At this point, we can mention a procedure, usually named after F. P. Ramsey (Ramsey, 1954; see also Sneed, 1971, Chaps. 3 and 4). We should try to discuss only the gist of the procedure rather than all technical details. Every time a theoretical term appears in a theory one should substitute a variable x for it, and the theory should be preceded by the expression "there exists a certain thing x, such that." According to R. Carnap:

> The Ramsey sentence continues to assert, through its existential quantifiers, that there is something in the external world that has all those properties that physicists assign to the electron. It does not question the existence – the "reality" – of this something. It merely proposes a different way of talking about that something. The troublesome question it avoids is not, "Do electrons exist?" but, "What is the exact *meaning* of the term 'electron'?" In Ramsey's way of speaking about the world, this question does not arise. It is no longer necessary to inquire about the meaning of "electron," because the term itself does not appear in Ramsey's language (Carnap, 1966, p. 252).

What Carnap says about the electron could be repeated word for word about the magnetic charge. But it seems to me that in agreement with the opinion that there are no purely observational terms, exactly the same facts can be said about such terms as "iron bar," "stone," "leaf," which are usually claimed to be observational.[22]

Physicists generally think of the electron in a way very close to that described. Hence Ramsey's procedure may appear only too natural to them. Nevertheless, their idea of a magnetic charge is certainly different. Apart from the still controversial magnetic monopole, physicists are likely to say that magnetic charges are *fictitious*. I think that *to a certain extent* it is only a matter of language.[23] Conveniently adhering to scientific language, we shall agree, without further comment, that magnetic charges are fictitious entities.

Incidentally, let us take this opportunity to comment on the use of analogy, which represents a powerful tool of investigation. As far as electrostatic phenomena are concerned, scientists would presumably have hit, sooner or later, on Coulomb's law, even independently of Newton's law of gravitation. But what of magnetic phenomena? It is hard to believe that the analogy with a great and universal law, which had only met with success, did not induce scientists to strain experimental evidence a little and to imagine magnetic charges.

With equations (2.26) and (2.27) we have left the province classically thought to belong to mechanics. Indeed, electricity and magnetism belong traditionally to another *chapter*. This historical heirloom, however, is definitely obsolete today and does not correspond to anything conceptually important.

As described in Chapter 4, material bodies consist primarily of electrically charged particles; the interaction between electric charges is mainly responsible for the mechanical properties of bulk matter. In what sense, then, can one say that electric forces are not mechanical, whereas gravitational forces are mechanical? It is merely somewhat of an outdated convention, which should be dropped, for the sake of clarity.

2.9. The field concept

The idea of a pure action at a distance has never fully satisfied the mind of all scientists. In the eighteenth century many scholars were already wise enough to follow Newton's example and to refrain from asserting facts that were not supported by experience. Nevertheless, many of them were considering a sort of tension or pressure in the medium, which could transmit the force. But how does this make sense, if there is only empty space between two interacting bodies?

Today, with the benefit of hindsight, we recognise that if it is a metaphysical prejudice to assume as necessary the presence of an ordinary material medium between bodies, it is also a metaphysical prejudice

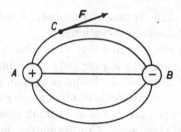

Figure 2.14

to allow empty space to have only *geometrical* properties and to discard
the possibility that it may also have *physical* properties. It took nearly
a century of theoretical and experimental work before physicists began
to see this possibility.

M. Faraday is mainly credited with having taken the first and decisive
step in the right direction (for a good historical and philosophical pres-
entation, see, e.g., Agassi, 1971).

We know that iron filings when placed on a cardboard and shaken
gently in the presence of a magnet, align themselves in characteristic
patterns. Every iron fragment becomes magnetized and orients itself
like a compass needle. Chains are formed, the positive end of one
fragment facing the negative end of the next fragment. In this way the
fragments attract one another and the chain becomes capable of transmitting a force.

Faraday assumed that this process of *polarization* should take place
even in empty space when magnetic phenomena occur, and that a similar process should take place for electric phenomena.

It is easy to account for the polarization of a material body, as will
be explained shortly. In a vacuum, however, if we do not want to
fabricate hypotheses, we must confine our statements to the fact that
there are things such as the *lines of force,* along which action is transmitted. What these lines of force precisely consist of, from a physical
point of view, is very hard to say; scholars conceived them for a long
time as being analogous to some sort of threads stretched between two
bodies (Fig. 2.14).

Nowadays the lines of force are not much more than a convenient
expression, a matter of language. But it would be a mistake to believe
that language could not have a profound influence on the development
of science. The case of the lines of force is typical and very instructive.
The language and the simple representation created by Faraday were
of paramount importance for later developments.

Today we see things in the following way. Consider two electric

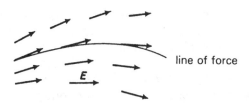

Figure 2.15

charges A, B of opposite sign as in Figure 2.14. If we place a positive charge C at any point of the intervening place, C will be subject to a force F, being the resultant of the repulsion by A and the attraction by B. The fact that a charge placed at any point of the space surrounding A and B experiences a force is expressed by saying that in that portion of space there is an *electric field*.[24]

Because, in this way, we are introducing a novel physical quantity, we have to state how it can be measured. To this end, one refers to a positive unit charge. The intensity of the electric field (or the electric field, for short) at a given point, is defined as the force experienced by a positive unit charge when located at that point.

The electric field, generally denoted by E, is plainly a vector. Visually, a region of space containing an electric field can be represented as in Figure 2.15. At every point, vector E is defined. A line of force is a line having at all points the same direction as E, or being everywhere tangent to E.

With some obvious changes, one can also define the magnetic field and the gravitational field, along with their lines of force.

At this point, one can restate the usual considerations about the operational definition, and can also be prepared to rebut the usual criticism. The electric field is a physical quantity, for we have given an unequivocal way to measure it.[25] We do not intend to deny anyone the right to think of a profound and unexplained reality on which the operational definition does not shed light. We only want to say that one can start to do serious physics without having to wait for the glorious day in which someone is capable of explaining what the field really is. If we had to wait for that day, we would not have radio and television today!

The *behavior* of the electric and magnetic fields was completely clarified by J. C. Maxwell by means of his celebrated equations. Maxwell was much influenced by Faraday's ideas. He was not in the least a supporter of agnostic formalism, and for a long time sought to cast his theory into a mechanical framework. However, as it turned out, posterity judged as really important (and extremely important) only the abstract theory of the electromagnetic field, not the vaguely intuitive and mechanistic concepts to which Maxwell appealed in order to justify his theory. H. Hertz said it explicitly: "Maxwell's theory consists of Maxwell's equations" (1832).

The last statement has often been criticized by those for whom equations are not enough. To repeat, at the risk of boring the reader, they have every right not to be satisfied. To be sure, it is worthwhile to go on digging, with a view to discovering what is hidden under the equations. However, it must be sincerely recognized that Maxwell's more or less mechanistic attempts are not in the least conducive to this end. Maxwell's equations represents the only firm starting point. Therefore Hertz was not altogether wrong.

The field can be described by either assigning at each point the vector *E* or assigning the *potential*. The potential at a point is defined as the work that the field does when a unit positive charge passes from that point to infinity. This definition makes sense, because it can be shown from Coulomb's law that the result does not depend on the path followed by the charge. The potential is measured in CGS electrostatic units of potential; more practically, one can use the Volt (*V*), which is a unit 300 times smaller.

The electric field *E* is always parallel to the direction of the steepest decrease of potential, and the faster the decrease the more intense it becomes.

Today we know that the electric charge is an intrinsic property of some of the fundamental particles. There exist positive, negative, and neutral particles. The first charged particle to be discovered was the *electron*, which carries a negative charge $e = (4.80325 \pm 0.00002) \times 10^{-10}$ electrostatic CGS units. A remarkable fact, not yet clearly explained is that all the elementary particles isolated so far have this same amount of positive or negative charge.

Due to their comparative mobility, the electrons are responsible for all the usual electric phenomena; they can even pass from one body to another. A neutral macroscopic body contains as many positive as negative charges. If it acquires or loses a certain number of electrons, it becomes negatively or positively charged. The production of electric charge by rubbing, known since antiquity, rests just on this process.

As long as the charges remain fixed, they give rise to the phenomena of electrostatics. A much richer class of phenomena occurs when the charges are set into motion. Naturally, such phenomena become more conspicuous when the number of the charges involved are greater and their motion is faster.

There are some materials, called *metallic conductors* (silver, copper, iron, and so on), within which the electrons are virtually free to move.

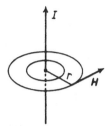

Figure 2.16

When a conducting wire carries a stream of electrons, one says that there is an *electric current*. The intensity of the current is defined as the total amount of charge passing through a cross section of the wire in a second. According to *Ohm's law*, the intensity of the current is proportional to the difference of potential between the two ends of the wire.

By convention, the current has the direction in which the positive charges move or would move if they were uniquely responsible for the current. As a matter of fact, only the electrons, which are negative, usually move. But a stream of electrons moving, say, to the right, produces exactly the same effects as a stream of positive charges moving to the left. Hence the direction of the current is, by convention, opposed to the motion of the electrons.[26]

An effective contrivance to generate a steady current was first built by A. Volta in 1799 with his electric pile. There is no point here to describe that device in detail. We shall only mention the decisive role it had in passing from pure electrostatics to electromagnetism.

2.10. Electromagnetism

From information about electric current it was soon possible to establish that electricity and magnetism are not separate phenomena, but are strictly connected, so that it is convenient to talk about *electromagnetism* (for history see, e.g., Whittaker, 1960).

The first important discovery was made in 1820 by H. C. Oersted who showed that an electric current gives rise to a magnetic field about it. A precise law for this phenomenon is that named after Biot and Savart. Let a straight wire of infinite length[27] carry a current *I* (Fig. 2.16). The lines of force of the magnetic field are circles lying on planes perpendicular to the wire and centered on the wire.

At a distance *r* from the wire the intensity *H* of the magnetic field

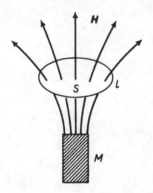

Figure 2.17

is given by

$$H = \frac{2I}{cr} \tag{2.28}$$

where c represents a constant. The direction of H is found by the corkscrew rule. A corkscrew advancing in the direction of the currrent must turn as is indicated by H.

A second phenomenon of importance is electromagnetic induction, discovered by Faraday in 1831. Let a magnet M (Fig. 2.17) be moved in the presence of a wire loop l. One can verify that a current arises in the wire.

In order to arrive at a quantitative law, we should introduce the concept of *magnetic flux*. The lines of force of the magnetic field go through the surface S bounded by l. If the lines of force represented the paths followed by the particles of a fluid, it would make sense to talk about the *flux*, that is, about the amount of fluid that goes across S in one second. This flux can be obtained by multiplying each element of S by the component of the fluid velocity perpendicular to it and by adding all the products. By analogy, the same operation can be carried out with the magnetic field in place of the velocity of the fluid. One has to multiply each element of S by the projection of H on the direction normal to it and add all the products. The result is termed the magnetic flux and will be denoted by Φ_H.

When Φ_H varies because the magnet is brought near it or is removed from it, a current arises in the wire l. As the current is a motion of charges, we can infer that an electric field E arises all along l.

The work done by the electric field E when a unit positive charge travels the whole loop l, is known as the *electromotive force* acting in the loop. If we denote the electromotive force by $\mathcal{E}$, the quantitative law of electromagnetic induction reads

$$\mathscr{E} = -\frac{1}{c}\frac{d\Phi_H}{dt} \tag{2.29}$$

First, we note that $\mathscr{E}$ turns out proportional to the time derivative of Φ_H. When the magnetic flux does not vary, no electromotive force arises, even if the flux is very strong. What only matters is the speed with which the flux is either increasing or decreasing.

Second, the minus sign on the right side of equation (2.29) means that we have to apply a countercorkscrew rule. The electromotive force is in the opposite direction to that of the rotation of a corkscrew advancing in the direction of increasing flux. As to c, it represents the same constant as in equation (2.28).

At this point, it is expedient to introduce the *magnetomotive force,* as an analogy with the electromotive force. Obviously, the magnetomotive force $\mathscr{M}$, acting along a loop l, is the work done by the magnetic field when a positive magnetic charge goes around the entire loop.

Let us return to the case of Figure 2.16, where a magnetic field H is engendered by an electric current I. By definition, a unit positive magnetic charge is acted on by a force H, tangent to the circle. If it completes a turn of the circle of radius r, the work will equal H times the length $2\pi r$ of the circumference. Hence by utilizing equation (2.28), we find

$$\mathscr{M} = \frac{4\pi}{c}I \tag{2.30}$$

One should show that the results do not depend on our having taken the loop l coincident with a circle. No matter what shape the loop may have, provided only that it encircles the current, one obtains equation (2.30).

The picture of the classical relations holding between electricity and magnetism can be completed by one more law, due to A. M. Ampère. Consider a current I carried by a straight wire of length l, embedded in a constant magnetic field H. If the wire is perpendicular to H (Fig. 2.18), one notices that it is acted on by a force F, perpendicular both to H and to the wire, given by

$$F = \frac{1}{c}IHl \tag{2.31}$$

If the wire instead is parallel to H, no force arises. More generally, if the wire and H are at an angle α, equation (2.31) must be multiplied by $\sin \alpha$. This means that the force is proportional to the projection of H on to the direction perpendicular to I in the plane of I and H.

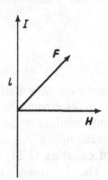

Figure 2.18

2.11. Maxwell's equations

There is no denying that the laws of electromagnetism, even cast into the simple and synthetic form just described, may appear somewhat distasteful to the beginner. Above all, one fails to perceive a harmonic connection between them.

J. C. Maxwell is credited with discovery in 1873 of the admirable synthesis that renders equations (2.29) and (2.30) virtually symmetric to each other. However, one had to wait for Einstein and his relativity theory in order to understand the perfect justifiability of equation (2.31).

The central point of Maxwell's contribution can be obtained in the following way. Consider a straight wire (Fig. 2.19) interrupted by a condenser, which is formed by two parallel metallic plates P_1, P_2. If a current I is flowing in the wire, plate P_1 gradually acquires positive charge, whereas plate P_2 acquires negative charge.

These charges give rise, inside the condenser, to an increasing electric field E, directed from P_1 to P_2. In a loop l encircling the current very far from the condenser we have a magnetomotive force $\mathcal{M}$, given by equation (2.30). But what will happen in a circuit l' wound around the condenser? Should we say that no magnetomotive force will arise there because no current goes across l'? Maxwell correctly surmised that the situation had to be exactly the same for l' and for l. This hypothesis has since been confirmed in a thousand ways, through the results derived from it. Note that the area inside l' is not crossed by a current but by an electric field whose flux is increasing. If this increasing flux were responsible for the generation of the magnetomotive force in l', we would get a fascinating analogy with equation (2.29), according to which an increasing magnetic flux generates an electromotive force.

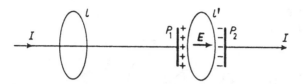

Figure 2.19

A little computation shows that the rate of increase of the electric flux Φ_E across l' is given by

$$\frac{d\Phi_E}{dt} = 4\pi I \tag{2.32}$$

Solving this equation for I and substituting into equation (2.30), we obtain

$$\mathcal{M} = \frac{1}{c}\frac{d\Phi_E}{dt} \tag{2.33}$$

Maxwell rightly guessed that the same magnetomotive force $\mathcal{M}$ should arise in l and in l' alike; in l it is caused by the true current I, whereas in l' it is caused by the displacement current. The latter is nothing but a variation of the electric flux across l'.

More generally, we can consider a circuit across which both a true and a displacement current are passing. Their effects can be added. By combining equations (2.30) and (2.33), we obtain

$$\mathcal{M} = \frac{1}{c}\frac{d\Phi_E}{dt} + \frac{4\pi}{c} I \tag{2.34}$$

As stated previously, there are no true magnetic charges. Hence if on the basis of the analogy of the true electric current we define a true magnetic current K, it will always be $K = 0$. As a result, no error is made if one rewrites equation (2.29) as

$$\mathcal{E} = -\frac{1}{c}\frac{d\Phi_H}{dt} - \frac{4\pi}{c} K \tag{2.35}$$

Equations (2.34) and (2.35), which show a high degree of symmetry, are known as *Maxwell's equations*.

As a matter of fact, for a reason to be explained shortly, Maxwell's equations are usually written for infinitesimal circuits, in a form that requires some familiarity with the differential calculus. It is not worthwhile here to tackle such complications, for it is merely a matter of mathematical form (although most suitable for many applications), whereas the substance of the underlying physics is perfectly embodied

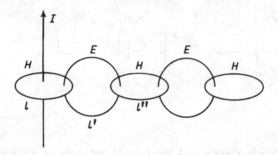

Figure 2.20

by equations (2.34) and (2.35). One says that a true or displacement current generates a magnetic field about it, such as that shown in Figure 2.16, whereas a true or displacement magnetic current generates an electric field about it, in a direction opposite to that of the former case.

The most important consequence of Maxwell's equations is that a varying electromagnetic field must propagate in space. To see this, consider an electric antenna, that is, a straight wire (Fig. 2.20) carrying an *alternating* current I, or a current going back and forth. Along a circuit l around the current there arises an alternating magnetic field H. Hence the magnetic flux across circuit l' varies all the time and gives rise to an alternating electric field E. As a result, there is a varying electric flux across l'' which engenders a magnetic field, and so on. Thus the electromagnetic field propagates away from the antenna.

The idea is very simple; however, a more precise and quantitative analysis of the process requires some calculus. Let us consider circuit l. In order to evaluate the magnetic field H around it, one cannot use the simple Biot and Savart law, equation (2.28), because it is not sufficient. For the area enclosed by l is not crossed by the true current I only, but also by the displacement current due to the electric field E around l'. Analogous considerations must be made for all the other circuits. It is evident that this interconnection by which the effects influence the causes, complicates the issue very much and calls for more sophisticated mathematics. One has to write Maxwell's equations for infinitesimal circuits and to introduce partial derivatives.

The most important result arrived at in this way is that the electromagnetic field about the antenna follows the oscillations of the current, however, with a retardation proportional to the distance from the antenna. At distance r each oscillation occurs only after time t, given by $t = r/c$. This evidently means that the electromagnetic field is propagated with velocity c.

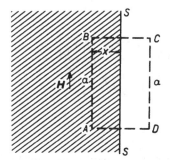

Figure 2.21

   With a simplified, but instructive, argument, we can derive the velocity of propagation in empty space also from equations (2.34) and (2.35). Suppose that the electromagnetic field is constant in a half space (shaded area of Fig. 2.21) and is zero in the other half space. No true current is present. Let the field H be parallel to the boundary SS of the half space and consider the ideal circuit represented by the rectangle $ABCD$, with $AB = a$. The magnetomotive force around it is readily found to be $\mathcal{M} = Ha$.
   By equation (2.34), a current must of necessity go through the area of the rectangle. Since by hypothesis no true current I is present, there must be a component E of the electric field perpendicular to the plane of the figure; suppose it is directed away from the reader. The electric flux is evidently given by $\Phi_E = Eax$. If x is constant, $d\Phi_E/dt$ vanishes and equation (2.34) cannot be satisfied. But if we admit that the electromagnetic field travels from left to right so that SS shifts to the right with a constant velocity v, we have $x = vt$ and there readily follows $d\Phi_E/dt = Eav$. Hence by applying equation (2.34), we get $H = E\, v/c$. Similarly, by reasoning, on a rectangle perpendicular to the plane of the drawing and applying equation (2.35) one would find $E = H\, v/c$. The last two equations readily yield $v = c$. Hence the field is propagated to the right with velocity c.
   The value of c can be derived from experiments of classical electromagnetism and turns out to be $c = 3 \times 10^{10}$ cm/s. As this value coincides with that of light speed in a vacuum, measured directly in different ways, Maxwell was led to set forth the hypothesis that light consists of oscillations of the electromagnetic field. Consequently, there arose the electromagnetic theory of light, which has had a huge number of confirmations.
   Today the value of c is derived by measuring the speed of electro-

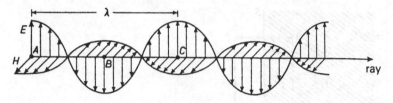

Figure 2.22

magnetic waves rather than by performing measurements on stationary electromagnetic phenomena. An enormous precision is obtained in this way and results in $c = (2.997924562 \pm 0.000000011) \times 10^{10}$ cm/s.

2.12. The electromagnetic waves

Every variation of the electromagnetic field taking place at a given point is propagated from there and reaches farther and farther away. An important kind of variation in time is represented by *sinusoidal*, or *harmonic*, motion. Harmonic motion is a back and forth motion, typically represented by the oscillations of a pendulum or of a tuning fork. If one plots the quantity a which varies harmonically, as a function of time, one obtains a curve, known as sinusoid.

When the current in the antenna oscillates sinusoidally, one obtains the simplest kind of electromagnetic waves. Along a *ray*, that is, along a line perpendicular to the antenna, the field takes the shape shown in Figure 2.22. The electric field and the magnetic field are perpendicular both to each other and to the ray. The configuration shown in the figure travels in bulk along the ray with velocity c. It is evident that at any fixed point of the ray, both the electric and the magnetic fields oscillate harmonically. The correspondence with the usual notion of wave is clear (e.g., of ocean waves).

The number of oscillations carried out in 1/s at a fixed place is called the *frequency* and will be denoted by ν. The duration of a complete oscillation, or period T, is evidently given by

$$T = \frac{1}{\nu} \tag{2.36}$$

The spacing between two successive maxima of the wave (Fig. 2.22) is known as the *wavelength* and is usually denoted by λ. A little reflection leads to the conclusion that the wavelength represents the distance covered by the wave during one period. Hence

$$\lambda = cT = \frac{c}{\nu} \tag{2.37}$$

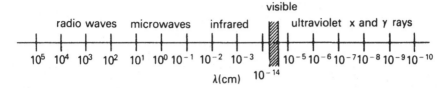

Figure 2.23

As c is constant, the wave can be characterized by either its wavelength or its frequency.

The wavelength range of the electromagnetic waves known today is amazingly wide (Fig. 2.23). Starting from the longest radio wavelengths (many kilometers), one can arrive with continuity to the most energetic γ rays (even less than one billionth billionth centimeter).

Light waves, that is, those waves that are capable of exciting our sense of sight, occupy an incredibly small portion of this entire *spectrum*. Roughly speaking, they range from $\lambda = 7 \times 10^{-5}$ cm (red) to $\lambda = 4 \times 10^{-5}$ cm (violet).

Two points such as A and C (Fig. 2.22), which reach the maximum of the field at the same time, are said to be in phase. They will also reach the minimum or, more generally, be at the same distance from the nearest maximum at the same time. By contrast, two points such as A and B, which at the same instant are found on the maximum and on the minimum, respectively, are said to have opposite phase.

In wave propagation one can define a *wave surface* as a surface whose points vibrate all in phase; they represent the spatial analog of the circles generated, say, by a stone thrown into a pond. As an example, a few wave surfaces generated by a finite length antenna, are shown in Figure 2.24 (dotted lines). At a great distance the wave surfaces tend to become spherical and concentric. The lines normal to the wave surfaces are called the rays. In the case of spherical waves the rays are evidently straight lines through the common center.

Sometimes plane waves are considered. In most cases they are simply spherical waves, generated by a faraway source, so that their curvature can be neglected. In the case of plane waves the rays are parallel lines; hence describing a beam of parallel rays is the same as describing plane waves (Fig. 2.25).

The electromagnetic field is propagated along the rays. However, this is not merely a kinematic phenomenon. A result of Maxwell's equations is that electromagnetic waves also carry some energy, in the sense that one must expend energy to generate them, but exactly the same amount of energy can be recovered at a distance where the waves

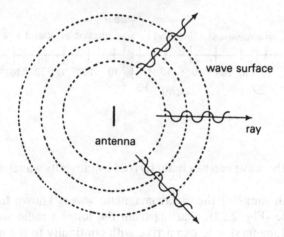

wave surface

ray

antenna

Figure 2.24

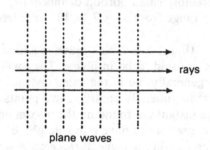

rays

plane waves

Figure 2.25

are received. The conservation principle for energy remains valid, provided that only *electromagnetic energy* is added to mechanical energy. When we generate an electromagnetic wave, we transform some energy, say, mechanical, into electromagnetic energy. This energy travels with the wave and can be recovered at a distance. Energy travels along the rays.

2.13. The polarization of material media

In general, we know that material bodies consist of atoms and molecules. These concepts will be discussed in some detail in Chapter 4. Here we only mention that an atom consists of a central nucleus, with positive electric charge and a cloud of negative electrons surrounding it.

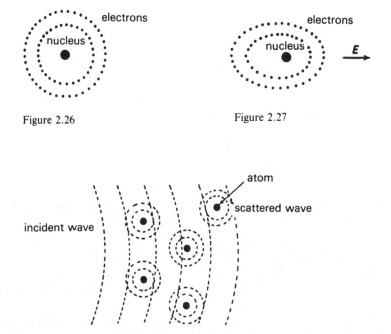

Figure 2.26

Figure 2.27

Figure 2.28

When the electron cloud is undisturbed, it takes roughly a spherical shape, centered on the nucleus (Fig. 2.26). However, when the atom is embedded in an electric field, the nucleus is acted on by a force in the direction of the field, whereas the electrons are pulled in the opposite direction. As a result, the atom becomes somewhat distorted (Fig. 2.27) and the nucleus no longer occupies the center. Altogether, the positive charge shifts a little toward the right, whereas the negative charge shifts to the left. The atom is said to be *polarized* and behaves like an *electric dipole* (the electric analog of a small magnetic needle).

When the field oscillates, the displacement of the charges follows the oscillations. As an electric current is nothing but a displacement of charges, the atom carries an alternating current and becomes a tiny antenna, which radiates a spherical wave.[28]

Suppose now that an incident electromagnetic wave impinges on to a material body (Fig. 2.28). Each atom is forced into oscillation and radiates a spherical wave (*scattered wave*). In this situation one observes the overall field, consisting of the incident, plus all the scattered waves. The process seems very complicated.

As a matter of fact, a mathematical investigation shows that the overall field behaves like one single wave, identical to the incident

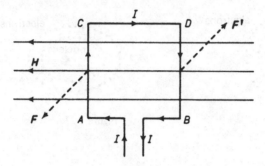

Figure 2.29

wave, except that it travels more slowly. Precisely, whereas the velocity of the incident wave is c, the velocity of the resultant wave is c/n, where n represents a positive number, called the *refractive index* of the medium. For glass, in a typical case, we have $n = 1.55$; for water, $n = 1.34$; for air, $n = 1.00029$. As the last value of n is very close to 1, propagation in the air is virtually identical to that in empty space.

Also, the magnetic field can give rise to polarization phenomena, although with a different mechanism. Consider a square loop of wire (Fig. 2.29), carrying current I, embedded in a uniform magnetic field H, parallel to the sides AB and CD of the square. Recalling what was said about equation (2.21), we infer that the sides AC and BD will be subjected to the opposite forces F and F', perpendicular to the plane of the drawing, whereas no force will act on the sides AB, CD. As a result, the loop will tend to turn and to place itself in a plane perpendicular to H. Arriving at this position, it will stop, because the electrodynamic forces will tend to widen it, not to turn it. Thus we see that the loop tends to orient itself in a magnetic field, like a compass needle. Exactly, the loop is equivalent to a magnetic needle perpendicular to its plane, and one can show that the loop's shape (whether or not square) is immaterial. Furthermore, the magnetic field generated by the current in the loop is also perfectly similar to that generated by a magnetic needle, or *magnetic dipole*.

This result led Ampère to set forth a fascinating hypothesis, according to which magnetic materials have a large number of tiny loops inside, which carry steady currents. These atomic loops are usually oriented at random and destroy the effect of each other. However, when the body is placed in a magnetic field, the loops orient themselves and become a system of aligned magnets, whose effects are added. The body is then magnetically polarized or magnetized.

Today we know that Ampère's hypothesis comes very near to the

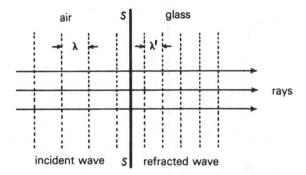

Figure 2.30

truth, even though it does not mirror it in detail, as will be seen when we introduce the spin (§4.11).

The magnetic polarization has in most cases a very modest intensity; consequently, the behavior of a material body with respect to electromagnetic propagation is virtually conditioned only by the electric polarization. An important exception is that of *ferromagnetic* materials, where the magnetic polarization can attain huge values and can also be conserved when the external field is removed. In the last case one has the well-known permanent magnets.

2.14. Reflection, refraction, dispersion

When an electromagnetic wave, say, a light wave, impinges on the smooth surface of a homogeneous body, coming from a vacuum, some interesting things occur. The incident energy is partly sent back, carried by a *reflected* wave, partly enters the body as a *refracted* wave, and partly is *absorbed* or disappears as light energy (whereas the body, of course, is heated). When the absorption is so strong that not even a small part of the energy goes through, the body is *opaque*; otherwise, it is *transparent*.

These facts have been known for a long time and their theory can be developed, starting with Maxwell's equations. We shall assume that they are known from experience and shall investigate some of their quantitative features.

Let SS (Fig. 2.30) represent the smooth surface of a transparent medium, say, glass; SS divides empty space or air from glass. Let a plane wave parallel to SS, or a beam or rays all perpendicular to SS, impinge on the glass surface. The reflected wave will recoil back in the air, hence it will have the same velocity and the same wavelength λ

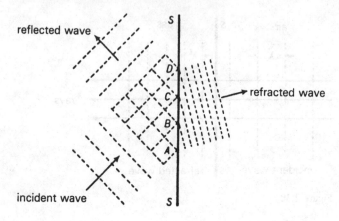

Figure 2.31

as the incident wave. The reflected wave will instead travel with the velocity c/n; from equation (2.37) we see that its wavelength λ is given by

$$\lambda' = \frac{c}{n} T = \frac{\lambda}{n} \qquad (2.38)$$

where n denotes the refractive index.

The wavelength in the glass therefore will, be smaller than in the air. We are now in a position to predict the result when the incident wave is inclined with respect to SS (Fig. 2.31). It may appear obvious that at A, B, C, D, where there are maxima of the incident wave, there are also the maxima of the reflected and refracted waves. If this were the case, the configuration would be that shown in the figure. As a matter of fact, this is not entirely correct; one should say instead that such points as A, B, C, D, which are in phase for the incident wave, should also be in phase both for the reflected and for the refracted waves. However, because the result would not change, let us give up argument for the sake of simplicity.

We see, then, that the reflected wave leaves at the same angle as the incident wave, whereas the refracted wave leaves at a smaller angle. Usually, one refers to the rays rather than to the waves. It is customary to consider only one of the incident rays, its point of incidence I, and the normal NN to SS through I (Fig. 2.32). The incidence angle i and the reflection angle r are equal, and the refraction angle r' is given by

$$\frac{\sin i}{\sin r'} = n \qquad (2.39)$$

This law is credited to W. Snell.

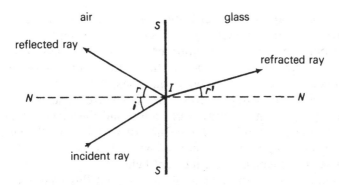

Figure 2.32

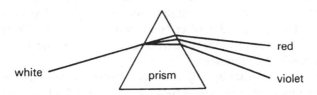

Figure 2.33

Refraction is a reversible process in the sense that if light comes from the glass, the incident and refracted rays exchange their roles.

Optics, when developed by means of rays, is called *geometrical optics*. It agrees perfectly with experience, as long as the wave surfaces are of infinite extent.[29] On the contrary, when the wave surface meets an obstacle of finite extent, some phenomena not predictable by geometrical optics arise. We shall deal with them presently.

Let us now carry out the refraction experiment with sunlight. As far as the reflected ray is concerned, there is nothing new. But the refracted light appears iridescent; it is as though several colors were present in the incident light, each one having a different refractive index.

This is precisely the conclusion that was drawn by Newton. To this end, he made use of a prism (Fig. 2.33) in which two successive refractions take place, one from air to glass and the other from glass to air. The incident light is white, whereas the light coming out of the prism shows a color *spectrum,* from red to violet. Red is the least deflected and violet is most deflected, from the initial direction.

The modern interpretation of these facts is as follows. Sunlight contains all frequencies of vibration (or all wavelengths). As a rule, the refractive index is an increasing function of frequency. As a result, the prism produces an angular separation of frequencies, deflecting less the lower and more the higher frequencies (*dispersion*).

When our eye is hit by radiation of a single frequency, or *monochromatic* radiation, we experience the sensation of a *pure color* or *spectral color*. The lowest visible frequencies ($\lambda \simeq 7 \times 10^{-5}$ cm) produce the sensation of red, whereas the highest frequencies ($\lambda \simeq 4 \times 10^{-5}$ cm) produce the sensation of violet, and the intermediate frequencies produce the other colors. The prism represents the simplest type of *spectroscope*, an apparatus capable of analyzing the different frequencies represented in the incident light.

The human eye is not capable of carrying out this analysis. When stimulated by light of different frequencies, it always responds with the sensation of *one single color*. As a result of experience, one can obtain any desired color by mixing, in suitable proportions, the lights of only three colors (e.g., red, green, and blue). *Colorimetry* is the science that studies the rules of composition of colors. Obviously, it can only refer to a conventional and average eye, with respect to which there are individual differences.

What does it mean that an object is red? It means that when hit by sunlight, the object absorbs some frequencies and reflects some other frequencies, which together give the sensation of red. These considerations can readily account for the change of color shown by the objects when instead of being illuminated by sunlight, they are illuminated with artificial sources. Furthermore, they may render obvious a concept that has long been obscure. When we say that an object *is* red, we should mean by this the entire process just described. As a matter of fact, "red" is in our mind, not in the object.[30] The object is just a filter of frequencies.

2.15. Lenses and images

Frequently in applied optics the incidence and refraction angles are very small (not greater than a few degrees). In this case, because the sine of an angle is virtually equal to the angle (expressed in radians), the law of refraction can take the approximate form

$$\frac{i}{r'} = n \tag{2.40}$$

This law is the basis for the elementary theory of optical instruments.

Let us first consider a thin prism, or a prism whose angle α (Fig. 2.34) is very small. In this case an easy argument shows that the deflection δ of a ray going through the prism is given by

$$\delta = (n - 1)\alpha \tag{2.41}$$

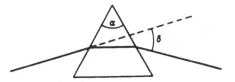

Figure 2.34

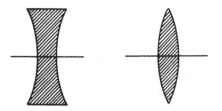

Figure 2.35

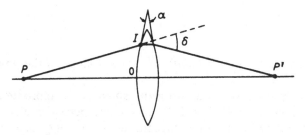

Figure 2.36

independent of the (small) incidence angle. As a particular case we have $\alpha = 0$, or a plane parallel plate. Here the deflection is zero.

A piece of glass limited by two spherical surfaces is called a lens. Figure 2.35 shows, in cross section, two examples representing a *convergent* lens and a *divergent* lens, respectively. The straight line through both centers of the spheres is called the *optical axis*. We shall deal with thin lenses, or with lenses whose thicknesses are very small (relative to the radii of the spheres).

Let us consider a light ray PI (Fig. 2.36) issued at point P of the optical axis of a convergent lens. Relative to it, the lens behaves as a thin prism of angle α, whose sides are represented by the tangent planes to the spheres at the points of incidence of the ray. The ray is deflected by the angle $\delta = (n - 1)\alpha$ and meets the optical axis at P'. Because α turns out to be proportional to the distance from I to the optical axis, the same is true of δ, and a simple geometric argument shows that *all* the rays issued by P (*object point*) will meet, with a good

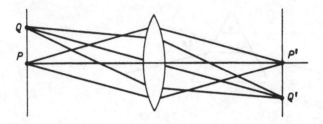

Figure 2.37

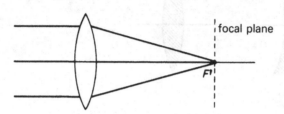

Figure 2.38

approximation, at P' (*image point*). One says that P' is the image of P.

Further analysis, always based on the small angle approximation, shows that every point Q (Fig. 2.37) of a plane perpendicular to the optical axis (*object plane*) has its image formed at a point Q' of a another plane perpendicular to the optical axis (*image plane*). An easy argument shows that in the case of Figure 2.36, if the object approaches the lens, the image recedes from it.

When the point object, located on the optical axis, is very far, its rays arrive all virtually parallel to the axis (Fig. 2.38); it is customary to say that the point is *at infinity*. The rays meet at the image point F', which here is called the *focus* of the lens. The name originates obviously from the case when the object is the sun. The distance from the lens to the focus is called the *focal distance* and the perpendicular plane through the focus is called the *focal plane*. Sometimes F' is called the rear focus, whereas the front focus is the point F symmetric with respect to the lens (Fig. 2.39). If a point source of light is located at F, the rays emanating from the lens form a parallel beam; the image is at infinity. We shall not describe the behavior of a divergent lens in detail. Obviously, it will tend to make the rays diverge instead of converge.

After these brief hints about geometrical optics let us now turn to vision. All ancient times, the middle ages, and part of the modern period

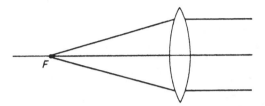

Figure 2.39

have grappled with the problem of vision, with little success. How do we manage to see objects?

Part of the difficulty arose from the failure to make a clear-cut distinction between what is an outer phenomenon and what is really within us, not only in the form of sensation or perception, but also as a subconscious data processing and a subsequent projection to the exterior. In this sense, images are within us and not on the objects.

The term *image* can have different meanings. Here we want to discuss the subjective representation of objects. Sometimes the word *phantasm* is used (from Greek: what appears). It will be stressed that phantasm is radically different from the image formed by a lens, as just described.

The naïve assumption that images are on the objects, led to a theory, according to which something like a peel or skin leaves the object and travels while decreasing in size until it penetrates into the eye. The correct approach was first found by Kepler (1604), who understood that one had to take each point of the object separately and to follow the path of each ray coming out of it.

We have already discussed what happens when light impinges on a smooth surface. When, on the contrary, the surface is rough, as is mostly the case, light is not regularly reflected, but is diffused. A single light ray impinging on the surface at *I* (Fig. 2.40) gives rise to an infinity of rays scattered in all directions. Each point of the surface of an illuminated object behaves as a point source of light.

Let us consider an object whose points are all close to a vertical plane (Fig. 2.39). A convergent lens forms an (upside down) image of it on another plane. If on the latter we place a photographic film or plate and enclose everything except the lens within opaque walls, we get a *camera*.

If we start from the situation of Figure 2.41 and make the object come closer to the lens, the image will recede from it. *Focusing* is a process by which we control the distance from the lens to the plate so as to bring the image back to the plate.

The human eye, considered an optical instrument, is essentially sim-

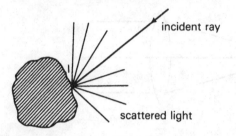

Figure 2.40

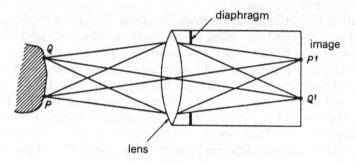

Figure 2.41

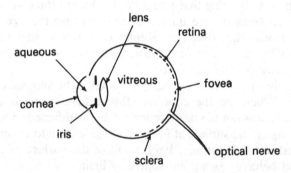

Figure 2.42

ilar to a camera. It consists of a sphere of about 2 cm in diameter (Fig. 2.42), enclosed by an opaque membrane (*sclera*), except in the front part, which is transparent (*cornea*). Behind the cornea we find a liquid (*aqueous*) and the lens, then a transparent sort of jelly (*vitreous*) until we come to the *retina* which covers most of the inner wall of the sphere. The retina is sensitive to light and sends its messages to the brain through the *optical nerve*. Situated before the lens there is a diaphragm

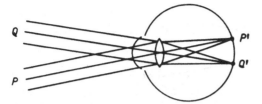

Figure 2.43

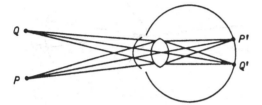

Figure 2.44

(*iris*) with a circular hole (*pupil*) whose diameter can vary (about 2 to 8 mm), according to lighting conditions.

When the eye is at rest, the images of objects at infinity are formed on the retina[31] (Fig. 2.43). As the object gets nearer, the image shifts behind the retina and focusing is needed. This is not accomplished in the same way as in a photographic camera, but by means of the lens which becomes more curved and more convergent (*accommodation*), as shown in Figure 2.44.

The retina is covered with cells sensitive to light, called *cones* and *rods*. The cones operate mainly in daylight conditions and are very thickly distributed in a small central spot of the retina (*fovea*). The rods are mainly in charge of night vision.

2.16. The physical theory of vision

From these facts concerning the operation of the human eye one can be led to formulate the physical theory of vision. This theory does not ignore *psychological* factors, inasmuch as it assumes that the observer associates an inner representation with the physical information collected by the eye. But it maintains that this representation is based uniquely on physical data. We shall first describe this theory, which in many cases is an excellent approximation and is, at any rate, a good starting point. Later we shall see that it does not represent the entire picture.

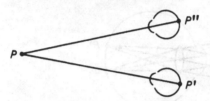

Figure 2.45

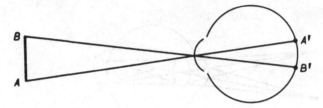

Figure 2.46

According to the physical theory, when a point source is presented to our eye we can tell the direction in which it is located from the point hit on the retina, and assess its distance from the effort of accommodation required to see it clearly.

As a matter of fact, accommodation can help in this assessment only at comparatively small distances. Beyond a few meters (say 5 m) the normal eye can see clearly without any accommodation; *infinity* begins at that distance.

But nature has endowed us with two eyes. When we fix our eyes on a point *P* (Fig. 2.45), we manage to bring its images on to two corresponding points *P'*, *P''* of both retinas (the foveas, in this case). This is achieved by turning our eyes so that their axes converge. The muscular effort of *convergence* gives us the necessary information to assess the distance of *P*. This mechanism is effective up to distances of about one hundred meters; beyond that there is infinity also for binocular vision.

Let us restrict ourselves to distances less than 100 m. When an object *AB* (Fig. 2.46) is presented to the eye, its retinal image *A'B'* is larger, the closer the object is. Being in a position to judge the distance, we can infer the size of the object from this. Moreover, as both retinal images of the complete object will not be equal because they were taken from two different points of view, we also have some elements by which to determine the *relief* of the object. A well-known application of this has been made to *stereoscopic* photography.

According to physical theory, we have all the elements needed to determine the distance, the size, and the relief of the object at which we are looking. The capability to make this determination does not seem to be innate, but is acquired with experience. The newborn baby is ostensibly not able to give a meaning to its visual sensations. In time, the infant starts little by little to connect the visual sensations with the tactile sensations received from surrounding objects and to adjust sight by touch. In this way we learn that a given luminous sensation, connected with a given effort of accommodation and convergence, corresponds to a given object at a given distance which we can touch by properly stretching our hand. When at this stage, we start to project our sensations out into the surrounding space. We get the impression of *seeing* things in space, definitely outside of us.[32]

In order for this piecing together of objects in space to be possible, the observer must receive all the information he needs, although the form in which such information is presented to him is immaterial, provided that it is a constant form. For he learns little by little from experience what kind of spatial object corresponds to what kind of information.

In this way one can easily clarify a number of doubts that have in turn arisen about the process of vision. For example, it was asked: Why do we see an object large, whereas the retinal image is small? or Why do we see it upright, whereas the retinal image is upside down? The explanation is that we do not see the retinal image! The image is only one link in the chain of physical events leading from the object to our brain; in this respect, its size is immaterial.

Everyone knows that a shortsighted person must wear divergent glasses. The first time the person puts on the glasses, all objects seem to be smaller than they really are. But in a short time, there is an adjustment of sight and everything comes back to normal size. In an eighteenth century experiment, a subject was supplied with special glasses that invert the images. The first days, with the glasses worn steadily, are a real disaster. Then the subject adjusts and starts to see normally, but is again in trouble when the glasses are finally removed!

An interesting kind of adjustment made in modern times is that of vision through the rearview mirror of an automobile. Expert drivers do not perform complicated works of interpretation; they simply *see* behind them. It is as though there were eyes in the back of their heads!

If we piece together the outer world by means of the physical data reaching our eye, a change of the data will entail a change in our representation of the objects. This is what is done by an *optical instrument* that transforms the light coming from the objects in order to render it more suitable to our purposes. The simplest optical instrument is obviously the *plane mirror*. Its operation is well known and no further

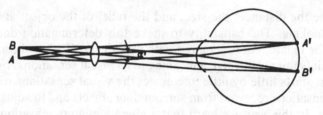

Figure 2.47

description is necessary. Through a plane mirror we see an object where it actually is not, but its size is unaltered (only left and right are inverted). We shall now briefly describe a few instruments that give magnification and thereby help us to distinguish details which the naked eye cannot see.

Our eye has limited resolving power. This means that when two points are too close to each other, the eye cannot separate them, but instead sees one single point. For example, this obviously must happen when both images fall on the same cone of the retina. As an order of magnitude one can say that two points are seen as separate by a normal eye only if their angular separation exceeds one minute of arc (one-sixtieth of a degree). Details finer than this lower limit cannot be perceived.

If an observer, looking at an object at a given distance, is interested in details finer than those he can distinguish, he can simply get closer to the object. In this way the angle under which any two points are seen, increases (Fig. 2.46). But there is a limit to this, represented by the limited power of accommodation of the eye lens. We cannot focus on a point closer than a certain distance. This limit varies with age, but is present for everyone.

In this case, one can resort to a *magnifying glass*. This is essentially a convergent lens of short focal length, which supplies us with the additional converging power required. But we need not put the lens and the object very close to our eye, as might be expected. The distance AB from the object to the eye (Fig. 2.47) is rather immaterial, and can remain the normal one. We only have to put the lens between the eye and the object so that the latter falls in the front focal plane of the lens. Thus the rays emitted by any one point of the object leave the lens as a parallel beam and the eye can see the point sharply, without accommodation (i.e., at infinity). But the beams coming from A and B are at an angle α' larger than the angle α in which both points were seen by the naked eye.

The magnifying glass is sometimes also termed the *simple microscope*. By combining two or more lenses, one can build a more powerful

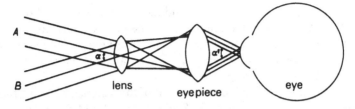

Figure 2.48

compound microscope, which will not be described here; usually, this is simply called the *microscope*.

Sometimes the details of an object are not sharp, because the object is too far away and, for some reason, we cannot get closer to it. In this case one can resort to the *telescope*. In its simplest version a telescope consists of two convergent lenses (Fig. 2.48), termed the *objective* and the *eyepiece*. The rear focal plane of the objective coincides with the front focal plane of the eyepiece. The latter is more convergent, that is, has a shorter focal length than the objective. As shown in the figure, two points A, B at infinity, which would be seen by the naked eye under angle α, are seen through the instrument, still at infinity, but under a wider angle α'. The ratio α'/α is called the *magnification* of the telescope.

The telescope was discovered in Holland at the beginning of the seventeenth century, but no reasonable theory of it was known at the time. Galileo heard about it, and built several samples of the telescope. As a matter of fact, the Galilean telescope is different from the one we described, in that its eyepiece is divergent. However, this is not of conceptual importance. It is to Galileo's credit that he was the first to believe in the real capabilities of the instrument, and to use it for observing the sky. The results he obtained were of the utmost importance.

In this respect, Galileo's work can shed some light on the discussion made about theoretical and observational terms (§1.7). Of course, Galileo did not talk in these terms but essentially understood that the things he was seeing through the telescope were observational (perhaps we would say theoretical), exactly to the same extent as those seen with the naked eye and, consequently, deserved exactly the same confidence. Instead his opponents said that they were fallacies, entities whose correspondence with directly observable things was deceptive.

2.17. How do we really see?

As already anticipated, we must now subject the physical theory of vision to serious criticism. The theory certainly represents a first and necessary step toward an understanding of the process, but it does not consider a number of essential factors. The first to set forth a clear and convincing analysis of the subject was G. Berkeley (1709) (see Berkeley, 1948, p. 141). One might say that his analysis gave rise to what today is called *psychological optics*.

We shall state the fundamental result of this science by using modern concepts and terms, as follows. *Seeing* means building an inner phantasm and projecting it into outer space by combining the information coming through the eyes with other information stored (mostly unconsciously) in the mind of the observer.

All this will become clearer when we understand some concepts of information theory (§3.11); but there should be no difficulty in comprehending the main idea. We know the structure of the visual world and apply this general knowledge to the interpretation of the data in each particular case.

Of course, should this process take place through conscious reasoning, the situation would be trivial and there would be no point in dwelling on it. But as a matter of fact, the process occurs mostly unconsciously, and so quickly, that we get the impression that all the data, including those inserted by ourselves, come from the object. This can be illustrated by a few examples.

First of all, we ask: "How do we assess the distance of an object farther than 100 m?" If the object is of a familiar kind so that we know, at least approximately, its real size, the task is easy.[33] From the angular size we can immediately infer the distance. This occurs when the object is, say, a man, a house, a car. We see it immediately as being far away, without even thinking about it.

When the size of the object is scarcely (if at all) known, the situation gets more complicated. Nevertheless, we often have some clues. For one thing, the transparency of the atmosphere is never perfect; there is always some dust or haze which makes objects in the distance appear fuzzy. We rely on this clue so much, even to the point of being deceived. Who has not noticed that on a very clear day, mountains seem closer?

Finally, let us consider the case of heavenly bodies. First, because no element enables us to make distinctions, we see all of them attached to one and the same surface, the *vault of heaven*. We place this surface simply beyond everything we have seen in the different directions. That is why the vault is flattened! Heavenly objects at the zenith seem closer than those at the horizon. Clouds are largely responsible for this. For when they are above us, their speed can be conspicuous, whereas when near the horizon, they move extremely showly; we infer from this that clouds near the horizon are farther away.

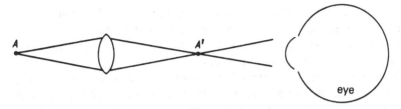

Figure 2.49

This same phenomenon is responsible for the fact that the sun and the moon appear much larger near the horizon than on the meridian. If their angular size is measured with an instrument, it is found to be identical in both cases! The effect is merely psychological, and depends on the fact that when near the horizon, the sun and the moon are thought to be farther away.

As Berkeley says: "This phaenomenon of the horizontal moon is a clear instance of the insufficiency of lines and angles for explaining the way wherein the mind perceives and estimates the magnitude of outward objects." (Berkeley, 1948, p. 203).

So far we have dealt with the case when the clues used in the physical theory are unknown and the observer cannot but resort to what he knows or believes he knows about the visual world. Things become extremely interesting when the physical data are available, but clash with the common experience of the observer. How does the observer behave in this case? More often than not, he decides against the physical data!

Consider the case of a magnifying glass, forming the image at infinity. The observer is not deceived and sees the image at the same distance as the object. In the case of a telescope, the situation goes the other way around. The object, which is virtually at infinity, should be seen still at infinity, although magnified. However, the lay person does not say that the telescope magnifies the object, but instead says that the telescope brings the object nearer.

The most striking thing happens with a convergent lens, which forms a real image A' of an object A (Fig. 2.49). The observer who sees the rays coming from A' should locate his phantasm at A'. In this experiment, no one ever sees an object floating in the air. The observer locates the image where he knows that there is something material, that is, either at A or on the lens. This experience is very instructive.

One could go on and on with geometrical examples, but the concept should be clear by now. Let us instead mention a case regarding color vision. When a scene is illuminated with monochromatic light, that is, with light of a single wavelength, the objects can only reflect that same wavelength, therefore they should all have the same color. This is not exactly what happens. One can perform the experiment, for instance, with a sodium lamp. The colors will be somewhat altered, but someone will still report seeing different colors.

The foregoing discussion should help convince the reader that vision includes a stage of data processing, which takes place immediately and unconsciously in the mind of the observer. In this connection it is of interest to mention that histology and embryology show that the retina is a very complex tissue, virtually an appendix of the brain tissue. If based uniquely on elementary sensations the theory is much too simple to be right.

If this is so, we must once again criticize the naïve assumption of those who want to base epistemology on the assumed immediacy of visual data and on their perfect correspondence with reality.[34]

2.18. Interference and diffraction

The hypothesis that light consists of waves has a long history. It was clearly postulated by C. Huygens at the end of the seventeenth century but was irrefutably established only at the beginning of the nineteenth century through the successful efforts of T. Young and A. Fresnel. The entire eighteenth century was dominated by the *corpuscular* hypothesis, entertained by Newton, according to which light consisted of tiny projectiles whose paths were the rays.

However, G. Grimaldi had shown already in 1690 that the paths were not straight lines and that light could turn obstacles around. He gave the name of *diffraction* to this phenomenon. The simplest case of diffraction occurs when one seeks to isolate one ray by placing an opaque screen with a tiny hole on the light path. It turns out that the outgoing light, far from approaching one single ray, spreads out into a cone. The smaller the hole is, the wider the cone.

A further phenomenon that cannot be reconciled with the corpuscular hypothesis is that called *interference*. The simplest device to produce interference is the two-hole interferometer, used by Young in 1804. A point source S of monochromatic light (Fig. 2.50) sends light to an opaque screen with two holes F_1, F_2. Due to diffraction, the light from each hole diverges in a cone and illuminates a spot on a second screen HK. Both spots partially overlap. The illumination in the common zone is not uniform and equal to twice that produced by a single hole, as might be expected. One perceives instead a number of fringes, that is, of parallel and equidistant lines, alternatively bright and dark.

The wave theory explanation given by Young is as follows. If S is equidistant from F_1, F_2, light oscillations on both holes are *in phase*

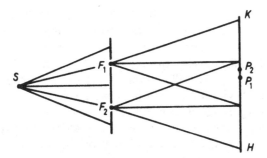

Figure 2.50

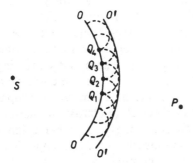

Figure 2.51

and will arrive in phase at a point P_1 of HK, also equidistant from F_1, F_2. At P_1 both oscillations will therefore reach maxima and minima at the same times and their effects will add. Hence P_1 will be a bright point. If we go to a neighboring point P_2, such that the path A_2P_2 is a half wavelength longer than A_1P_2, both oscillations will arrive there with *opposite phases*. When one reaches its maximum, the other will be at a minimum (i.e., a maximum in the opposite direction). Thus their sum will steadily vanish and P_2 will be dark. The bright fringes are the locus of points in phase, whereas dark fringes are the locus of points with opposite phases. With this device Young was also able to measure the wavelength of the light used.

Now we consider diffraction. This was accounted for by A. Fresnel, by digging out an old principle, due to C. Huygens, and adding something essential to it. Huygen's principle of elementary waves states that each point Q_1, Q_2, . . . , (Fig. 2.51) attained by a light wave emitted by a source S, becomes in turn the source of a spherical wave (elementary wave). According to Huygens, the next wave surface $W'W'$ is the envelope of all the elementary waves (i.e., a surface tangent to all of them).

Fresnel preserved the hypothesis of the emission of elementary waves by Q_1, Q_2, . . . , but benefiting from Young's experience, inferred that in order to evaluate the effect of all the elementary waves at a point P, one should consider the phase with which each wave arrives at P. All waves in phase will add their vibrations, but one must subtract the effect of all vibrations with opposite phase.

By applying this Huygens–Fresnel principle,[35] and carrying out some computations, it is found that when a wave propagates in free space, the results are those predicted by ray optics. If on the contrary, part of the wave surface is intercepted by an opaque obstacle, light can, to some extent, turn around it and invade the shade zone. As a result, the passage from the lit area to the shade is never sharp; it is somewhat fuzzy and presents a number of maxima and minima of brightness.

Fresnel's calculations turned out quite in agreement with the results of experience. After Young and Fresnel, no one dared any longer to doubt the wave nature of light. One could be sure that something was oscillating. But *what* was oscillating? The best people could do at the time was to postulate a very odd elastic medium, which propagated the waves. One had to wait for J. Maxwell to clarify the matter.

A diffraction process of great consequence occurs when a light wave is collected by a lens, such as the objective of a telescope or microscope, or the lens of the eye. If the incoming wave is plane, it is concentrated onto a point focus. However, the wave surface is not free and of infinite extent. It is of necessity limited by the rim of the lens or by the eye pupil. This limitation gives rise to diffraction, with the result that not all light goes exactly through the focus. The light turns out to be distributed in a little *disk,* surrounded by a few rings, alternatively bright and dark. As very little light goes through the rings, one can ignore the rings and consider only the disk. If the lens diameter is denoted by D and the focal length by f, the radius of the disk is approximately expressed by

$$r = \frac{\lambda}{D} f \tag{2.42}$$

Because every point of the object is represented in the image by a disk of finite extent and not by a single point, the image is not sharp. The effect is similar to that produced by poor focusing.

Quantitatively, the effect can be described by introducing the concept of *resolving power*. The resolving power of an instrument is defined as the minimum spacing at which two points are still perceived as separate. Experience shows that two points are just at the limit of the resolving power when their images are spaced by a distance about equal to the radius r of the diffraction disk. Hence from equation (2.42) we may conclude that the smallest angular distance α between two points seen as separate through a telescope, whose objective has diameter D, is given by

$$\alpha = \frac{\lambda}{D} \qquad (2.43)$$

In the case of the eye, if D represents the diameter of the pupil (about 2 mm), the result is that α is about one minute of arc. This is just what experience gives for a normal eye. One should not conclude, then, that the resolving power of the eye is actually limited by diffraction; rather, one should say that better refractive media and finer cones would not help the eye, because the wave nature of light would render them useless anyway. Nature likes economy.

If we want to resolve the details of a distant object (say, a heavenly body), ten times or a hundred times better than with the naked eye, we must resort to a telescope whose diameter is ten times or a hundred times that of the eye pupil.

In the case of the microscope the object is near, and virtually in the front focal plane of the objective. In this case one does not consider the angular separation, but the linear separation. The resolving power is simply equal to the radius r given by equation (2.42), where f denotes the focal length of the objective (or virtually the distance from the object to the objective).

2.19. The Galilean relativity

Maxwell's theory of electromagnetism and the discovery of the electromagnetic nature of light brought again to the surface a problem that seemed to have been solved once and for all by Galileo.

In a celebrated page of the *Dialogue Concerning the Two Chief World Systems*, Galileo writes:

Shut yourself up with some friend in the main cabin below decks on some large ship, and have with you there some flies, butterflies, and other small flying animals. Have a large bowl of water with some fish in it; hang up a bottle that empties drop by drop into a wide vessel beneath it. With the ship standing still, observe carefully how the little animals fly with equal speed to all sides of the cabin. The fish swim indifferently in all directions; the drops fall into the vessel beneath; and, in throwing something to your friend, you need throw it no more strongly in one direction than another, the distances being equal; jumping with your feet together, you pass equal spaces in every direction. When you have observed all these things carefully (though there is no doubt that when the ship is standing still everything must happen in this way), have the ship proceed with any speed you like, so long as the motion is uniform and not fluctuating this way and that. You will discover not the least change in all the effects named, nor could you tell from any of them whether the ship was moving or standing still. In jumping, you will pass on the floor the same spaces as before, nor will you make larger jumps toward the stern than toward the prow even though the ship is moving quite rapidly, despite the fact that during the time that you are in the air the floor under you will be going in a direction opposite to your jump. In throwing something to your companion, you will need no more force to get it to him whether he is in the direction of the bow or the stern, with yourself situated opposite. The droplets will fall as before into the vessel beneath without dropping toward the stern, although while the drops are in the air the ship runs many spans. The fish in their water will swim toward the front of their bowl with no more effort than toward the back, and will go with equal ease to bait placed anywhere around the edges of the bowl. Finally the butterflies and flies will continue their flights indifferently toward every side, nor will it ever happen that they are concentrated toward the stern, as if tired out from keeping up with the course of the ship, from which they will have been separated during long intervals by keeping themselves in the air. And if smoke is made by burning some incense, it will be seen going up in the form of a little cloud, remaining still and moving no more toward one side than the other. The cause of all these correspondences of effects is the fact that the ship's motion is common to all the things contained in it, and to the air also. That is why I said you should be below decks; for if this took place above in the open air, which would not follow the course of the ship, more or less noticeable differences would be seen in some of the effects noted. No doubt the smoke would fall as much behind as the air itself. The flies likewise, and the butterflies, held back by the air, would be unable to follow the ship's motion if they were separated from it by a perceptible distance. But keeping themselves near it, they would follow it without effort or hindrance; for the ship, being an unbroken structure, carries with it a part of the nearby air. For a similar reason we sometimes, when riding horseback, see persistent flies and horseflies following our horses, flying now to one part of their bodies and now to another. But the difference would be small as regards the falling drops, and as to the jumping and the throwing it would be quite imperceptible (translated by S. Drake).

In this passage, Galileo essentially states that the physical laws appear to be the same in a fixed laboratory as in a laboratory traveling with uniform and straight motion (*Galilean relativity*). Naturally, Galileo refers to those physical processes that were known in his days, that is, to mechanical processes.

How can one verify a similar statement? In order to do this in a quantitative way, one must, first, be able to measure with comparative precision spatial coordinates and time.

Let us compare a fixed laboratory (or reference system or reference frame) K with a laboratory K' in uniform and straight motion. The observer located in K will make use of a system of rectangular Cartesian coordinates x, y, z whose origin O is fixed (Fig. 2.52), whereas the observer located in K' will refer to a system x', y', z', with origin at O', fixed relative to him, hence moving with a constant velocity v relative to K. Without any loss of generality, we can assume that the

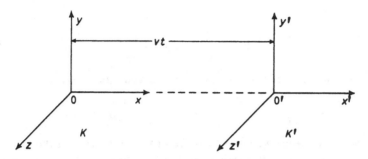

Figure 2.52

axes x and x' coincide (i.e., x' glides along x) and that y, y' and z, z', respectively, are maintained parallel. Let each observer be provided with a perfect clock. The clocks are set in such a way that both are at zero hours at the instant when O' passes through O.

Let an instantaneous event (as the blowing up of a bomb) take place at point P. The observer in K will say that the event has taken place at x, y, z, at the instant t, and the observer in K' will say that the event has taken place at x', y', z, at the instant t'. For all we know, before 1900 the two identical clocks represented two instruments belonging to one and the same operational class as described in §1.8; they should yield identical results, independently of the velocity v. Hence we should have $t = t'$.

Now since the displacement of K' with respect to K is given by vt, we obtain the obvious relations

$$x' = x - vt$$
$$y' = y$$
$$z' = z \tag{2.44}$$
$$t' = t$$

This system of equations is known today as *Galileo's transformation*. As a matter of fact, this name can be justified only by concepts developed three centuries after Galileo. In the seventeenth century it would have seemed silly to write down obvious equations such as equation (2.44), and to give them a pretentious name.

Let us examine the bearing of Galileo's transformation on the laws of mechanics.

Let a point P be in motion so that its coordinates x, y, z are functions of t. The K observer will measure a velocity V, whose components are expressed by

$$V_x = \frac{dx}{dt}, \qquad V_y = \frac{dy}{dt}, \qquad V_z = \frac{dz}{dt} \qquad (2.45)$$

whereas the K' observer will measure a velocity V' with

$$V_x' = \frac{dx'}{dt'}, \qquad V_y' = \frac{dy'}{dt'}, \qquad V_z' = \frac{dz'}{dt'} \qquad (2.46)$$

In these equations we can substitute t for t', by virtue of the last equation (2.44). Next, by considering the first three equations (2.44), we get

$$V_x' = \frac{d(x - vt)}{dt} = \frac{dx}{dt} - \frac{d(vt)}{dt}$$

$$V_y' = \frac{dy}{dt}, \qquad V_z' = \frac{dz}{dt} \qquad (2.47)$$

It is readily seen that $d(vt)/dt = v$,[36] hence on comparison with equation (2.45), we obtain

$$V_x' = V_x - v, \qquad V_y' = V_y, \qquad V_z' = V_z \qquad (2.48)$$

These are the equations describing the transformation of velocity in going from K to K'. Both observers do not obtain the same result for the velocity of P. Thus velocity is not invariant under Galileo's transformation. The result is trivial. In order to obtain V', one has to subtract v from V.

As v is a constant, there follows that when V is constant, V' is constant, too. Hence the law of inertia, if valid in K, is also valid in K'. A body not subject to any force moves uniformly and in a straight line, both in K and in K'.

Let us denote by a the acceleration in K, and by a' the acceleration in K'. One has to evaluate the time derivative of the velocity by taking into account equation (2.48). As v is a constant, its derivative vanishes, as is readily seen. Therefore we obtain

$$\frac{dV_x'}{dt} = \frac{dV_x}{dt}, \qquad \frac{dV_y'}{dt} = \frac{dV_y}{dt}, \qquad \frac{dV_z'}{dt} = \frac{dV_z}{dt} \qquad (2.49)$$

All the components of the acceleration are invariant. As a result, acceleration is an invariant; both K and K' measure the same acceleration of P.

Now let a force F act on a body of mass m. If F is measured by a spring dynamometer, both K and K' will notice the same extension of the spring and, consequently, will get the same value for F. Hence the second law of dynamics has the same form $F = ma$, or $F' = m'a'$, in both systems; there also follows that $m = m'$, whereby mass is invariant.

If F and F' are measured by the dynamometer, the law of action and reaction is also valid for K and K' alike. Therefore the three laws of mechanics (i.e., all of classical mechanics) are invariant under Galileo's transformation. As a result, it is impossible to determine by means of mechanical experiments conducted within one system whether the system is at rest or in uniform motion. What physical meaning can the statement have that one system is at rest as the other is moving? Within the scope of mechanics, there is no means to verify such a statement.

Nonetheless, Newton remained attached to the notion of absolute space and time. According to his conception, which was shared by many scientists,[37] there exists an immovable space, independent of the observer; bodies move or are at rest with respect to this space. Likewise, time flows outside of us, at an inexorable pace, independent of the system in which it is measured.

Once again, Newton's position suggests some interesting remarks. Newton insisted on the existence of absolute rest and absolute motion, at the same time when he was discovering his laws which are the same for an observer at rest as for an observer in uniform motion. Such laws render it impossible to ascertain whether the observer is at rest or is moving.

Scholars have long debated about the effect metaphysical prejudice may have had on Newton's choice.[38] Undoubtedly, his ideas were accepted and largely shared by his successors, in spite of the very serious arguments to the contrary, brought out by scientists such as C. Huygens and G. Leibniz. Leibniz, in particular, had quite a modern conception of space – as being a relation between material things.

2.20. Einstein's relativity

The advent of electromagnetism and Maxwell's synthesis seemed to prove the absolutist vision of Newton as right.

Note that Galileo's relativity does not affirm the invariance of all physical quantities. We have already stressed, for instance, that velocity (hence energy, momentum, and so forth) is not an invariant. But physical laws are invariant, because mechanics is based on three axioms that remain invariant.

The situation is different in the case of Maxwell's equations. They represent general laws; nevertheless they contain, in an essential role, the constant c, the speed of light. Hence if Galileo's transformation is valid, Maxwell's equations are not invariant.[39] In particular, by measuring the speed of light in a reference frame, one could infer whether the frame is at rest or is moving.

The mathematical problem concerning the invariance was parallel to the physical problem of the *ether*. Maxwell had shown that light was an electromagnetic wave propagation; the analogy with the propagation of mechanical waves (seismic waves, sound waves, surface waves in a liquid, and so forth) was only too obvious. Mechanical waves always occur in a material medium whose particles oscillate. Hence there is a material substance, relative to which it makes sense to say whether we are moving or are at rest. It should therefore be legitimate to state whether the observer is at rest or is moving with respect to the system in which the waves are propagated.

At this point, it was only natural to ask: In what medium do electromagnetic waves travel? The usual material media are ruled out because light propagates also in a vacuum.[40] A special substance, the ether, was considered, endowed with properties very different from those of all other material substances. It had to be so thin as to penetrate all bodies, and had to be capable of transmitting light vibrations. And because light vibrations are transversal to the direction of propagation, the ether had to behave as a solid body.

The problem of detecting the observer's motion relative to ether was tightly connected to that of deciding whether the velocity of light was added to that of the observer. As the speed of light is enormous relative to that attainable by any observer, one resorted to the motion of the earth in space. The earth's velocity in the solar system is about 30 km/s. This is still ten thousand times less than the speed of light, nevertheless there was some hope of measuring it.

A. A. Michelson set out to do this by conducting an interferometric experiment, first carried out with E. W. Morley in 1887; the experiment was later repeated many times, with ever increasing precision. The result is that the velocity of light does *not* add to that of the earth.[41] Whatever the direction of propagation relative to the earth's motion, the speed of light is always found to be the same.

A solution to the problem was indicated by Einstein in 1905. However, it is worthwhile to note two facts which have more than an historical importance. First Michelson many years after his experiment, still believed in the ether. Second Einstein, according to his own declaration, was not prompted to elaborate his theory by Michelson's experiment. Yet that experiment was considered by many people as the crucial experiment in favor of relativity! For the time being, let us confine ourselves to merely recording these two facts whose relevance will become clearer later (see, e.g., §2.28 and §2.29).

The Galilean relativity does not apply to any motion, but only to uniform motion. If the ship described by Galileo were sailing on a rough sea and were subjected to pitching and rolling, the observer would obviously realize his movement. In such a frame the simple laws of Newtonian mechanics are plainly not valid.

We must conclude that there are some reference frames in which the laws of Newtonian mechanics are valid, and other frames in which they are not valid. The former are called *inertial frames*.[42]

Galileo's relativity consists of the statement that a frame moving relative to an inertial frame is also inertial. The earth represents an inertial frame only as long as we do not exceed a certain precision in the measurements and certain values for the parameters. This is due to the rotation of our planet about its axis, as can be shown, for instance, by *Foucault's pendulum*. It is a result of Newtonian mechanics that a pendulum, once set into motion, oscillates in a fixed plane. A pendulum on the earth appears instead to oscillate in a plane that rotates slowly in opposite direction to the terrestrial rotation. One usually says that an inertial frame must preserve its orientation relative to the fixed stars.[43] In any event, the observations suggested by Galileo were not so precise as to reveal that the earth is not an inertial reference frame.

Einstein,[44] first of all, extended the relativity principle to the electromagnetic phenomena by postulating that "the same laws of electrodynamics and optics should be valid for all reference frames for which the equations of mechanics are valid." Today this first postulate may be worded as follows:[45]

1. In all inertial frames the same physical laws are valid.

To this, Einstein added as a second postulate the invariance of the velocity of light.

2. Light travels in empty space always with velocity c, independent of the motion of the source.

It is easy to see that if this second postulate holds, the notion of *simultaneity* becomes relative and depends on the observer's frame. Consider again Galileo's ship. Let a lamp be lit at a point of the deck, at middistance between the bow and the stern. To a passenger on the ship, light will appear to employ the same time to reach both the bow and the stern. Therefore to that observer, both events: A = light's arrival at the bow and B = light's arrival at the stern are simultaneous. For an observer on the shore, however, as light travels, the bow is running away from it, whereas the stern is advancing toward it. As a result, event B will occur before event A. From this we can draw the conclusion that two events simultaneous for the passenger, may not be simultaneous for the person on the shore.

Einstein inferred from similar considerations that the concept of time had to be revised and that the absolute Newtonian time had to be given up. One had to analyze with operational methodology how time is measured and in what way clocks can be synchronized in different inertial frames. This was the key that opened the door to relativity and which allowed the construction of the theory.

At this point, however, we shall leave Einstein's original procedure. It is more convenient to pass to a more general representation, considered a few years later by H. Minkowski.

2.21. The Lorentz transformation

Space and time are physical entities of a particular kind, in a sense, privileged with respect to all others. For in order to describe any physical process, one must first specify places and times in which it takes place. This peculiar nature of space and time has always interested scientists and philosophers who have considered space and time as either a substrate in which reality is cast (Newton) or a substrate into which we cast reality (Kant).

Newtonian space and time are absolute and do not depend on the observer. For space the absolute reference frame was thought for some time to be the ether. A body moves or is at rest relative to the ether. Two events distant in time can occur at the same place of the ether.

When the ether was put aside, it started to become evident that (as had been previously surmised by some people) considering two events occurring at the same place at different times did not make sense, unless with respect to a given reference frame. If I make a date with someone for tomorrow in this room, do I mean absolutely in this same place? No, because the earth travels along its orbit and tomorrow this room will be faraway from where it is now. I can make the date for exactly a year from now when the earth will have resumed the position in the solar system that it has today. However, the solar system moves within our galaxy, our galaxy moves with respect to the other galaxies, and so on! Hence to the best of our knowledge, absolute identity of place at different times is a physical absurdity.

When Einstein showed, conversely, that simultaneity at different places does not have an absolute meaning either, people began to realize that (1) there is a certain symmetry between space and time and (2) they are more intimately connected than was previously thought.

From this one can be led to surmise, after Minkowski, that space–time is a unique continuum, not separable into space and time, unless with respect to a given observer. In place of the *points* of space, one should consider the *events* of space–time, each one specified by four coordinates x, y, z, t. Hence space–time represents a four-dimensional continuum.

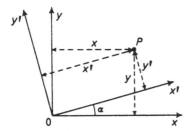

Figure 2.53

What transformation of the four coordinates must be carried out in passing from the inertial frame K to the inertial frame K' of Figure 2.52? Galileo's transformation equation (2.44) is a *linear* transformation, in that its coordinates appear raised to the first power. This may remind one of a very important kind of linear transformation of coordinates, precisely that existing when the reference frame is turned by a given angle.

For simplicity, let us refer to the two-dimensional space (plane) of Figure 2.53, related to the rectangular axes x, y. Each point P of the plane is specified by its coordinates x and y. Let us now consider a new reference frame x', y', obtained from the first by rotation of angle α counterclockwise. Point P in the new frame will have the coordinates x', y' given by[46]

$$x' = x \cos \alpha + y \sin \alpha \qquad (2.50)$$

$$y' = - x \sin \alpha + y \cos \alpha$$

Hence x' and y' are both linear combinations of x and y.

It is useful to note that transformation equation (2.50) leaves invariant the expression $x^2 + y^2$. In other words, as readily verified, $x'^2 + y'^2 = x^2 + y^2$. This is obviously a necessary consequence of the fact that $x^2 + y^2$ represents the square of the distance from O to P and such a distance is not varied by the rotation of the axes.

If from a two-dimensional space we pass to a three-dimensional space, we find on rotation of the reference frame, that the new coordinates x', y', z' are again linear combinations of the old coordinates x, y, z. We shall not write these general formulas but instead shall limit ourselves to a particular case. Let us suppose that the x, y axes are those indicated in Figure 2.53 so that the z-axis is perpendicular to the drawing. Let the system turn by an angle α about z. The new coordinates x', y' will obviously be those indicated in the figure, whereas z' will coincide with z. In other words, the transformation will be ex-

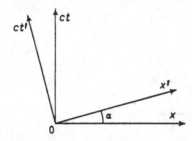

Figure 2.54

pressed by equation (2.50), in conjunction with a third equation $z' = z$.

If we now pass to a four-dimensional space, we shall have to add a fourth axis t, orthogonal to x, y, and z. Of course, we cannot form an intuitive visualization of the system but can easily generalize the mathematics. In particular, we can derive the result that when the frame turns by α in the xy plane, the transformation will still be given by the two equations (2.50) plus the two equations $z' = z$ and $t' = t$.

After recalling these elements, let us return to Galileo's transformation equation (2.44) and ask: "Could the transformation that must replace it be a rotation of the frame x, y, z, t? The first equation (2.44) suggests that in this case the rotation should occur in the plane x, t. Let us represent this plane in Figure 2.54, the axes y and z being both perpendicular to the drawing.

The reader will notice that t has been replaced by ct. This has been done for two reasons: (1) It is convenient that all coordinates should be *homogeneous*, which requires the fourth coordinate to have the dimension of length, like the others. (2) It turns out that with respect to what may be called a natural scale, we usually measure time with an enormous unit.[47] Everything becomes more harmonious and symmetric if we take as unit of time the time employed by light to travel 1 cm. This is precisely what we do when we replace t by ct.

By applying equations (2.50), we have

$$x' = x \cos \alpha + ct \sin \alpha$$
$$ct' = -x \sin \alpha + ct \cos \alpha \qquad\qquad (2.51)$$
$$y' = y$$
$$z' = z$$

Let us assume that this represents the right transformation. Next, consider the motion of point O' of Figure 2.52. In K' it represents the origin where $x' = 0$. Relative to K point O' moves according to the law $x = vt$. On the other hand, from the first equation (2.51), by putting $x' = 0$, we obtain $x = ct \tan \alpha$. Then we have

$$-\tan \alpha = \frac{v}{c} \qquad (2.52)$$

from which the angle α is determined.

Let us now examine the reason that we want to change Galileo's transformation, that is, Einstein's postulate.

Suppose that when O' passes through O, a lamp is lit at O. Light will propagate in all directions, in particular, along the x axis. For the observer K, light at time t will have reached the points $x = ct$ and $x = -ct$. Altogether, we can write $x^2 - c^2t^2 = 0$; indeed, solution of this equation gives $x = \pm ct$. By Einstein's second postulate, light for the observer K' will have reached $x' = \pm ct'$, and we can write $x'^2 - ct'^2 = 0$. Then in order for the transformation to be the right one, it should leave invariant[48] the expression $x^2 - c^2t^2$.

Unfortunately, equations (2.51) do not leave invariant the difference $x^2 - c^2t^2$, as is readily verified. But let us remember that because they represent a rotation, equations (2.51) leave invariant the sum $x^2 + c^2t^2$, or the square of the distance from the origin to the point x, ct. Therefore it would, be convenient to change the sign of t^2. But this cannot simply be obtained by changing the sign of t because $(-t)^2 = t^2$.

The difficulty is analogous to that faced by Renaissance mathematicians in connection with the solutions of higher order equations. There is no real number whose square is negative. The difficulty was overcome by introducing, besides the real unity 1, the *imaginary*[49] unit i, such that $i^2 = (-i)^2 = -1$ and $i(-i) = (-i)i = 1$. This imaginary unit is also of help in our case.

Let us replace ct by ict and take this expression as the fourth coordinate.[50] Thus equations (2.51) become

$$x' = x \cos \alpha + ict \sin \alpha$$

$$ict' = -x \sin \alpha + ict \cos \alpha \qquad (2.53)$$

$$y' = y$$

$$z' = z$$

In place of equations (2.52), we find $-i \tan \alpha = v/c$, then by multiplying both sides by i

$$\tan \alpha = i\frac{v}{c} \qquad (2.54)$$

Expressing $\sin \alpha$ and $\cos \alpha$ in terms of $\tan \alpha$, we obtain

$$\sin \alpha = \frac{i(v/c)}{\sqrt{1 - v^2/c^2}}, \quad \cos \alpha = \frac{1}{\sqrt{1 - v^2/c^2}} \quad (2.55)$$

After substituting into equations (2.53) and making some obvious simplifications, we have

$$x' = \frac{x - vt}{\sqrt{1 - v^2/c^2}}$$

$$y' = y$$

$$z' = z \quad (2.56)$$

$$t' = \frac{t - (v/c^2)x}{\sqrt{1 - v^2/c^2}}$$

These equations represent the celebrated Lorentz transformation.[51]

When v is negligible compared to c, the Lorentz transformation gives back Galileo's transformation (as is readily verified). This is necessary because for ordinary velocities Galileo's transformation perfectly agrees with experience.[52] Moreover, equations (2.56) ensure the invariance of the expression $x^2 - c^2 t^2$, or, more generally, $x^2 + y^2 + z^2 - c^2 t^2$, as required by Einstein's second postulate. We shall see later (§2.24) how Einstein's first postulate can also be satisfied.

2.22. Length contraction and time dilation

Consider a rod fixed in K' and parallel to the x axis. The rod is moving with velocity v relative to K. Let an observer in K' measure the positions of both ends of the rod, finding the abscissas x'_1 and x'_2. He will infer that the length of the rod is $L' = x'_2 - x'_1$.

The observer in K can conduct the same operation, finding $L = x_2 - x_1$. But be careful! Because the rod is moving with respect to K, both abscissas x_1 and x_2 must be measured at the same time. This, of course, will mean at the same time for K, that is, at the same value for t. Let this value be, say, $t = 0$. By putting $t = 0$ in the first equation (2.56), one has $x'_1 = x_1/\sqrt{1 - v^2/c^2}$ and $x'_2 = x_2/\sqrt{1 - v^2/c^2}$. By subtracting, one obtains

$$L = L' \sqrt{1 - v^2/c^2} \quad (2.57)$$

Thus the K observer will say that the rod has a length equal to L' multiplied by the factor $\sqrt{1 - v^2/c^2}$, which is always less than 1. The observer will get the impression that when in motion, the rod shrinks and becomes shorter. This is the contraction that had been postulated by H. A. Lorentz and G. F. Fitzgerald as an ad hoc hypothesis, and it had been used to explain the result of Michelson's experiment. After

Einstein's analysis the contraction no longer represents a very odd property of material bodies, but instead a result of the operational definition of length and duration.

Next, let two events take place in K', say, at O' at times t_1' and t_2'; their time separation, as measured by K', will be $T' = t_2' - t_1'$. At O', there exists $x' = 0$. Substituting this into the first equation (2.56), we get $x = vt$ (as was obvious). Putting this value of x in the last equation (2.56), and carrying out a few simplifications, we arrive at $t' = t \sqrt{1 - v^2/c^2}$. Therefore if t_1 and t_2 represent the instants measured by K, we have $t_1 = t_1'/\sqrt{1 - v^2/c^2}$ and $t_2 = t_2'/\sqrt{1 - v^2/c^2}$. On subtracting, we obtain for the duration $T = t_2 - t_1$, as measured by K,

$$T = \frac{T'}{\sqrt{1 - v^2/c^2}} \qquad (2.58)$$

Thus to the observer at rest in the system where the events occur, the duration appears to be shorter. Conversely, to the observer with respect to whom the system in which the events occur, is moving, time appears longer. This is the celebrated phenomenon of the relativistic *time dilation*.

The events we have been discussing may simply represent the arrival of the hands of the K' clock at two well-determined positions on the dial. Hence the K observer will infer that the K' clock is *slow*, compared with his own clock.

This fact, which at first might seem incredible, has received a considerable number of qualitative and even quantitative confirmations. One of the most fascinating is the effect shown by the μ particles or *muons* produced by cosmic rays. A primary high energy particle arrives from outer space and encounters the atmosphere. At the height of about 10 km, it clashes with an atomic nucleus of the atmosphere. The nucleus disintegrates and gives rise, among other things, to a number of π mesons or *pions*. These particles are unstable and their (average) lifetime, as measured at rest in the laboratory, is 2.6×10^{-8} s. They cannot go very far during this time (a few meters at most). Pions disintegrate and generate muons. Muons are also unstable and their lifetime, measured when they are at rest in the laboratory, averages to 2.2×10^{-6} s. Even in the limited case when muons travel at light speed, they would cover only a few hundred meters during their lifetime, hence only very few would be able to reach the ground.

However, experience shows that the muons reaching the ground are many more than would be expected according to this argument. It is as though muons traveling at high speed had a longer lifetime than muons at rest. By carrying out experiments at different heights above sea level and by counting the arriving muons, one finds that their life-

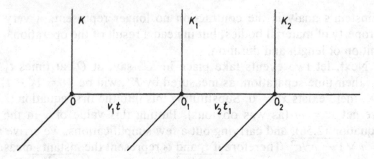

Figure 2.55

time appears to the observer at rest exactly dilated by the factor equation (2.58).

Time, measured by a clock at rest in K', is called the proper time of K'. It represents the shortest time (among those measured by different observers) that can be measured between two events occurring at one and the same place of K'. By equation (2.58), the proper time τ of K' can be derived from the time t measured by K by means of the relation

$$\tau = t \sqrt{1 - v^2/c^2} \qquad (2.59)$$

Proper time, obviously, represents an *invariant*.

2.23. The limiting velocity, the past and the future

Let us consider three inertial frames of reference K, K_1, K_2 (Fig. 2.55) and assume that K_1 is moving with velocity v_1 relative to K, whereas K_2 is moving with velocity v_2 relative to K_1. We ask: "With what velocity v does K_2 move with respect to K?" In Galileo's and Newton's mechanics the answer would plainly be $v = v_1 + v_2$.

According to equation (2.54), K_1 is turned in space–time with respect to K by the angle α_1, such that $\tan \alpha_1 = i(v_1/c)$, whereas K_2 is turned with respect to K_1 by the angle α_2 such that $\tan \alpha_2 = i(v_2/c)$. We have to find the angle $\alpha = \alpha_1 + \alpha_2$ by which K_2 is turned with respect to K, from which v can be derived by $\tan \alpha = i(v/c)$.

In elementary trigonometry one can prove the relation $\tan(\alpha_1 + \alpha_2)$ $= (\tan \alpha_1 + \tan \alpha_2)/(1 - \tan \alpha_1 \tan \alpha_2)$. By introducing the immediately preceding expressions into this equation, one readily arrives at

$$v = \frac{v_1 + v_2}{1 + v_1 v_2/c^2} \qquad (2.60)$$

This important relation tells us how velocities are added in relativity.

It is evident that when v_1 and v_2 are small compared with c, we again have Galileo and Newton's formula. But as the velocities increase, the denominator of the right-hand side expression increases, too. The result is always less than $v_1 + v_2$. It is readily verified that by combining two velocities smaller than c, one never attains or exceeds c. Precisely, one obtains $v = c$ only when either $v_1 = c$ or $v_2 = c$, or both.

If we assume that a material object can attain any velocity only by gradually adding new velocity, we must infer that c represents a *limiting velocity*, never attained or exceeded by any material object. Actually, no object or particle has ever been found traveling faster than light.[53]

Let us return to the discussion about time. We had mentioned that simultaneity of two events is relative to the reference frame in which they are observed. We can now ask: Given two events A and B, can we always find a frame in which A and B are simultaneous? If in a frame A is later than B, does there exist a frame in which A is earlier than B?

As is customary in these arguments, we can limit our description to the plane x, t (this is not a loss of generality, because y and z do not vary). Let us denote by x_A, t_A and x_B, t_B the coordinates of both events measured in K and add a prime to the corresponding quantities measured K'. From the last equation (2.56), we get

$$t'_A = \frac{t_A - (v/c^2)x_A}{\sqrt{1 - v^2/c^2}} \, , \qquad t'_B = \frac{t_B - (v/c^2)x_B}{\sqrt{1 - v^2/c^2}} \qquad (2.61)$$

Let us assume that B is later than A in K, so that

$$t_B > t_A \qquad (2.62)$$

Is it possible for B to be earlier than A in K', so as to have $t'_B < t'_A$?

Expressing t'_A and t'_B by equations (2.61), our problem readily becomes that of establishing whether it is possible to have $t_B - (v/c^2)x_B < t_A - (v/c^2)x_A$ or

$$c(t_B - t_A) < \frac{v}{c}(x_B - x_A) \qquad (2.63)$$

As $|v/c| < 1$, we see that this is impossible when $c(t_B - t_A) > |x_B - x_A|$, or when the spatial distance from A to B is smaller than the distance covered by light during the interval $t_B - t_A$. Because light travels at limiting velocity, we can say that if a signal emitted at event A can reach event B, temporal inversion is impossible. If we have instead $c(t_B - t_A) < |x_B - x_A|$, the inversion is possible and there is even a frame in which A and B are simultaneous.

Sometimes it is said that if A can be the cause of B, then A will be earlier than B, in any frame. This assertion is evidently in agreement

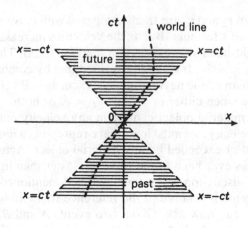

Figure 2.56

with the common concept of cause. We shall discuss this concept in more detail in §4.20. But taken at face value, the concept requires that if A is the cause of B, something must leave A and reach B. This is why it seems inconceivable that the effect may precede the cause.

Let O (Fig. 2.56) represent the event here and now, or the place where I am at the present instant. The two straight lines $x = ct$ and $x = -ct$ divide the plane x, ct into four regions. The shaded area above O represents my potential future, for each one of its points represents an event I could witness, provided that I moved in the right direction with a suitable velocity $v < c$. The shaded region below O represents my possible past, because I might have witnessed any one of its events. The unshaded regions (elsewhere) are absolutely out of my present reach, because I cannot influence any one of their events nor can those events influence me now.

As an example, consider the star Alpha Centauri (the nearest to the solar system). Its distance is about four light years; that is, its light takes four years to reach us. I have no means (not even theoretical) to influence an event that will occur on Alpha Centauri two years from now. By the same token, an event that occurred on Alpha Centauri two years ago can in no way influence me now.

If we add a third dimension y to x and ct, the shaded region becomes a *cone* about the ct axis. If we add a fourth dimension z, we obtain a *hypercone* which cannot be visualized. However, it is customary to talk about the *cone of the past* and the *cone of the future*. During my lifetime I travel along a *world line*, coming from the cone of the past and going through O into the cone of the future. The conical surface bounding the shaded regions is called the *light cone* and represents the locus of the events attained by light arriving to O or emitted by O. Its equation is $x^2 + y^2 + z^2 = c^2 t^2$.

The past, the future, and the light cone are invariant concepts in that they do not change under a Lorentz transformation, or when from an observatory K we pass to another observatory K'.

2.24. The invariance of the laws of physics

Let us now turn to the principle of relativity, which requires that all the laws of physics should be the same in all inertial frames. Consider, in three-dimensional space, a body of mass m, to which force F is applied. The body acquires an acceleration, such that $F = ma$. In a rectangular reference frame x, y, z a vector can be specified by its three components, hence we get the three equations $F_x = ma_x$, $F_y = ma_y$, $F_z = ma_z$. If we now change the orientation of the rectangular axes about the origin, the components F_x, a_x, and so on of the vectors will change according to well-determined laws, depending on the angles by which the axes have rotated (laws of *covariance*). We shall not describe these laws in detail, but it is evident that they are the same transformation as obeyed by the coordinates of a point, because the coordinates of point P coincide with the components of vector OP joining the origin O with P. Vector components vary, but the law $F = ma$, being a natural law, cannot depend on the orientation of the axes and does not change.

Let us consider this curious problem. Let us build a vector F' whose first two components are equal to F_x and F_y, whereas the third component is equal to the temperature T of the body, on which the force is acting. Can the equation $F' = ma$ represent a physical law? No, because whereas F_x and F_y obey the covariance laws of the components of a vector, and thus vary in the same way as the components a_x and a_y of the acceleration, T remains unchanged when the reference frame turns and cannot vary as a_z. The law $F' = ma$ would change, depending on our choice of the reference frame. But this is absurd, because the behavior of the physical world cannot depend on how we choose the reference axes. As a conclusion, the physical laws must be expressed in terms of *true* vectors, whose components obey the covariance transformation laws.

Conversely, if a physical law is expressed in terms of true vectors, we may be certain that it remains unchanged, whatever the orientation of the observer may be. These considerations are trivial as long as we deal with three-dimensional space, because a vector in that space can be visualized as an oriented line without any reference to components.

The situation, however, is different in the case of space–time. We can well imagine that there are vectors also in space–time, but we cannot *see* them. All we can do is to study them by means of their four components; accordingly, they are called four-vectors. In order to ascertain whether four numbers, derived by measurement, represent a four-vector, we have to verify whether they have the proper covariance under rotation of the frame of reference. Only in this way can we recognize the true vectors of space–time. A law, expressed as an equality between four-vectors, is invariant under frame rotation.

We know that the passage from an inertial frame or reference to another corresponds simply to a rotation of the axes in space–time (Lorentz transformation). We can therefore conclude that the laws of physics, when written in four-vector form are invariant, when we pass from one inertial frame to another. Hence the relativity principle will be satisfied if we succeed in writing all the laws of physics in that form. This program can be fully realized. We cannot, however, develop it here in any detail but shall hint at a few important results that can be derived from it.

A four-vector of paramount importance in physics can be obtained in the following way. Let us consider the momentum $p = mv$ of a mass m particle moving with velocity v. We may wonder whether the three components P_x, P_y, P_z can represent the first three components of a four-vector. It is readily verified that this is not possible. We evidently have

$$P_x = m \frac{dx}{dt}, \qquad P_y = m \frac{dy}{dt}, \qquad P_z = m \frac{dz}{dt} \qquad (2.64)$$

where dx, dy, dz represent the increments of the coordinates of the particle during dt, so that $v_x = dx/dt$, and so on. Now dx, dy, dz vary as x, y, z, or as the first three components of a four-vector in space–time. Hence P_x, P_y, P_z could represent the first three components of a four-vector, only if dt were invariant (assuming m to be invariant). However, dt varies as the fourth component of a four-vector and cannot be invariant.

We can overcome the difficulty, replacing dt by the increment of the proper time of the particle, which by equation (2.59) is $d\tau = dt \sqrt{1 - v^2/c^2}$, and is an invariant. Hence the three quantities

$$P_1 = \frac{m \, v_x}{\sqrt{1 - v^2/c^2}}, \qquad P_2 = \frac{m \, v_y}{\sqrt{1 - v^2/c^2}},$$
$$P_3 = \frac{m \, v_z}{\sqrt{1 - v^2/c^2}} \qquad (2.65)$$

represent the first three components of a four-vector. When v/c is very small, that is, in all the cases of classical mechanics P_1, P_2, P_3 coincide with the components P_x, P_y, P_z of ordinary momentum. We are therefore allowed to redefine momentum as

$$P = \frac{m\,v}{\sqrt{1 - v^2/c^2}} \tag{2.66}$$

In this way, we virtually do not alter the results of classical mechanics, whereas we are about to build a four-vector to be used in the invariant laws of relativistic mechanics.

The problem is now to find the significance of the fourth component of this four-vector. The fourth coordinate of space–time is ict. Thus the fourth equation to be added to the three equations (2.65) is plainly

$$P_4 = m\,\frac{d(ict)}{d\tau} = m\,\frac{ic}{\sqrt{1 - v^2/c^2}} = \frac{i}{c}\,E \tag{2.67}$$

where for convenience we put

$$E = \frac{mc^2}{\sqrt{1 - v^2/c^2}} \tag{2.68}$$

What does E mean?

When the particle is at rest, the value of E becomes

$$E_0 = mc^2 \tag{2.69}$$

If the particle is set into motion, E increases, due to the denominator of equation (2.68). When v/c is small, we can write[54] $1/\sqrt{1 - v^2/c^2} = 1 + v^2/2c^2$, so that

$$E = mc^2 \left(1 + \frac{1}{2}\frac{v^2}{c^2}\right) = E_0 + \frac{1}{2}\,mv^2 \tag{2.70}$$

Hence when the particle, starting from rest, acquires velocity v, the quantity E increases by $mv^2/2$, that is, by the value of the kinetic energy, equation (2.18). This suggests that E may represent the energy of the particle. If this is true, there follows that the particle at rest has a rest energy E_0 given by equation (2.69), proportional to the mass; when the particles are in motion, the kinetic energy is added to the rest energy. It is to Einstein's credit that he soon derived this daring result, much before any experimental proof of this concept was possible.

What does it mean that the particle at rest has energy E_0? This means that the particle must have, in one way or another, the capacity to yield energy or to perform work equal to E_0.[55] This is verified in a large number of ways in modern microphysics. The particle can vanish and thereby can furnish energy (thermal, radiant, kinetic of other particles) exactly equal to mc^2. Mass must therefore be conceived as a sort of frozen energy, which under suitable conditions, can be set free.

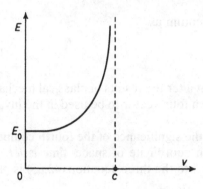

Figure 2.57

The kinetic energy $E - E_0$ equals $mv^2/2$, only as long as v/c is very small. As a matter of fact, E (and, consequently, the kinetic energy) grows to infinity as v tends to c. This is illustrated in Figure 2.57, which shows the behavior of E as a function of v according to equation (2.68).

We can now understand better why velocity c cannot be attained. If a particle is accelerated in a modern accelerator, it acquires some energy on each turn. Velocity increases, but the rate of increase gradually slows down when approaching c. In order to reach c, we should confer an infinite amount of energy to the particle and, of course, we shall never succeed.

We have found that the three components of p, together with iE/c, form a four-vector $\wedge$. We must therefore think that the momentum and the kinetic energy are both parts of a unique entity of space–time, invariant under frame rotation. We perceive the spatial components of $\wedge$, which we call the momentum, as separate from the temporal component, which (apart from a factor) we call the energy. However, this decomposition depends on how our frame of reference is oriented, or what amounts to the same, on what inertial frame we are in. The square of the length of $\wedge$

$$\wedge_1^2 + \wedge_2^2 + \wedge_3^2 + \wedge_4^2 = p^2 - \frac{E^2}{c^2} \tag{2.71}$$

will obviously be an invariant, because it cannot depend on the orientation of the axes. Its value for the K observer, who sees the particle in motion, will be the same as for the K' observer, who sees the particle at rest. For the latter, because $p = 0$ and $E = E_0$, the value of the invariant is $-E_0^2/c^2$. Hence we can write $p^2 - E^2/c^2 = -E_0^2/c^2$. Substituting the value equation (2.69), for E_0, we readily arrive at

$$E = c \sqrt{p^2 + m^2c^2} \tag{2.72}$$

This remarkable equation, which expresses energy as a function of momentum, is fundamental in modern physics.

In the apparently absurd hypothesis of a vanishing mass, a particle can still have energy and momentum, which according to equation (2.72), are connected by the relation

$$p = \frac{E}{c} \tag{2.73}$$

This is exactly the relation that exists for the momentum and energy of electromagnetic radiation. No wonder for as we shall see in due course (§4.6), electromagnetic radiation consists of zero mass particles (*photons*).

In order to apply the relativity principle to the electromagnetic field, one has to consider mathematical entities more complex than simple vectors. Physical laws are expressed by means (1) of invariant numbers not associated with any direction (called *scalars*); (2) numbers associated each with one direction (*vectors*); and (3) numbers associated each with several directions (*tensors*). The last two kinds of entities can represent quantities independent of the orientation of the axes, provided that their components obey suitable covariance laws.

It is found that all of the laws of electromagnetism, in the first place Maxwell's equations, can be expressed by means of equations involving only scalars, vectors, and tensors of space–time. The content of these equations does not depend on the orientation of the frame of reference. Thus the principle of relativity is satisfied. In this theory the electric and magnetic fields are only the projections on to the coordinate planes of a special tensor of space–time. What appears to us as electric or as magnetic field therefore depends on the inertial system we are in.

One can explain in this way a phenomenon that for a long time had seemed very odd. We have seen that a wire carrying a current, when embedded in a magnetic field perpendicular to it, experiences a force perpendicular to both. This happens because although for an observer at rest in the laboratory there is only a magnetic field, for an observer moving with a charged particle inside the wire, the electromagnetic tensor also presents an electric component. In other words, the particle will also see an electric field, perpendicular to the wire and to the magnetic field, and therefore will be acted on by a force in that direction.

Inversely, let us place a wire in a magnetic field perpendicular to it, and give it a lateral motion perpendicular to the lines of force (Fig. 2.58). A charged particle of the wire will see an electric field parallel to the wire and will be pulled in that direction. Hence an electromotive force arises in the wire.

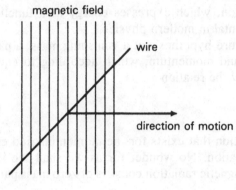

Figure 2.58

This is one of the phenomena that Einstein believed needed an explanation. He surmised that the electromotive force had to be produced by an electric field so as to restore the symmetry between this case and the case when the wire is at rest and the magnet is moving. In the latter case the magnetic flux is varying and the electric field arises in agreement with Maxwell's equations. These considerations were largely responsible for inspiring Einstein to enlarge on relativity.

2.25. Gravitation

The theory presented so far is called *special relativity* and accounts for the equivalence of all inertial frames, in the sense that the laws of physics are identical in all of them. As one of its main results, there is no way to verify, by means of experiments conducted within a laboratory K, whether the laboratory is at rest or in uniform motion. The notion of rest is therefore void of an absolute physical significance.

But is this sufficient in order to eliminate absolute space? What about noninertial frames? If the laboratory is accelerated or rotates, with respect to what does it accelerate or rotate?

E. Mach suggested an intriguing principle, according to which a mass moves (and is accelerated or rotates) relative to all the other masses of the universe.[56] The mass of a body, or its resistance to acceleration, should be due to its gravitational interaction with all the other bodies of the universe. If this is true, there exists, in a sense, a privileged system, specified by the distribution of all masses in the universe (the fixed stars system).

A famous argument in favor of absolute space, due to Newton, was based on the experience of a water pail turning about a vertical axis. The water surface becomes curved (a *paraboloid*), when the water

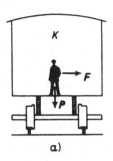 

Figure 2.59

turns "in absolute terms," and not when it is merely turning relative to its container (e.g., at outset of the motion, when the pail turns and the water is still at rest). In order for this experiment to be able to decide in favor either of Newton's ideas or of Mach's principle, one should try to turn the fixed stars around the water at rest! This is impossible, hence Newton's argument is powerless. Einstein was profoundly influenced by Mach's ideas; however, he followed a somewhat different path in approaching his general relativity.

Let us ask: "In what way does a noninertial frame differ from an inertial frame?" "By what means can an observer, conducting experiments only inside K, realize that K is accelerated?"

Consider a passenger on a railroad carriage K in uniform motion on a perfect straight track. No experiment carried out within K, can tell him that the carriage is moving. But if the train arrives at a turn, the passenger experiences a force F (Fig. 2.59(a)) pulling him sideways, hence he realizes that the train is moving. Generally, this effect is normally counteracted by suitably tilting the track plane (Fig. 2.59(b)). The ideal condition is obtained when F, combined with the weight P, gives rise to a resultant force P', perpendicular to the floor of the carriage. In this way one reverts to a normal situation.

However, starting uniquely from the evidence he has inside K, could the passenger construe the situation the other way around and think that the motion is still straight, the track being tilted outward in Figure 2.59(a) and returning horizontal in Figure 2.59(b)? This seems to be possible.[57] Further, he could entertain the daring idea that at the outset of the situation, Figure 2.59(a), it is the earth's center that shifts to the right, so that in Figure 2.59(b) the track simply returns horizontal!

Suppose now that the train, running on a straight track, stops. When the brakes are applied the passenger feels projected forward. Could he not construe this situation as being due to a new planet, suddenly placed in front of the train and pulling him with its gravitational force? He certainly can, for the following reason. If M denotes the mass of the planet, a mass m at distance r will experience the force $F = GmM/r^2$, hence an acceleration a such that $ma = GmM/r^2$. On dividing by m, we have $a = GM/r^2$. As a result of the equality of inertial and gravitational mass, a does not depend on m. Thus all the free bodies of the passenger's laboratory will undergo the same acceleration, and the result is exactly the same as when it is the laboratory to be accelerated in the opposite direction.

All this is absurd, one will say; the passenger knows very well that there is no new planet in front of him. Of course, but he knows this fact only owing to evidence collected *outside* of K. Making experiments only *inside* of K, he will never be able to tell a true acceleration from a gravitational effect. Indeed, the acceleration of space vehicles is measured in g, as is well known.[58]

We are thus led to accept the *principle of equivalence*, according to which the presence of a gravitational field is equivalent in all respects to an acceleration of the reference frame.[59] Therefore it follows that gravity must be so intimately connected with the space–time characteristics of the reference frame that it must be taken into consideration when developing relativity.

2.26. General relativity

As a result of the foregoing considerations, the omnipresence of gravity can cast serious doubts on the existence of inertial frames.

It is sometimes believed that within a system falling freely in space the effects of gravity fully disappear. As generally known, an astronaut on board an artificial earth satellite is "weightless," hence the system should appear to be perfectly inertial. As a matter of fact, it is inertial only *locally*, that is, in a limited region. For each object in free motion inside the satellite represents only a different satellite of the earth and moves along its own Keplerian ellipse. No one of these ellipses can appear as a straight line in the satellite frame, as is proved by the fact that the mutual distance never does go to infinity. Nevertheless, it is true that in a very small region of space–time as that accessible to the astronaut, a free object appears to move with great precision, in a straight line.

The situation vaguely recalls that obtained in a limited portion of the earth's surface, such a portion appears close by to be a plane. A stone flung along the surface of a frozen lake seems to slide in a straight line. Yet we know that if its motion can go on indefinitely, the path would be a great circle of the earth's surface. The paths of two stones flung in different directions would eventually meet, and thus cannot be straight lines.

The great circle has a very important property in common with the straight line. The latter is the shortest line between two points of a plane (or of space), whereas the former is the shortest line between two points on a spherical surface. A line drawn on a surface, having this property, is called a *geodesic line* (or a *geodesis*, for short) of the surface. The geodesics of a plane are straight lines, whereas those of a spherical surface are great circles.

One can accept from intuition (and it could be proved in a rigorous way) that a body free to move on a surface, moves along a geodesic of the surface.

A triangle, whose sides are geodesics is an ordinary triangle if the surface is a plane; but it can have very odd properties if the surface is curved. For example, the sum of its internal angles is, in general, different from 180°. To convince ourselves of this, let us consider a geodesic triangle drawn on the earth's surface, whose sides are two meridians leaving the pole at right angles and one-fourth of the equator. Such a triangle has three right angles! It is clear that the geometry on the earth's surface is not strictly Euclidean as that of a plane, but only locally Euclidean. Only within a suitably limited region around us can we consider the earth to be a plane and be able to use Euclidean geometry.

On the earth's surface, we obviously cannot use the ordinary straight axes of Cartesian coordinates. We must resort to *curvilinear coordinates*, and normally the *latitude* θ and the *longitude* φ are used. In this case the *coordinate lines* are represented by the parallels and the meridians, which meet at right angles.

If R denotes the radius of the earth, the distance between two points on the same meridian, whose latitudes differ by $d\theta$, equals $R\,d\theta$, whereas the distance between two points on the same parallel at latitude θ, whose longitudes differ by $d\varphi$, equals $R\cos\theta\,d\varphi$ (for the radius of the parallel is $R\cos\theta$). Because the geometry is locally Euclidean, one can apply Pythagoras' theorem and obtain for the distance ds between two points P_1, P_2, whose latitudes and longitudes differ by $d\theta$ and $d\varphi$, respectively (Fig. 2.60), the expression

$$ds^2 = R^2\,d\theta^2 + R^2\cos^2\theta\,d\varphi^2 \tag{2.74}$$

The square of the line element ds turns out to be a *quadratic* expression in $d\theta$ and $d\varphi$. If the meridians and the parallels did not meet at right angles, the triangle of Figure 2.61 would not be a rectangular triangle.

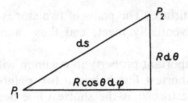

Figure 2.60

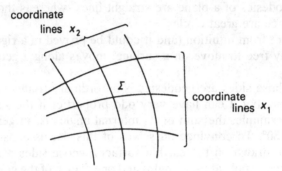

Figure 2.61

As a result, equation (2.74) would contain, besides the terms in $d\theta^2$ and $d\varphi^2$, also a term with the mixed product $d\theta \, d\varphi$ (Carnot's theorem of trigonometry).

The spherical surface referring to its meridians and parallels represents only a particular case. In general, on a curved surface Σ (Fig. 2.61) one can establish (in an infinite number of ways) a system of curvilinear coordinates x_1, x_2. The square of the distance separating two very close points will be given by

$$ds^2 = g_{11} \, dx_1^2 + g_{12} \, dx_1 \, dx_2 + g_{22} \, dx_2^2 \tag{2.75}$$

where g_{11}, g_{12}, g_{22} represent three well-determined functions of the x_1, x_2 variables.[60] In the case of equation (2.74), one has $x_1 = \theta$, $x_2 = \varphi$, $g_{11} = R^2$, $g_{12} = 0$, $g_{22} = R^2 \cos^2 \theta$.

As was shown by K. Gauss, the three functions g_{11}, g_{12}, g_{22} completely determine the metric of the surface and its curvature. In the case of equation (2.74) one would find that the curvature is constant, with a positive radius R. But there are also surfaces whose curvatures vary from point to point, and even surfaces with negative curvature. The plane represents a surface with zero curvature.

A curved surface embodies a non-Euclidean space with two dimensions and can easily be visualized. But in three or more dimensions, a non-Euclidean space loses this useful property; however, the mathematical generalization, introduced by B. Riemann, does not offer great conceptual difficulties.

Following Einstein's idea, we consider space–time as a four-dimensional *Riemannian manifold*, represented by

$$ds^2 = g_{11} dx_1^2 + g_{22} dx_2^2 + g_{33} dx_3^2 + g_{44} dx_4^2$$
$$+ g_{12} dx_1 dx_2 + g_{13} dx_1 dx_3 + g_{14} dx_1 dx_4 + g_{23} dx_2 dx_3 \quad (2.76)$$
$$+ g_{24} dx_2 dx_4 + g_{34} dx_3 dx_4$$

where x_1 x_2, x_3, x_4 denote any system of curvilinear coordinates and the g's are functions of them. The g's determine the space–time geometry and, in particular, the curvature. The g's are in turn determined by the distribution of masses in space–time, according to ten very complicated differential equations discovered by Einstein.

According to general relativity, the masses present in the various parts of the universe determine its curvature and, more generally, its geometry. A body left to itself follows a geodesic of space–time. In the solar system, space–time is curved by the mass of the sun.[61] In this way the motion determined by gravitation is brought back to an inertial motion.

2.27. Consequences of general relativity

The theory of general relativity represents, without any doubt, a majestic construction and an admirable synthesis of the laws of dynamics with the law of gravitation. It meets a fundamental demand of the human mind for harmony and unity. But apart from these merits, in what way does the theory improve our grasp of physical phenomena? What experimental facts does it predict, beyond those already understood by classical physics and special relativity?

With the advent of general relativity, a fairly peculiar situation has arisen in physics. Newtonian mechanics, the wave theory of light, Maxwell's electromagnetism, and even special relativity have each in turn conspicuously broadened the class of the predictable phenomena. From abstract theories, they have very shortly become indispensable components of applied physics and engineering.[62] Nothing of this kind has occurred on the first appearance of general relativity. Physicists have been compelled to consider very refined and difficult ad hoc experiments in order to "prove" its validity.

Only one of the tests consists of something that had already been known–the precession of Mercury's perihelion, that is, the fact that the ellipse described by the planet about the sun is slowly turning in

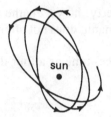

Figure 2.62

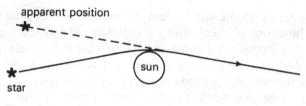

Figure 2.63

its own plane (Fig. 2.62). The perturbations due to the other planets cannot account for the whole amount of the precession. General relativity instead could give an explanation, valid within very reasonable limits of precision. Of course, the theory predicts that the phenomenon should take place for all planets, but with different quantitative features. Only Mercury is close enough to the sun to give a measurable result (and then, very small: 49 seconds of an arc per century!)

A second effect, predicted by general relativity and tested only after its appearance, is the deflection of a light ray passing through a strong gravitational field. This can be verified during a solar eclipse when the rays from a star pass very close to the sun. The star is seen shifted a little outward (1.75 seconds of arc) with respect to its normal position (Fig. 2.63). The effect was first verified during an eclipse in 1919; but for a long time some doubts remained about the exact amount of the deflection. Recent precision measurements, conducted with radio-astronomical methods, have fully confirmed the amount of deflection predicted by general relativity and have shown that in this respect, general relativity is still preferred, compared to other rival theories of gravitation which were developed in the last decades. Moreover, by means of radar techniques, a collateral effect predicted by the theory has been verified, that is, the retardation of an electromagnetic wave passing very close to the sun; of course, the effect is minute, but detectable.

The third classic phenomenon predicted by general relativity, and afterward verified, is the slowing down of a clock when embedded in a strong gravitational field or, what amounts to the same thing, the shift toward longer wavelengths of the radiation emitted by atoms at the surface of a star. Today physicists master some techniques (the Mössbauer effect) of astounding precision, which allow them to verify this effect even in the earth's gravitational field.

General relativity has shed new light on a puzzling result already found by special relativity. A subject on which there is an enormous amount of literature is the *twin paradox*. There are twin brothers A and B. Twin A leaves on board a spaceship, as twin B remains on earth. If A travels for some time at a very high speed, his proper time will be slower than that measured by B. Hence when A comes back to earth, he will be younger than his twin brother B.

A hasty reasoning might lead to the conclusion that the situation is paradoxical. For one might argue that in A's frame it will be B who leaves and comes back; hence it will be B who turns out younger. Be careful! The A and B reference frames are not at all equivalent. Whereas B's frame is virtually inertial, A's frame will be violently accelerated, at least on departure, inverting the direction, and arrival. Because special relativity does not apply to noninertial frames, we cannot very well tell how the situation appears to A. But if one carries out the computations according to general relativity, one finds that both in A's and B's reference frames, it is A who eventually turns out to be younger than B, and by the same amount.

The test of this effect, which is comparatively easy when one observes a beam of unstable particles traveling in a cyclic accelerator, has been performed also at macroscopic level by J. C. Hafele and R. E. Keating in 1971. Two airplanes A and B carrying two indentical atomic clocks orbit the earth in opposite directions. Their motions combine with the earth's rotation, so that A, which is flying, say, eastward, has a higher speed than B, which is flying westward. When A and B come back to land, it is possible to verify that A's clock is slow relative to B's clock. Naturally, it is a question of an extremely small fraction of a second (about 3×10^{-7} s); however, the result is in excellent agreement with general relativity.

Another important phenomenon that must be mentioned is that of *gravitational waves*. One can ask the question: Are gravitational actions transmitted instantaneously, as seems to be implied by Newton's theory,[63] or at a speed not greater than c, as is required by special relativity?

To begin with, let us clarify the meaning of the question. A naïve interpretation could be the following. If at a distant place a new star is suddenly born (from nothing), how long will it take before we experience its gravitational pull? In this wording, the question is over-simplified. As far as we know today, the birth of a star from nothing is an impossible event. It is very dangerous, or even nonsensical, to ask what would happen, if a fundamental law of nature were violated!

However, we can assume that the star is moving so that the gravitational action gradually changes its direction. One can think of a periodic process, such as that shown by a double star whose components rapidly rotate about one another. In such cases, according to general relativity, the variations of the gravitational field should be transmitted by special waves at the speed c.

The experimental detection of gravitational waves is extremely difficult and requires a huge and very sensitive apparatus. A number of laboratories throughout the world are tackling the problem. Although no direct confirmation of the existence of gravitational waves has been obtained as yet, there is at least an important *indirect* confirmation. J. H. Taylor has recently discovered a system of two very dense stars (see §5.11) revolving about each other with a very short period (about eight hours). The period is slowly increasing in a measurable way, which means that the system is gradually losing energy. The energy lost per year corresponds fairly well to the energy that the gravitational waves should carry away according to general relativity. No other plausible explanation can be found. Note incidentally that the precession of the *periastron* (about 4° per year) is many orders of magnitude larger, consequently, is measured much better than the precession of Mercury's perihelion. This double system has raised considerable excitement in the scientific community, because it has confirmed a considerable number of effects predicted by general relativity, all at the same time.

As to the main object of general relativity, one cannot help but ask: "Why should we incorporate gravitational forces into the geometrical framework of space–time and omit other types of forces, in particular, electromagnetic forces?" Einstein devoted several years of his life in replying to this objection, by means of a *unitary* theory of the gravitational and electromagnetic fields. His attempt, however, was not fully successful. Moreover, particle physics has shown that there are other forces, besides gravitational and electromagnetic. Should we not unify all of them?

An additional challenge is represented by the quantization of the field of general relativity, which no one so far has been able to achieve in a satisfactory way. In conclusion, one can say that general relativity is still an open theory, probably subject to change and development (see, e.g., Brillouin, 1970; Synge, 1970; Cowperthwaite Graves, 1971).

2.28. Physical theories

In the first chapter we discussed several preliminary questions regarding the method of physics. However, no thorough analysis of the concept of *theory* was developed. Such analysis was deferred to a later time, so as first to provide readers with some concrete material on which to base their reasoning. We begin the discussion of this important subject now and shall come back to it from time to time in the following chapters.

An examination of the history of classical physics, from Galileo to Newton, from Maxwell to Einstein, seems to show the tendency to go from simple and isolated laws of type such as equation (1.6) to complex theoretical assertions, from which a great number (hopefully, all) of those laws can be derived.

Many people in the past believed, and some today still believe, that physicists are simply carrying out Francis Bacon's program, starting from sense data and particular facts and then climbing with gradual ascent to more and more general assertions. As a matter of fact, this conception is neither true to history nor capable to stand the test of critical analysis.

Already as we saw with Galileo, discussed in Chapter 1, some form of theory always precedes experience, otherwise one could not have sensible experiments but only a collection of disconnected and useless data. At the same time, it was Galileo's contention that a physical theory has no value if it is not supported by experience.

Later on the idea that one cannot derive theory from experience by carrying out, as it were, a synthesis of a great number of results, but can only compare the theory with experience (sometimes by means of one single experiment), was gradually accepted and understood better.

First, I shall set forth my personal point of view which, I believe, is shared by many physicists, although with a number of slightly different opinions. Later, I shall survey some other kinds of philosophical views, which have been proposed and are currently maintained.

A physical theory T consists of a number of general assertions, or axioms, from which we can derive by logic, one or more laws of the kind such as equation (1.6). As is obvious, one or more laws of this type can also be found among the axioms of T.[64] A physical theory must always have an associated *domain of validity*. The domain of validity D of T is embodied by the class of all physical phenomena for which the predictions of T are known to be in agreement with experience. The domain must be specified by describing the physical situations in which to perform the experiments, along with the precision of the apparatus used and the range of admissible values for the parameters involved.

The specification of D has necessarily an historical character. This means that at a given epoch it is known[65] that certain experiments are in agreement with T; these are the experiments of domain D. Other experiments of the same kind, perhaps carried out with better precision, lie outside D, consequently, one must reserve judgment about them. If later it is possible to establish that some of these experiments are also in agreement with T, domain D will be widened to include them.

Extrapolation, that is, the assumption that T is valid also outside of D, is not allowed in physics, unless for *heuristic* purposes.

The physicist will constantly strive to perform experiments outside of D (by either improving on his precisions or widening the class of the physical situations considered) and to compare the results with those predicted by T. This is an excellent methodological prescription, and it represents the main road along which science is progressing today. However, those who watch science from outside, are often likely to make a mistake. They may believe that the physicist entertains the faith that T is valid outside of D. Starting from this assumption, they can hardly resist the temptation to criticize (very easily) the physicist's position.

This misunderstanding arises only because modern physicists always have in mind a finite domain D. This enables them to affirm with certainty the validity of T within D. Those who are not very familiar with physicists' work, may mistake this certainty for a naïve faith in the unlimited validity of T.[66] Any student having properly followed physicists' curricula, should be immune from this faith.

Experimenting outside of D, one may find a phenomenon E that clashes with T.[67] At this point, those who think that a theory is either absolutely valid or false have no choice. They must say that T has been disproved or falsified by E. And a false theory simply has to be dropped.

To understand how that attitude is unreasonable, one only has to recall the example of Lavoisier's law, already mentioned in §1.9. A law that had an immense influence on the birth of chemistry, and is still indispensable today, should be false![68]

A much more reasonable conclusion is that T is still valid within D, whereas the fresh evidence E merely determines the boundary of D in that direction. Naturally, one would wish to have a theory capable of accounting for E. This is the task of the theoretical physicist who begins working as soon as the result E is known. He seeks to build a novel theory T' whose validity domain should include both D and E.

Incidentally, one should remark that the chronological sequence described, which might appear to be the only one possible, is often reversed in reality, particularly in modern physics.[69] First T' is born, and only later is the new experiment E performed. If E agrees with T' (and clashes with T), one says that T' is confirmed. Otherwise, T' is discarded.[70]

How and why does a new theory arise? Ordinarily, it is arrived at by sheer intuition or analogy; it can largely be due to a desire for unity and symmetry, that is, to an intellectual need.

It is interesting to know that the actual chronological order of E and T', although very important for history, has only little (if any) significance for the logical reconstruction of the development of physics.[71]

We shall schematize the development of physics, by introducing a historical ladder of theories $T_1, T_2, T_3, \ldots$, which are valid in the wider and wider domains $D_1 \subset D_2 \subset D_3 \subset \cdots$.[72] It is an *internal* history, or a rational reconstruction, which does not necessarily coincide in detail with actual history.[73] This view will be illustrated by means of two extreme cases; but it should be emphasized that all intermediate cases are possible.

Case I. In a given epoch a theory T_n is known to agree with all the experiments $E_n{}^1, E_n{}^2, E_n{}^3, \ldots$ of a certain domain D_n. Some fresh evidence $E_{n+1}^1, E_{n+1}^2, \ldots$ that clashes with T_n is found. Someone proposes a novel law, an ad hoc hypothesis H, which can account for the new evidence, without contradicting that of D_n. Then the new theory $T_{n+1} = T_n + H$ is accepted, with a validity domain of $D_{n+1} = D_n + E_{n+1}^1 + E_{n+1}^2 + \cdots$.

As an example of case I, let us take (in a very schematic way) the situation of electricity and magnetism in 1820. Scientists were acquainted with the laws of mechanics, with both Coulomb's laws and with the possibility of producing electric current by a chemical device (Volta), and so forth. Oersted discovered new evidence, not predicted by the existing T_n. The hypothesis H was added, stating that a current gives rise to a magnetic field according to a law such as that of Biot and Savart equation (2.28), and the theory $T_{n+1} = T_n + H$ was accepted. Its domain D_{n+1} also included the effect of a current on a magnet. Next, in 1831 M. Faraday discovered electromagnetic induction, which had not been predicted by T_{n+1}. Accordingly, one had to add the hypothesis H' – that a variable magnetic flux generates the electromotive force, equation (2.29); thus the theory $T_{n+2} = T_{n+1} + H'$ was adopted. And this can go on and on.

Case II. The theory $T_n = T_m + H + H' + \cdots$, consists of many assertions, some of which are detached from and independent of the others, as in the example of case I. Someone works out theory T'_n, which is valid within the domain D_n of T_n, but contains fewer independent hypotheses. This T'_n might be preferred to T_n, for the sake of economy, unification, harmony. Nevertheless, if there is no other advantage, scholars tend to be conservative and to retain T_n for D_n. It may happen, however, that T'_n is able to predict one or more novel phenomena $E^1_{n+1}, E^2_{n+1}, \ldots$, which lie outside of D_n (and clash with T_n). If we find by experience that these phenomena really occur in the way predicted by T'_n, we say that T'_n is confirmed and can be accepted to represent T_{n+1}, with the validity domain $D_{n+1} = D_n + E^1_{n+1} + E^2_{n+1} + \cdots$.

As an example for case II, we can again refer to electromagnetism. Before Maxwell one had only a collection of separate laws. Maxwell unified almost all of them with his equations. They predicted the existence of electromagnetic waves. When Hertz discovered these waves, Maxwell's theory was confirmed. Einstein was confronted with that theory and the additional hypothesis that an electric charge, moving in a magnetic field, experiences a certain force. Moreover, there was the ad hoc hypothesis of Lorentz and Fitzgerald, which was necessary to account for the result of Michelson's experiment.[74] Einstein worked out a special relativity theory that unified everything. The theory also predicted some novel phenomena, such as time dilation, the equivalence of mass and energy, and so forth. The experimental verification of these phenomena represented a confirmation of the theory. In mechanics, however, there remained as an independent hypothesis, the equality of inertial and gravitational masses. With general relativity the independent hypothesis disappears. The new phenomena predicted are the precession of Mercury's perihelion, the deflection of light rays in a gravitational field, the red shift of spectral lines in a gravitational field. Once these facts were experimentally verified, the theory was confirmed.

We have said that between both extreme cases, several intermediate cases are possible. Obviously, the case that physicists like best is the extreme case II. The extreme case I with its ad hoc hypotheses, independent of all the rest, is accepted with some uneasiness, and is considered to be only a provisional stage.

I believe that the uneasiness is largely to be attributed to psychological reasons. One has only to think that the ad hoc quality of a hypothesis can depend uniquely on that chronological order to which very little conceptual bearing should be attributed. The independent hypothesis H is called ad hoc if it is adopted merely in order to explain new experimental evidence. However, the hypothesis can sometimes precede the facts it accounts for. Can it then be called ad hoc?

As illustrations of hypotheses that were independent of preexisting science, one may mention the law of universal gravitation (introduced in order to unify the free fall of bodies, the motion of the moon about the earth, and the motion of the earth about the sun), the displacement current (introduced for symmetry and by analogy with electromagnetic induction), the wave nature of particles (introduced for symmetry and analogy, as will be explained in §4.9). If the new facts that these hypotheses predicted (perturbations of planetary motions, electromagnetic waves, electron diffraction) had been discovered before their introduction, one would have discussed ad hoc hypotheses.

The unification of all science in a single theory with as few independent hypotheses as possible represents an excellent methodological prescription. The essential result, however, is the construction of the historical ladder of theories.

2.29. The richness of the man–nature relation

After having set forth the general outlook that I believe to agree best with the ideas of many physicists, I would like to discuss briefly some other interpretations of the meaning of physical theories which are of interest. I do not have a strong desire to give an impossible completeness of information, nor do I intend to refute the ideas of those who entertain a view about science different from my own. It is only a question of taking cognizance of the different aspects of an extremely rich and complex reality, such as that represented by the man–nature relation. Each one of these aspects has an interest and a validity of its own, which is worthwhile knowing.

I am not an *eclectic*, satisfied with putting together a number of disconnected, perhaps contradictory, assertions on the same subject. I hope to have shown in the foregoing pages, that I have my own viewpoint; this is, I believe, indispensable. Nevertheless, when the very nature of the subject has many facets, noting a few alternative points of view can only be beneficial to our knowledge. Naturally, there are many disadvantageous points of view from which one cannot see anything of interest; but those from which one *can* see something of interest are undoubtedly more than one.

There is a very common attitude, even among good scholars, that always surprises me and which I cannot help but disapprove. This becomes evident sometimes when two persons are discussing the foundations of science and one of them bursts out: "But this is *x*-ism!" Here *x* can stand for any one of the roots: ideal-, rational-, positiv-, neopositiv-, Kant-, neo-Kant-, reduction-, historic-, phenomen-, physical-, Platon-, aprior-, illumin-, empiric-, logic-, nominal-, instrumental-, conventional-, operational-, psycholog- . . . !

Sometimes the tone is exactly the same as when one charges: This is theft, swindle, murder, bigamy, cattle rustling![75] Serious scholars, however, are at times more reasonable and refrain from issuing a foolish moral censure. They confine themselves to suggesting (or even explicitly asserting) that x-ism has been refuted and, consequently, is false. If one is eager to know where this refutation resides, one will more often than not learn that x-ism meets somewhere with some serious difficulty, which prevents it from furnishing a complete interpretation of reality.[76] Fancy that! If I knew a doctrine which did not show this kind of difficulty somewhere, I would look at it with much suspicion. I would be afraid of having been deceived by a sleight-of-hand trick. The doctrine would probably be so superficial and unsubstantial as to tell very little about reality.[77]

Of course, I am not stating that all x-isms are equivalent. I only mean that it is sometimes instructive to consider one side of reality, for a moment, from the standpoint of an x-ism different from our own. Very often we can discover some truth, which we could not otherwise have discovered.

However, we should be careful not to put aside too hastily some important aspect of the complex relationship man–nature merely because we are unable to fit the *whole* of reality into it.

An aspect of science that would be ridiculous to deny is the *instrumental* aspect. Undoubtedly, science can be disinterested but it can also *serve some practical purpose*. The zeal with which numerous philosophers and physicists tend to belittle this aspect of science could certainly be devoted to a better cause. Today more and more people refuse to recognize the traditional superiority of abstract speculation, over the analysis of science in relation to society. In spite of many verbal battles, the interplay between science and its technical applications has not yet been sufficiently understood. Probably, we still have not found the right method to tackle the problem.

Returning to epistemology, we recall that according to *instrumentalism* (see, e.g., Popper, 1965, p. 107), physical theories are nothing but instruments, or recipes, which enable us to predict some facts, starting from the observation of other facts. The origin of this conception is very old. One can think, for instance, of Cardinal Bellarmino who sought to save Galileo by maintaining that the Copernican theory was a useful tool for calculating the positions of heavenly bodies, but did not regard reality. Among those who, in one way or another, can be called instrumentalists are G. Berkeley, M. Mach, H. Hertz, J. Dewey, W. Heisenberg, P. Dirac.

In my opinion, a great mistake is often made regarding the interpretation of the expression: "are nothing but instruments." A hammer is nothing but an instrument for driving nails. But what is important for epistemology is that the person using the hammer *knows* that he can use it to drive nails. The *cognitive* value, thrown out of the window, comes back through the main door; one knows something about the real world. As was already emphasized (§1.10), the true assertions of physical science are *true* in the sense that they actually correspond to the facts; a sense that is not only intuitively acceptable, but has also been formalized and rendered precise by A. Tarski.

It is largely believed today that our ancestor of the order of the primates started to become human when he learned how to use instruments, perhaps simple stones. It seems that language, symbolic representation, and abstract thought, developed together with the capacity to use instruments in a relation of mutual dependence. If this is true, the epistemologist should be very careful when talking about "nothing but instruments."

Instrumentalism sometimes goes hand in hand with a strict *phenomenism*, which maintains that only phenomena and their interconnections are of importance. This view is worth knowing. In my opinion, it even had a beneficial influence when it induced some people to give up building unnecessary castles in the air about a deep, but unverifiable reality. However, if one clings uncritically to the phenomenist dogma, one may fail to realize that certain theoretical constructions just represent the best connection between the phenomena and that, in any event, all connections are theoretical. One may even arrive at the unfortunate position of Mach who could not bring himself to accept atomism.

A close relative of instrumentalism is *conventionalism* (see, e.g., Wisdom, 1971; Lakatos, 1971). Also this view contains a great deal of truth. It is impossible to deny that the structure of science is to some extent conventional. More than once we have called attention to those points of physics that are definitely conventional, and have warned against the temptation to hypostatize as reality what merely exists as a convention. But there are some not so trivial issues, such as the following.

K. Gauss, one of the great precursors of non-Euclidean geometry, once performed some measurements on a triangle formed by three mountains, a few tens of kilometers apart from one another. He wanted to verify whether the sum of the angles of the triangle was different from 180°. Obviously, he did not find any such difference which, as we know today, may exist only on a cosmic scale. Later, H. Poincaré, taking part in the discussion on non-Euclidean geometries, remarked that even if a triangulation such as that performed by Gauss, should yield a result different from 180°, one would not be authorized to conclude that "real geometry" is necessarily non-Euclidean. For such a conclusion evidently implies having taken for granted that light travels in a straight line. This is only a convention. By relinquishing it, we could still retain Euclidean geometry; the experiment would then tell us that the light rays are curved. We have only to decide what geometry is most convenient for our purposes and, consequently, adopt it by convention.

Pursuing this line, one can arrive, like P. Duhem, at the affirmation that all theories are nothing but conventions; as such, they can be good or bad, but not true or false. When we realize that a given convention is bad, we give it up, and make a better convention. Naturally, what should cause perplexity is the affirmation that theories are *nothing but* conventions. Conventions about what? And why can they be good or bad?[78]

Additional relatives of instrumentalism are *pragmatism* (C. S. Peirce, W. James, F. Schiller, J. Dewey) and *operationalism* (P. W. Bridgman). The former has had some important influence on scientific thought, but only for its methodological component (instrumentalism), not for its metaphysic and fideistic component. As to the latter (Bridgman, 1927), we have already considered it (§1.5) in connection with the definition of physical quantities. In spite of the biting criticism that has been directed against the operational procedure, I do not believe that such a procedure can be dispensed with if physical quantities are to be defined in a reasonable way, corresponding to what is actually done in real physics. But I am well aware of the danger of transforming operationalism into a general philosophy of which one risks becoming prisoner. One can come, for example, to Bridgman's criticism to general relativity (Bridgman, 1936), which frankly appears to be misguided and not to the point.

All these doctrines have in common the concern not to receive in science any unprovable assertion or any statement of doubtful meaning. This attitude was made precisely and assumed as a norm in the antimetaphysical battle fought by the *neopositivist* school. The pioneer in that direction was L. Wittgenstein, with his assertion that metaphysical statements are void of *meaning*. A sentence has a meaning only if it is a truth function of a number of elementary assertions, each one of which expresses an observable fact.[79] This means that once the experimental observation has established which elementary assertions are true and which are false, one can deduce by the laws of logic whether the main sentence is true or false. Any other kind of statement is void of meaning and is therefore inadmissible in science.

Neopositivism (or logical positivism or logical empiricism or neoem-

piricism), is a school of thought born about half a century ago from the
Vienna circle (M. Schlick, O. Neurath, R. Carnap, K. Gödel, F. Waismann, and others) and the *Berlin circle* (H. Reichenbach, C. G. Hempel, R. von Mises, and others), having interaction with other scholars
such as L. Wittgenstein, K. Popper, C. Morris, W. V. Quine, E. Nagel,
J. Wisdom, G. Ryle, A. Tarski, and several others (see, e.g., Weinberg,
1948; Kraft, 1968). The scientific influences most strongly felt were
those of C. Mach and A. Einstein in physics, and those of D. Hilbert,
G. Frege, and B. Russell in logic and mathematics. Many of the foun-
ders of the school either came directly from science or had a solid
preparation in science. Not all physicists agree or disagree on the same
points of the doctrine, of course; but one cannot deny that the general
outlook of the logical positivists reflects very closely an attitude largely
diffused in the community of modern physicists. Their remarkable
willingness to accept criticism, to receive the results of colleagues, to
revise one's own ideas (of which R. Carnap was a model), is just the
kind of professional standard that is popular among physicists.

The antimetaphysical outlook is the point that connects neopositivism with nineteenth century's positivism and justifies its name. The
only source of meaningful knowledge is experience. From it, one obtains the basic assertions which are elaborated on later by the rigorous
methods of formal logic (logical empiricism).

In my opinion, a most important step forward, from a foundational
point of view, was that scholars began to wake up from what can be
called the *dogmatic sleep* of pure empiricism. Empiricists had maintained that the experimental facts can by themselves verify or falsify
a scientific assertion. But a little reflection should reveal that a statement can be contradicted only by another statement,[80] and not by an
experimental fact, that is, by an extralinguistic entity. In this way,
language becomes the protagonist of science. Science consists of a set
of noncontradictory statements, deducible from one another in turn in
a given, possibly formal, language. The language can be conventional;
but everything must check on the *syntactic* level, and one must be sure
to respect the fixed points to which the framework is anchored, consisting of the basic assertions (or *protocols*) furnished by experience.

Neopositivist scholars passed in turn from a purely *phenomenist* language (where the basic assertions concern sense data) to a *physicalist*
language (where the basic assertions concern physical objects); from
a purely *syntactical* analysis (i.e., from the formal properties) of language, to a *semantical* approach (i.e., reference to the objects about
which one speaks); from the criterion of *verification,* to the much
weaker one of *confirmation.* There would be no point here to discuss
all the details.

In my opinion, the most difficult point still to be clarified is in what way basic assertions are derived from experience. How is an extra-linguistic fact actually translated into a linguistic fact? Of course, one can merely refer to psychology and take the fact as primitive and not susceptible to analysis. It seems to me that this position is not so utterly unacceptable as some students maintain; there must be a starting point. But it is essential to realize that a basic assertion, just because it is not an experimental fact, is automatically "theory-laden." This was already stressed in connection with the distinction between theoretical and observational terms, proposed by the logical positivists, which in my opinion has to be discarded (§1.7).

Luckily, experience teaches us that the basic assertions, although theory-laden, can have *intersubjective validity*. This, I believe, is the primitive and not easily analyzed fact which can rescue what is usually called the *objectivity of science*.

According to K. Popper (see Popper, 1959, 1965, 1972), the basic assumptions are the result of a decision analogous to that of a jury. Hence they are, to a certain extent, intersubjective; however, they are always liable to being revised. Science is like a pile-dwelling built on a marsh. We drive the piles deep enough to make sure that they are sufficiently stable to support the entire framework.

The exclusion of metaphysics from science is also in Popper's program; however, he refrains from declaring void of meaning any assertion that cannot be derived from observable data. Rather, he proposes a criterion of demarcation between those propositions that can, and those that cannot, belong to empirical science. A proposition is of the former kind if there exists, at least conceptually, an experiment capable of falsifying it. For example, the law of universal gravitation is scientific, because two masses not attracting each other are perfectly conceivable; if they existed, an experiment could contradict the law. On the contrary, the statement: "Mr. So and So has the Oedipus complex" is not in any way falsifiable and does not belong to science (according to Popper, psychoanalysis is not an empirical science).[81]

Scientific theories can never be verified in absolute terms, but can be falsified even by a single counterexample. As a result, physical theories never go beyond the stage of hypotheses. This position might even seem trivial to anyone who believes that a theory can be either true or false in absolute terms (i.e., regardless of the validity domain). Because one can never prove that it is absolutely true, it must of necessity remain a hypothesis. But Popper goes on in a very ingenious way. Why is a given theory (i.e., hypothesis) accepted in a given historical period? The most common and naïve opinion is that we accept the most probable hypothesis. Popper maintains exactly the opposite.

In the spirit of the demarcation criterion, it is reasonable to assume that a hypothesis has the greater empirical content the more numerous and dreadful are its potential falsifiers. Yet the more numerous they are, the less probable is the hypothesis. As an example, let us take for granted that the planets move along plane curves, and consider the hypotheses: (1) that the planets move along circular paths and (2) that the planets move along ellipses. A group of four observations can falsify (1), but not (2); for in general, no circle can go through four arbitrary points of a plane, whereas an infinite number of ellipses (or, rather, second-degree curves) go through them. Hence hypothesis (1) is less probable; but it has a greater empirical content, because if verified, it restricts more severely the freedom of nature relative to our knowledge and yields more information.

Popper's prescription to the scientist is to try and imagine the most daring, hence most improbable, hypothesis and to submit it to experience. The latter will never be able to confirm the hypothesis. But the more numerous and severe are the experimental checks that the hypothesis has gone through, without being falsified, the more it will become verisimilar or, as Popper puts it, corroborated.

In this way, scientists build a succession of more and more daring, or improbable, hypotheses endowed with more and more empirical content. Although the language is diametrically opposed to the one we have used (falsification, instead of confirmation), the net result is surprisingly analogous to that illustrated by means of the historical ladder of theories.

According to I. Lakatos, one should not consider or assess isolated theories, but sequences of theories that show continuity and are subject to the same methodological rules (Lakatos, 1970). Each one of these sequences embodies a program of scientific research (e.g., Newton's program, Maxwell's program, Einstein's program). There are no genuine crucial experiments,[82] because each program builds up a protective belt which renders it impossible to hit its hard core. For example, to defend Newton's absolute space, one can imagine the ether dragging or the Lorentz–Fitzgerald contraction; to defend determinism against quantum mechanics, one can imagine the hidden variables (see §4.17); and so on. The program represents a progressive problem shift, as long as the continuous adjustments in the protective belt bring fresh empirical contents, that is, lead to predict novel experimental facts. It is instead a degenerating problem shift, when the new hypotheses are ad hoc and are unable to predict anything new. History of science consists of competition between rival programs, of which the progressive ones win, whereas the degenerating ones are rejected. However, victory may only be provisional; a new hypothesis of great empirical content can sometimes give new life to a degenerating program.

In this framework one could, by way of example, interpret the fight between the wave and particle hypotheses of light. Huygen's wave hypothesis did not bring about for a long time the discovery of new facts. When Fresnel, however, added an auxiliary hypothesis that afforded the prediction of a great number of novel experimental facts, the program outdid its rival and was resumed with great enthusiasm. It then became clear that Newton's program had accepted some ad hoc hypotheses of little empirical content in order to explain interference.

Not everyone selects in the same way and at the same time to which program to adhere. We have, for instance, the limiting case of Einstein who embarks on a progressive program, urged by reasons of consistency and universality, although virtually ignoring Michelson's experiment. But simultaneously, we have the protecting-belt position of Michelson, who many years later, in spite of the result of his own experiment, could not bring himself to give up his faith in the ether.[83] Several other cases, especially taken from microphysics, could be mentioned.

Lakatos's methodology of research programs still has a rational character. By contrast, the interpretation of the history of science given by T. S. Kuhn,[84] seems to introduce, at least to some extent, an irrationalistic component. There are periods of normal science during which scientists apply well-established theories and devote themselves to the solution of puzzles. All this takes place within a *paradigm,* dogmatically accepted by the scientists of that period. If *anomalies* relative to the paradigm start to pile up, the scientific community refuses to recognize them as such, in the same way as when we look at a figure and tend instinctively to put aside those details that do not check with our preconceived interpretation (recall §2.17). This process goes on until there occurs a scientific revolution and a change of paradigm. Borrowing a term from form psychology, Kuhn talks about a *Gestalt switch.* This term denotes a well-known phenomenon: sometimes when we are looking at a figure and are accepting a given interpretation of it, we suddenly realize that we see a different figure.[85] Then, it is difficult to return to our previous interpretation.[86]

The Kuhnian vision is very interesting and should be judged seriously. I think it unfortunate that this concept should have become so popular with a number of people who appreciate, above all, the irrationalistic component of it, with the unconscious, or even explicitly declared, purpose to deny the value of science.

In my opinion, one of the serious shortcomings of this vision is its tendency to imagine a nonhistorical space in which the same patterns are repeated again and again without variation. In such a view, scientists seem to be unaware of the fate that dominates them.

In politics the image of history as *magistra vitae* is indeed very far from reality; dictators of present times do not seem to study very much the history of the dictators of the past in order to avoid their errors. But scientists today are educated by long training, not to repeat the errors of their predecessors. Of course, they make errors all the same; but very seldom are they errors of the same kind. In particular, critical awareness is much more alive today than one or two centuries ago; nowadays physicists are very careful not to remain prisoners of a paradigm. Some assertions about the behavior of scientists in general (including scientists of today, and perhaps of tomorrow), derived from a discussion about *epicycles, caloric, phlogiston, electrical conflict,* and so forth, can be annoying. Modern physicists have in turn their *partons,* their *quarks,* their *gluons* (§4.27); but they are fully aware that one can talk about such objects only in a hypothetical manner. Today the physicist devoted to the sacred dogmas is largely a myth.

3   The physics of the irreversible

3.1. Reversibility and irreversibility

We have already remarked that macroscopic processes are usually divided into *reversible* and *irreversible* processes. Reversible processes can evolve equally well both ways in time, whereas irreversible processes can go only one way. Let us consider a process governed by one or more equations of the kind $f(a_n, t) = 0$, where t denotes time and a_n denotes, for short, all the other quantities involved. In the case of reversibility the process described by $f(a_n, -t) = 0$ is also perfectly possible, and does not clash with any physical law. In the case of irreversibility the process $f(a_n, -t) = 0$ is instead forbidden; it clashes with the second law of thermodynamics (to be illustrated shortly), and hence cannot be observed.

All the physical phenomena considered so far are reversible. A planet can revolve about the sun one way or the other,[1] an electromagnetic wave can either diverge from a point or converge on it, a light ray can pass from air to glass or follow the reverse path, from glass to air.

By contrast, if we put a lump of sugar into our tea and stir, we expect it to dissolve and (so to speak) to disappear; but no one would expect that by turning the teaspoon in the opposite direction, the lump of sugar should reappear.

Irreversibility represents something more essential and important than a mere property of some physical processes of the outer world. For we ourselves are irreversible systems, although in a somewhat different and more complicated way than the simple systems to be studied here. No one can expect to see a man grow younger, become a baby, and eventually go back to his mother's womb. Our life is one way; it is a long succession of partial deaths, until total death.

This fact, so often recalled with woeful words by poets, so often used by priests and philosophers to emphasize the smallness of man and to "prove" the need for a "being" *not subject to time*, is essential for our perception of the physical world. Psychological time flows always in one direction, but one may easily be led to think that this represents the essential feature of an outer physical entity, independent of us. Time appears to have an arrow of its own, pointing to the positive direction. All this can only surprise us, when we think that instead spatial dimensions do not show any privileged direction.

The problem of time, of its flow and its asymmetry, represents one

152

of the most profound and fascinating mysteries of the physical world as well as of our own nature. We are far from having completely solved it. However, the study of irreversible processes and of their interpretation in terms of atomic constituents sheds a good deal of light on it.

This study belongs traditionally to *thermodynamics*, because the most typical cases of irreversibility occur when there are heat exchanges or similar processes. Thermodynamics was first tackled from a merely phenomenological point of view, which led to the enunciation of a few very general laws. Later, about the middle of the last century, a statistical interpretation was discovered, based on the chaotic motions of the microscopic constituents of bulk matter.

At first, it was believed, somewhat uncritically, that the new interpretation could be derived uniquely from mechanics, with the addition of the notion of *probability*. That is not quite true and one needs at least one further postulate, as will be discussed later on.

Probability rapidly became one of the major tools of physics and of the whole of modern science. First of all, after its brilliant success in statistical thermodynamics, probability turned out to be essential for the construction of quantum mechanics. Further, it was largely used to build *information theory*, a new branch of science, arisen about the middle of our century. Information theory, which in a way is a relative of thermodynamics, has supplied us with an appropriate language in which to formulate a considerable number of problems in pure and applied science. Finally, it is worthwhile to mention that in the last decades, students of logic and methodology have analyzed the connection between probability and *inductive inference*.

We shall start by summarizing, briefly, the fundamentals of elementary thermodynamics. Next, we shall illustrate the problems of probability, information theory, and statistical mechanics. Eventually, we shall be in a position to tackle the problem of the time arrow.

3.2. Temperature and heat

The physical quantity called *temperature* owes its origin to the need to render objective our intuitive ideas of hot and cold. In order to measure temperature, we can take advantage of the observation that nearly all material bodies expand when heated. Due to several practical reasons, a thermometric substance largely used is mercury. Everyone is familiar with the *mercury thermometer* in its various forms.

But as soon as a reasonable precision of measurement is attained, one notices that the thermometers built with different substances (mercury, water, alcohol, and so forth) do not exactly coincide. For-

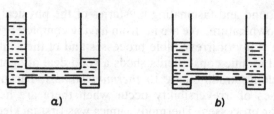

Figure 3.1

getting about this difficulty for the time being, we refer to a well-determined substance, say, mercury, and to the *Celsius scale*. This is obtained by first immersing the thermometer in a mixture of ice and water and marking 0°C, then in boiling water (at ordinary pressure) and marking 100°C; the whole interval covered by the mercury column is then divided into 100 equal intervals (*degrees*). Later, we shall introduce a more rational temperature scale, independent of the thermometric substance.

When two bodies at different temperatures are put into contact, the warmer one cools down, as the colder one warms up, until one and the same temperature is reached (*thermal equilibrium*).

The process is very similar to that observed in the case of connecting vessels containing a liquid (Fig. 3.1). If initially the levels are different, Figure 3.1(a), some liquid will pass from one vessel to the other until the levels are identical, Figure 3.1(b).

If the temperature is assimilated to level, one has to assume that some kind of fluid passes from the warm body to the cold body. It must be an imponderable fluid, because the weight of the body that is heated does not increase. Actually, the existence of such a fluid, called *caloric*, was postulated and maintained throughout the early nineteenth century. But it was rejected when scientists began to understand better the relations between heat and energy.

It is important to note that the fluid hypothesis was not so wide off the mark as is usually assumed and that it does not deserve the pitying look it usually gets. That something passes from the hotter to the colder body is perfectly true, even today. It is also true that this something is endowed with a few important properties of ordinary fluids (e.g., conservation in the conduction process). Talking about the quantity of heat as we do today instead of about the caloric is but a question of nomenclature. The naïve assumption consisted merely in granting the caloric *all* the properties and *only* the properties of ordinary fluids,[2] and in considering particles of caloric, in much the same way as one would consider molecules today.

Heat is measured in *calories*. The calorie is defined as the quantity of heat required to increase by 1°C the temperature of 1 g of distilled water about 15°C.

By means of the heating of a given mass of water or by making use of other processes, one can build various types of *calorimeters*, that is, of instruments measuring the quantities of heat exchanged between bodies. We shall not describe such instruments, because this would not shed much light on the conceptual framework. But we shall mention that the *specific heat* of a substance is defined as the quantity of heat needed to increase by 1°C the temperature of 1 g of that substance. The specific heat of water is obviously 1.

3.3. Perfect gases

Gases have quite an interesting thermal behavior.[3] The laws obeyed by a gas become especially simple when the gas is extremely rarefied. In this case one talks about a *perfect gas* or an *ideal gas*.

This peculiar concept can be explained as follows. We shall write down some general laws for the behavior of a perfect gas. It is found by experience that any real gas conforms to these laws as long as the accuracy of the measurements does not exceed a certain limit. If we attain a better accuracy, we shall show some discrepancies with respect to the behavior of a perfect gas. It suffices then to rarefy the gas further, that is, to increase the size of its container, to regain the behavior of a perfect gas. If we go again to more accurate measurement, we have to expand the gas still further, and so on.

It is known that gas exerts *pressure* on the walls of the container. Pressure is defined as the force exerted by the gas on a unit surface of the wall, and is naturally measured in dyne/cm^2.

We need not limit ourselves to the walls of the container, but can speak about the pressure acting on any (even imaginary) surface within the gas. A principle, named after B. Pascal, states that the pressure at a given point of the gas does not depend on the orientation of the surface. Hence if the container is not too large (has laboratory size),[4] one can speak simply of the pressure P of the gas, equal at all points.

When a perfect gas is maintained in *isothermal* conditions, that is, when its temperature is constant, whereas its volume V is varied, one finds that *Boyle's law* (1660) holds

$$PV = \text{constant} \tag{3.1}$$

The product of pressure and volume is a constant.

If the gas is instead maintained at constant pressure and its temperature t is varied, one will find the *Gay–Lussac* law (1801)

$$V_1 = V_0 (1 + \alpha t) \tag{3.2}$$

where V_0 denotes the volume at 0°C, V_1 the volume at temperature t and α is a coefficient equal for all (perfect) gases, whose value is about 1/273.

Thus the volume decreases when the temperature decreases. It is interesting to note that because $\alpha = 1/273$ the volume should vanish when t goes down to -273°C! Although equation (3.2) is valid only for a perfect gas,[5] this seems to indicate that the temperature $t = -273$°C should have a very peculiar significance. This temperature is referred to as the *absolute zero*.

Let P_0 represent the constant pressure at which equation (3.2) is valid. Multiplying both sides of the equation by P_0, we get

$$P_0 V_1 = P_0 V_0 (1 + \alpha t) \tag{3.3}$$

On the lefthand side we have the product of the pressure and the volume. By Boyle's law, such a product will remain constant if we pass to any new volume V and to the corresponding pressure P, although leaving the temperature unchanged. Hence for the product PV, measured at temperature t, we can write

$$PV = P_0 V_0 (1 + \alpha t) \tag{3.4}$$

where $P_0 V_0$ represents the same product, measured at 0°C.

The temperature counted from the absolute zero is called the *absolute temperature*. Denoting it by T, we get

$$T = t + 273 \tag{3.5}$$

If t is measured in °C, T is said to be measured in *Kelvin* (K).

Substituting $T - 273$ for t in equation (3.4) and remembering the value of α, we obtain

$$PV = \frac{P_0 V_0}{273} T \tag{3.6}$$

We now mention Avogadro's law, which states that one *mole*[6] of a gas at a given temperature and pressure occupies a given volume independent of the kind of gas.[7] As a result, the expression $P_0 V_0 / 273$, denoted by R, has the same constant value for all gases, as long as we refer to one mole. To find the value of R, we have to take a mole of any gas and to measure its volume V_0 at a given pressure P_0, at 0°C. In particular, it is found that at normal atmospheric pressure ($P_0 = 1.01 \times 10^6$ dyne/cm²) one mole of any gas occupies the volume

$V_0 = 2.24 \times 10^4$ cm^3. On computing, we find then $R = 8.31 \times 10^7$ CGS units.

Equation (3.6) when written for a mole of gas becomes

$$PV = RT \tag{3.7}$$

This fundamental law is known as the *law of perfect gases*.

3.4. Heat, work, and internal energy

The rough hypothesis of the caloric could not be maintained any longer when it was realized that work can be converted into heat and, conversely, according to a fixed ratio. This occurred about the middle of the nineteenth century.[8]

Then it became gradually clear that heat was only one form of energy. The precise conversion rate was measured by J. P. Joule, with a celebrated experiment, in which a given amount of work was converted into heat by friction. The mechanical equivalent of the calorie is found to be $J = 4.18 \times 10^7$ erg/calorie.

Having established that heat is a form of energy, and having measured the value of J, one can dispense with the *calorie* and express Q directly in *ergs*. This will be done from now on to simplify the writing.

The foregoing considerations should convince one that material bodies can store energy, for instance, in the form of heat. This energy is known as the *internal energy* and will be denoted by U.

Let us supply a body (or system of bodies) with a small amount of heat dQ.[9] This heat will partially be stored as internal energy, and will partially be utilized to do work on external bodies. Hence we write

$$dQ = dU + dW \tag{3.8}$$

where dU denotes the increase of the internal energy and dW the work carried out.

Let the system now under consideration undergo any kind of transformation, subject to the condition that, at the end, everything be returned to the initial condition. The system is said to have carried out a *closed cycle*. As the final condition is identical to the initial one, the internal energy will have returned to its initial value. Hence the increments dU, corresponding to the various portions of the cycle, will add up to zero (some will be positive and some negative).

If Q denotes the sum total of the increments dQ (positive when heat is given to the system, negative when subtracted from it) and W the sum total of the works dW (positive when the system does work, negative when work is done on the system), we have, from equation (3.8),

Figure 3.2

at the end of the cycle

$$Q = W \tag{3.9}$$

This equation states that every time a system goes through a closed cycle, the heat spent and work obtained are equal. This is the precise wording of the equivalence principle between heat and work, and is known as the *first law of thermodynamics*.[10] When the cycle is not closed, one must also take into account internal energy and make use of the complete equation (3.8).

In the case of a gas one can give dW an explicit form. Consider a gas at pressure P, enclosed in a container, whose walls will be denoted by S (Fig. 3.2). Suppose that the walls slowly recede and take the form S', very close to S. During the expansion the gas exerting a pressure (hence a force) on the walls does work. Each surface element is acted on by a force equal to P times its area a. The work will be obtained by multiplying that force by the normal displacement dl. The product $a\,dl$ equals the volume swept by the surface element (shaded region). Adding up the contributions from all surface elements of S, we get the work

$$dW = P\,dV \tag{3.10}$$

where dV represents the volume increase of the container in going from S to S'.

Thus for a gas we can rewrite equation (3.8) in the form

$$dQ = dU + P\,dV \tag{3.11}$$

It is evident that in case of contraction, instead of expansion, dV will be negative.

We have assumed that the displacement of the wall takes place sufficiently slowly so as to give the gas time to readjust and to exert the pressure P steadily. It is interesting to do the experiment in the opposite condition.

Let a gas be enclosed in the portion A of the otherwise empty container AB (Fig. 3.3). Suddenly, the dividing wall between A and B is removed; the gas rushes also to occupy B. This is obviously an expansion without work done by the gas, so that in equation (3.8) we set

Figure 3.3

$dW = 0$. Also, $dQ = 0$; hence $dU = 0$. As a result, in this experiment there is no variation of internal energy.

Let us perform the experiment with a perfect gas. This means, of course, taking a real gas and going to the limit of a high degree of rarefaction. We observe that the temperature T of the gas does not vary, that is, it is the same before and after the expansion. We infer from this that when the internal energy of a perfect gas does not vary, the temperature does not vary either, regardless of the changes in pressure and volume. Therefore to each value of U there corresponds one and only one value of T. Conversely, because T is found to be always an increasing function of U, to each value of T there corresponds one and only one value of U. We say that in a perfect gas the internal energy is a function of temperature only.

3.5. Specific heats of a gas

The specific heat of a gas depends on the conditions in which it is measured. It is customary to distinguish two extreme cases and to talk about the specific heat at constant volume c_V and the specific heat at constant pressure c_P.

If we heat 1 g of a perfect gas in a container with fixed walls (constant volume), we have by definition

$$dQ = c_V \, dT \qquad (3.12)$$

On the other hand, putting $dV = 0$, we get from equation (3.11) $dQ = dU$, hence on substituting in equation (3.12), we find

$$dU = c_V \, dT \qquad (3.13)$$

Because U depends solely on T, this expression of the increment of internal energy will be valid *for any process*, even for a process with varying volume.

Let us now turn to the case of constant pressure. The gas when heated tends to expand; hence to keep it at constant pressure, the container's volume will be properly increased. Accordingly, the gas will do work and the specific heat c_P will be greater than c_V, because part of the heat will be employed to do this work.

By definition, in the expansion of 1 g of the gas at constant pressure, we have

$$dQ = c_P \, dT \tag{3.14}$$

On substituting in equation (3.11), we obtain

$$c_P \, dT = dU + P \, dV \tag{3.15}$$

hence from equation (3.13)

$$c_P \, dT = c_V \, dT + P \, dV$$

or

$$c_P - c_V = P \frac{dV}{dT} \tag{3.16}$$

where one must remember that P is constant by hypothesis.

This equation refers to 1 g of the gas. If instead we want to refer to one mole, that is to M grams of the gas (where M denotes the molecular weight), dV will be M times greater and we have

$$c_P - c_V = \frac{1}{M} P \frac{dV}{dT} \tag{3.17}$$

From the gas equation (3.7) we evidently have, *at constant pressure*, $P \, dV = R \, dT$. We can derive from this the value of dV/dT and substitute it into equation (3.17), thus obtaining

$$c_P - c_V = \frac{R}{M} \tag{3.18}$$

This remarkable equation enables us to evaluate the difference between the specific heats for any perfect gas. Such difference varies from one gas to another only inasmuch as the molecular weight varies.

It is customary to consider the ratio γ of the specific heats

$$\gamma = \frac{c_P}{c_V} = \frac{R}{Mc_V} + 1 \tag{3.19}$$

which, of course, is always greater than 1. It is found experimentally that, to a good approximation, $\gamma = 1.67$ for a monoatomic gas, $\gamma = 1.40$ for a biatomic gas, and smaller values are found for more complex molecules (see §3.9).

Consider an ideal gas undergoing an *isothermal* expansion. The gas will be enclosed, say, within a cylinder, and provided with a piston, which is gradually receding. The gas pushes the piston and is thereby doing work. Because in the expansion the gas tends to cool, one must continuously supply it with heat. This heat is integrally turned into work, because by definition $dT = 0$, hence from equation (3.13) $dU = 0$ and, finally, by equation (3.8), $dQ = dW$.

We wish to evaluate the work performed by a mole of gas in isothermal expansion when its volume varies from V_1 to V_2. To begin with, from the gas equation (3.7) we have $P = RT/V$, so that the elementary work dW is expressed by

$$dW = P\,dV = RT\frac{dV}{V} \tag{3.20}$$

In order to evaluate the total work W obtained in going from V_1 to V_2, we have to add all the elementary works dW corresponding to the various increments dV, which go from V_1 to V_2. This is the same as *integrating*, which was explained in connection with equation (1.26). We write

$$W = \int_{V_1}^{V_2} RT\frac{dV}{V} \tag{3.21}$$

Mathematical analysis teaches us how to carry out this integration (with R and T constant), obtaining[11]

$$W = RT\ln\frac{V_2}{V_1} \tag{3.22}$$

Next, let us consider an *adiabatic* expansion, that is, an expansion during which the gas does not exchange any heat with the exterior, perhaps because the walls of the container are impenetrable to heat. Naturally, the gas now is cooled by the expansion and would be heated by a compression. Because $dQ = 0$, equation (3.11) yields $dU + P\,dV = 0$, hence, by equation (3.13), $c_V\,dT = -P\,dV$. This equation refers to 1 g, so that for one mole we obtain

$$Mc_V\,dT = -P\,dV \tag{3.23}$$

Expressing P with $P = RT/V$ (gas equation), we get

$$\frac{dT}{T} = -\frac{R}{Mc_V}\frac{dV}{V} \tag{3.24}$$

Recalling equation (3.19), we may also write

$$\frac{dT}{T} = -(\gamma - 1)\frac{dV}{V} \tag{3.25}$$

We can now add the values of both sides of equation (3.25) for all the dT that go from the initial temperature T_1 to the final temperature

T_2 and the corresponding dV that go from the initial volume V_1 to the final volume V_2. Thus we have to evaluate the integrals

$$\int_{T_1}^{T_2} \frac{dT}{T} = - (\gamma - 1) \int_{V_1}^{V_2} \frac{dV}{V} \qquad (3.26)$$

In a way analogous to that of the isothermal case, we obtain

$$\ln \frac{T_2}{T_1} = - (\gamma - 1) \ln \frac{V_2}{V_1} \qquad (3.27)$$

Considering some properties of logarithms,[12] we finally get

$$\frac{T_2}{T_1} = \left(\frac{V_1}{V_2}\right)^{\gamma - 1} \qquad (3.28)$$

As to the work done in an adiabatic expansion, we have from equation (3.23)

$$dW = - Mc_V \, dT \qquad (3.29)$$

Adding up all the elementary works and taking into account the minus sign on the righthand side, we readily get

$$W = Mc_V(T_1 - T_2) \qquad (3.30)$$

From equations (3.28) and (3.30) we conclude that in an adiabatic expansion ($V_2 > V_1$) the gas cools down ($T_2 < T_1$) and does positive work; and the other opposite occurs for an adiabatic compression.

3.6. The second law of thermodynamics

We have stated that mechanical and electromagnetic processes are reversible, whereas thermodynamic processes are not. But this is only a very simplified first approximation, and valid only in an abstract sense. In fact, no real physical phenomenon can take place without implying some thermodynamic process that renders it irreversible.

For example, let us consider the motion of the moon around the earth. The moon travels along its nearly circular orbit, the earth's gravitational pull on the moon being balanced by the centrifugal force. The earth, in turn, rotates about its axis in the same direction. Energy and momentum conservation seem to guarantee that the motion of the system should go on indefinitely, always in the same way. Moreover, there seems to be no reason that the motions could not be reversed.

However, if we look more closely, we notice that the moon's gravitational pull gives rise to tides on the earth (in the liquid as well as in the solid and gaseous parts). The tides dissipate a fraction of the mechanic energy and turn it into heat irreversibly. The earth–moon system is losing mechanical energy all the time, consequently, its motion varies. Precisely, the moon is slowly receding from the earth and the earth's rotation is slowing down. The reverse process is impossible, because we are not confronted with purely mechanical phenomena and because a thermodynamic process is involved.

What value, then, can the statement have that mechanical and electromagnetic phenomena are reversible? It has the same value as when we refer to a limiting case, whereas the limit is actually unattainable. It is very true that a purely mechanical process cannot be practically realized; but one can approach it better and better by taking special precautions (such as eliminating friction as much as possible, performing the experiment in a vacuum, screening external radiation, and so on). In other words, one can try to approach the case when all perturbations causing irreversibility are negligible compared to the main process under study. In this case the process can be considered to be reversible.

But once we have taken this point of view, we realize that even in thermodynamics, a process can be very near the ideal conditions of reversibility. This happens when the process takes place by successive steps, all very near equilibrium. For instance, when a gas expands very slowly in a cylinder, the force it exerts on the piston is obviously nearly equal (and opposite) to that acting on the piston from outside. A small variation on either one of these forces can make the process run in the opposite direction. In contrast, the expansion without external work, illustrated in Figure 3.3, is essentially irreversible. No one would deem that once the gas has invaded the whole volume AB, a small variation of the parameters may suffice to make it draw back and gather again in A.

In much the same way, we can discuss the case when a hotter body C_1 and a colder body C_2 are brought into contact. If the temperature difference is substantial, heat passes from C_1 to C_2 and a small variation of the conditions cannot reverse the process. If, however, the difference is infinitesimal, an infinitesimal variation of the initial conditions could render C_2 hotter than C_1, thereby causing heat to flow in the opposite direction. Hence the latter case can be considered to be reversible.

The isothermic and adiabatic expansions considered in the previous section can be thought to be as close as we want to a succession of ideal equilibrium states. Therefore they can be considered as reversible. But let us emphasize again that they are only limiting processes, or the result of an abstraction.

The observation that no real process is exactly reversible led in the last century to the conclusion that there must be a universal law demanding it.

It is of interest to note that the assumption of the irreversibility of even one single kind of physical processes, if properly stated, can entail the irreversibility of the physical world in general.

Accordingly, the second law of thermodynamics can be based on a suitable postulate such as one of the following, which only requires that each of a given class of phenomena are irreversible:

1. It is impossible that *the only result* of a process should be the passage of heat from a body to another body at higher temperature (Clausius' postulate).

2. It is impossible that *the only result* of a process should be the production of work from the heat supplied by one single source at fixed temperature (Lord Kelvin's postulate).

We have stressed in both cases the words *the only result*, because they are indispensable. As a matter of fact, the previous processes *can* very well occur, provided only that they are accompanied by some other change of the physical situation. Consider, for example, postulate 2, which denies that work can ever be produced simply by subtracting heat from the ocean. If besides the ocean, we have at our disposal a refrigerator, we can obtain work from the heat of the ocean, provided that a fraction of this heat passes to the refrigerator. This will be made clear presently.

Both postulates are not independent of each other. If one is valid, the other must be valid, too. As an example, let postulate 2 be false. In that case one could obtain work from the ocean water, next transform the work into heat by friction, and finally transfer the heat to a source at higher temperature than the ocean. In this way one could violate postulate 1.

Some results of significant generality can be drawn from the postulate of thermodynamics, by means of a particular reversible machine (in the sense already illustrated), which carries out a *Carnot cycle*.[13] This is a closed cycle performed by a perfect gas, say, one mole. The gas can be put into contact and exchange heat with two heat sources Σ_1, Σ_2, at the fixed temperatures T_1, T_2, with $T_1 > T_2$. Further, the gas can expand or be compressed reversibly, exchanging work with the surrounding.

The cycle is represented schematically in Figure 3.4, where pressure is plotted against volume. The gas is initially at 1, and has volume V_1 and temperature T_1. When kept in contact with Σ_1, it is made to expand isothermally to state 2, where its volume is V_2 and its temperature is still T_1. Next, the gas is isolated from the sources and undergoes an adiabatic expansion[14] until state 3, where its volume is V_3 and its tem-

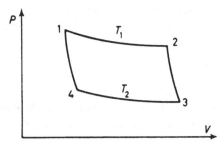

Figure 3.4

perature T_2. Then it is put into contact with Σ_2 and is isothermally compressed until it reaches state 4, where its volume is V_4 and its temperature T_2. Finally, the gas is isolated from the sources and is adiabatically compressed until it is brought back to temperature T_1. One can demonstrate (or derive by inspecting the diagram) that the volume V_4, where the isothermal compression is stopped, can be selected in such a way so that the adiabatic compression would return exactly to 1, that is, to the initial conditions. Thus the cycle is closed.

Let us evaluate the work produced by this peculiar thermal machine. We get positive work from the expansions $1 \to 2$ and $2 \to 3$ and negative work from the compressions $3 \to 4$ and $4 \to 1$. Note that in the first and second adiabatic processes the temperature varies from T_1 to T_2 and from T_2 to T_1, respectively. Because of equation (3.30), the corresponding works are equal and have opposite signs. There remains only the work due to the isothermal processes, and from equation (3.22) we obtain

$$W = RT_1 \ln \frac{V_2}{V_1} - RT_2 \ln \frac{V_3}{V_4} \tag{3.31}$$

Utilizing equation (3.28) for the adiabatic transformations, we have

$$\frac{T_2}{T_1} = \left(\frac{V_2}{V_3}\right)^{\gamma - 1} \quad \text{and} \quad \frac{T_1}{T_2} = \left(\frac{V_4}{V_1}\right)^{\gamma - 1} \tag{3.32}$$

Hence it is readily seen that $V_3/V_4 = V_2/V_1$, and, on substituting into equation (3.31), we finally get

$$W = R(T_1 - T_2) \ln \frac{V_2}{V_1} \tag{3.33}$$

The ratio between the work produced by the machine and the heat taken from the latter source (i.e., the heat that *costs* something), is called the *efficiency* of the machine. Because in an isothermal expansion the heat supplied to the gas is completely transformed into work, the efficiency e can be evaluated by dividing W by the work performed in the first isothermal expansion, that is, by $RT_1 \ln(V_2/V_1)$. Hence we have

$$e = \frac{T_1 - T_2}{T_1} \tag{3.34}$$

This remarkable relation shows that the efficiency depends solely on the temperatures of the sources and turns out to be greater, the greater their relative difference is.

Equation (3.34) has a *universal* significance. No machine performing a closed cycle between the temperatures T_1 and T_2 (e.g., taking heat from Σ_1 and yielding heat to Σ_2) can have an efficiency greater than that of the Carnot machine C.

To show this, let us imagine that a machine M has a greater efficiency than C. Since the Carnot machine is reversible, we can let M run and utilize its work to operate C in the reverse. If M has an efficiency greater than C, it will take from Σ_1 an amount of heat smaller than that given back by C. Hence the net (and only) result will be the transfer of heat from the lower to the higher temperature, which is absurd, by virtue of postulate 1.

It is also clear that every reversible machine must have the efficiency given by equation (3.34). If its efficiency is smaller, the machine is irreversible.

If there is only one heat source, we must put $T_1 = T_2$, hence by equation (3.34), $e = 0$. As a result, we cannot obtain work from one single source, in agreement with postulate 2.

In a real thermal machine we shall therefore have at least a high temperature source Σ_1 and a low temperature source Σ_2 (e.g., the atmosphere). The heat supplied by Σ_1 will partially be transformed into work, but will partially go to Σ_2, and will be virtually lost. The efficiency can never attain the value 1, unless T_2 is the absolute zero.

3.7. The entropy

It is of interest to note that the Carnot machine, originally introduced and discussed for economic purposes, has yielded results of essential significance for the whole of physics. Let us consider a Carnot machine running in the forward direction, taking heat Q_1 from Σ_1 and giving heat back to Σ_2. Recalling once more that in an isothermal expansion, heat is integrally transformed into work, we readily find that the efficiency can also be expressed as $(Q_1 - Q_2)/Q_1$. Equating this ratio to the right side of equation (3.34), we can readily get

$$\frac{T_1}{T_2} = \frac{Q_1}{Q_2} \tag{3.35}$$

Hence the temperatures are proportional to the amounts of heat exchanged. And this plainly holds for any machine that operates reversibly, exchanging heat with two sources.

All the properties derived so far have been obtained with reference to a particular scale of temperature, for instance, to that of the mercury thermometer, and are valid within certain precisions. We noticed that different thermometric substances yield slightly different results; it is therefore not to be expected that a particular substance would agree to any great accuracy with the universal laws.

By applying a procedure illustrated several times, we can take equation (3.35), not established very precisely, and turn it into a *definition*. We thus arrive at the *thermodynamic scale of temperature*.

As an example, let a Carnot machine operate between the temperature T_1 of boiling water (at standard pressure) and the temperature T_2 of melting ice. Let us measure both Q_1 and Q_2 and put $T_1 - T_2 = 100$ K. This relation, together with equation (3.35), forms a system of two equations with the two unknowns T_1 and T_2. We can thus derive the thermodynamic values of T_1 and T_2, hence the position of the absolute zero (which results to be defined by $T_2 = 273.16$ K). Any other temperature T can then be determined by means of a Carnot machine operating between T and a temperature already established (say, 0°C, or 273.16 K).

It will readily be surmised that all this is satisfactory in theory, but it does not lend itself to very accurate temperature measurements, because heat quantities can only be measured in a fairly coarse manner. Nevertheless, we have the great advantage that all the thermodynamic laws discussed become exact by definition.

A Carnot machine operating in reverse can serve as a refrigerating machine. It can remove heat from an enclosure at the lower temperature T_2 and pour heat into an environment at the higher temperature T_1. To this end, one must spend some work W. The quantity of interest in this case is the ratio r' between W and the mechanical equivalent of the heat Q_2 removed at temperature T_2. Economy will require that r' should be as small as possible. From this discussion, there results

$$r' = \frac{T_1 - T_2}{T_2} \tag{3.36}$$

We see by inspection that if T_1 is held fixed, as T_2 tends to vanish, r' tends to infinity. The amount of work needed to remove even a small quantity of heat from the cool enclosure becomes very large and can go beyond any limit. On the other hand, no refrigerating machine can have a value of r' smaller than that fixed by equation (3.36), otherwise it could be shown that one can violate postulate 1. We derive the conclusion that the absolute zero is unattainable. We can approach it, but the closer we get, the more expensive any further step becomes, until we are forced to give up.

The image of the physical world emerging from the discussion of the second law of thermodynamics is the following. Each heat quantity is *worth* more, the higher the temperature at which it is available, for the larger, then, is the fraction of the heat that can be converted into work.

Let us consider the case when heat goes from a higher temperature down to a lower one, producing work in the process. If the machine is reversible, we can hope to be able, by means of a second reversible machine, to take the heat at the lower temperature and bring it back to the higher one. But because no real process is reversible, a fraction of the heat necessarily remains at the lower temperature and cannot be retrieved. Thermal energy goes down step by step in this way and is *degraded*; that is, it becomes less and less capable of being utilized. Mechanical energy can be entirely converted into heat but cannot be reversed.

If the time comes when the world has reached the same temperature everywhere, and all energy, mechanical or other, has been converted into heat, *thermal death* will have been attained, that is, the end of any interesting or distinguishable process.

One would wish to characterize quantitatively the degradation of energy. This was accomplished by Clausius by means of the concept of entropy.

Suppose that a material system goes through a reversible process, during which it takes heat Q from a source Σ at temperature T. We shall say that its *entropy S* is increased by Q/T.

If the reversible process occurs by successive steps in which the system takes heat Q_i from source Σ_i at temperature T_i, the entropy increase will be represented by the sum $\Sigma Q_i/T_i$.

With infinitesimal notation, we say that every time the system receives heat dQ at temperature T, its entropy increases by

$$dS = \frac{dQ}{T} \tag{3.37}$$

In going from state A to state B the entropy increase will be

$$S_B - S_A = \int_A^B \frac{dQ}{T} \tag{3.38}$$

Let us now consider a system going through a reversible and closed cycle as that of Figure 3.4. The increase of the system's entropy is zero, because by virtue of equation (3.35), the entropy gain Q_1/T_1 equals the entropy loss Q_2/T. We know that this holds for any reversible machine operating between T_1 and T_2. One can fairly easily show (by

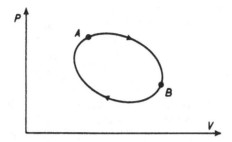

Figure 3.5

means of a number of Carnot machines operated *in series*) that the entropy does not vary even if the system exchanges heat (reversibly) with a number of sources Σ_i at temperatures T_i.

Thus referring, say, to a *PV* diagram as in Figure 3.5, we can conclude that if the system goes through a reversible and closed cycle, starting from A and coming back to A, its entropy does not vary. Let us then evaluate the entropy increase, going first from A to B along the upper line (in the direction of the arrow), then from B to A along the lower line (again in the arrow's direction). We have

$$\int_{A}^{B} \frac{dQ}{T} + \int_{B}^{A} \frac{dQ}{T} = 0 \qquad (3.39)$$
$$\underset{\substack{\text{upper} \\ \text{line}}}{} \quad \underset{\substack{\text{lower} \\ \text{line}}}{}$$

because the cycle is closed.

If the second path is followed in the opposite direction (i.e., from A to B along the lower line) the sign of all the dQ's, hence the sign of the whole integral, will be reversed. We have then

$$\int_{A}^{B} \frac{dQ}{T} = \int_{A}^{B} \frac{dQ}{T} \qquad (3.40)$$
$$\underset{\substack{\text{upper} \\ \text{line}}}{} \quad \underset{\substack{\text{lower} \\ \text{line}}}{}$$

Thus the entropy increase evaluated going from A to B along the upper line equals that evaluated going from A to B along the lower line. As is evident, this result has general validity, hence we can state that the entropy increase in going from state A to state B does not depend on the path followed as long as the process is reversible.

As a conclusion, once we have fixed the value of the entropy at a given initial point A,[15] the value of the entropy at any other point B is a function of B only; that is, it depends uniquely on where B is located and not on the particular reversible process with which it has been attained. We can therefore use the notation $S(B)$ for the value of the entropy at B. It is customary to say that entropy is a *state function*, in that it depends on the state of the system and not on the particular reversible transformation through which it has been attained. Thus the past history of the system becomes immaterial; when the system is at B, its entropy can be defined as $S(B)$ regardless of whether it has been reached in a reversible or irreversible way. Accordingly, one can make the convention that regardless of how B has actually been reached, $S(B)$ must be evaluated as though a reversible path had been followed.

Let us consider a mole of a perfect gas. From equation (3.11) and (3.13), we have $dQ = Mc_V dT + RT \, dV/V$, so that equation (3.37) becomes

$$dS = Mc_V \frac{dT}{T} + R \frac{dV}{V} \tag{3.41}$$

Carrying out the integration[16] between the initial state T_0, V_0 and a final state T, V, and utilizing integrals of a kind already seen, we get

$$S = Mc_V \int_{T_0}^{T} \frac{dT}{T} + R \int_{V_0}^{V} \frac{dV}{V} = Mc_V \ln \frac{T}{T_0} + R \ln \frac{V}{V_0} \tag{3.42}$$

Remembering now that $\ln(T/T_0) = \ln T - \ln T_0$ and $\ln(V/V_0) = \ln V - \ln V_0$, as $\ln T_0$ and $\ln V_0$ are constants, and entropy is defined up to a constant we can simply write

$$S = Mc_V \ln T + R \ln V + \text{constant} \tag{3.43}$$

This is the expression of the entropy of a perfect gas in state T, V, by whatever path that state has been reached.

When a system consists of several subsystems, which exchange heat with one another, the total entropy will be defined as the sum of the entropies of all subsystems.

Let us now consider an *isolated* system, or an ideal system that does not exchange energy of any kind with the environment. For example, we may refer to a Carnot machine C plus its sources Σ_1, Σ_2, the sources being isolated from the surrounding and being large enough to preserve constant temperatures in spite of their heat exchanges with C. It is readily seen that the overall entropy does never vary during the cycle. In fact, to every entropy increase or decrease of C there corresponds an identical decrease for Σ_1 or increase for Σ_2, respectively. But if we replace C with an irreversible machine, the entropy will increase.

For example, suppose that the first isothermal expansion $1 \to 2$ (Fig. 3.4) is replaced by an expansion without outer work, such as that of Figure 3.3. The entropy increase of C will be the same as though the process were reversible, because the final state 2 is the same (remember that the temperature does not change in an expansion of that kind), whereas no entropy decrease occurs for Σ_1 (because the source does not yield any heat). Alternatively, besides operating C in a normal way, let us connect Σ_1 directly with Σ_2, so that some heat Q passes from the former to the latter. The entropy of Σ_2 will increase by Q/T_2, whereas that of Σ_1 will decrease by Q/T_1, which is smaller. Thus there will be an overall increase of entropy.

More generally, if a portion l of the cycle is reversible and the other portion l' is irreversible, the total cycle is irreversible. Then its efficiency will be less than that of C and, consequently, we have $Q_1/T_1 < Q_2/T_2$, as is easily verified. This means that the overall entropy of the sources increases. On the other hand, the machine goes through the closed cycle and comes back to the initial state, so that its entropy is unchanged. The net result is an increase of entropy. But entropy cannot increase along l because that portion of the cycle is reversible, hence it must have increased along l'.

We can derive the conclusion that *the entropy of an isolated system is constant in a reversible process and increases in an irreversible process*. Because no real process can be exactly reversible, the entropy of an isolated system steadily increases. If it reaches a maximum, the system has attained its thermal death.

Following a famous statement made by R. Clausius in 1867, scientists have often been tempted to generalize and affirm that the entropy of the universe is increasing toward a maximum, after which there will be thermal death. But is the universe an isolated system? If it is infinite, what sense does it make to say that its entropy reaches a maximum?

No one can answer this and similar questions, hence one should be very careful not to affirm with certainty the results of unreliable extrapolations. However, it is undeniable that the entropy of that part of the universe that we are able to observe is increasing. Energy is actually being degraded. Everything induces us to believe that at least at the present time we live in a system that evolves irreversibly.

3.8. The nonlinear development of classical thermodynamics

Although we have stated at times that we are not concerned with the history of physics per se, we do not intend to ignore the historical perspective, or worse, to contend that history is immaterial to the comprehension of science and its epistemological implications. It is rather a question of distinguishing the various aspects of human knowledge (not ignoring their reciprocal relations).

Classical thermodynamics represents perhaps the best illustration of the difference between *internal* history as a rational and linear reconstruction of the development, and *actual* history, as read directly in the documents. For the greater part of the nineteenth century, the latter consists of an unbelievable sequence of attempts in divergent directions (nearly independent of one another), of reversions, of obscure or imprecise formulations, and of luminous anticipations (see, e.g., Truesdell, 1971; Bellone, 1973). There is much to be learned in following this intricate net. Above all, it is interesting to see some of the most important trends of modern methodology grow out of the difficult path of last century's thermodynamics.

It is very instructive to examine the attitude of the various authors toward the demand for an explanation of the phenomena, or for the discovery of what *underlies* them. Hypotheses and intuitive models were proposed and all possible views were represented.

First, the figure of J. Fourier (see, e.g., Herivel, 1975) stands out in a peculiar way. His approach in 1822 is in many respects an anticipation of some trends that became prominent in his and in our century. Fourier deals with heat conduction and for the first time states that what matters is the *equation of the process*, regardless of the explanation or of the *model* one wants to adopt. In this respect, Fourier can be considered the pioneer of what today would be called the *formalization* and the *axiomatization* of physical theories. Starting from a few general axioms about heat propagation, he gives a striking example of the *mathematical physics* of continuous media and derives analytically a number of precise results.

Giving up models and placing stress on the equations of the phenomena, Fourier starts an important tradition, which reaches up to Dirac. Two facts must be noticed, however, which mark a profound difference between the somewhat sterile approach of Fourier and the very fruitful one of Dirac. (1) One has to acknowledge that heat conduction is a phenomenon of comparatively modest physical interest, whose study brought about little progress in thermodynamics. (2) One cannot ignore that in Fourier's work the germs of an attitude, which later rendered physicists fairly critical toward a certain type of mathematical physics, are already present. We are referring to that procedure by which, starting from some general laws established with a *limited precision*, one derives, by refined and rigorous mathematics, a number of *extremely precise* results, in cases void of physical interest, which no one is willing or able to verify experimentally! Sometimes, one cannot blame physicists for their lack of enthusiasm toward that kind of activity; an activity that can be misleading and can induce us to believe that we know much more about the physical world than we actually do.[17] There is nonetheless an important circumstance that should not be forgotten. Sometimes the efforts of the mathematical physicist lead to an elaboration or a better comprehension of some mathematical tools, which turn out to be essential to further development of physics. Fourier himself offers a brilliant example of this: The *Fourier analysis*, which allows us to consider any process as the superposition of countless sinusoidal processes, today represents one of the fundamental tools for the understanding of physical reality. Heat conduction is only a minor application of it!

Apart from Fourier's unusual figure, the academic culture in the first decades of the last century remained firmly anchored to the idea that an explanation of the *nature* of heat was indispensable to the progress of science. In the light of the achievements of eighteenth-century science, it was only natural to seek this deeper nature in mechanics. In other words, scientists would imagine *mechanistic* models.

In this connection, I would like to make a general remark. I believe that the emphasis often placed on *mechanism*, its philosophical bearing, and its rise and *fall* is rather unwarranted. It is only too natural that when the "T_n-rung" of the historical ladder of theories has been attained (§2.28) one should seek to explain the phenomena external to D_n by means of T_n, before giving up and starting to look for T_{n+1}. At this point, one could talk about "T_n-ism." Granted that all this has a methodological importance, it is difficult to see why such importance should be felt exclusively in the case of mechanics. When H. Lorentz at the turn of the century sought to interpret the structure of matter by means of an electronic theory, one could have discussed *electronism*. Yet, for some psychological reason, we tend to give mechanism much more importance.

In any event, whereas the attempt to apply T_n-ism is a sound and very useful methodological approach, it is an error to think that it should represent a necessary step and that it is impossible to understand or explain something when the attempt does not succeed. This was already mentioned in connection with Maxwell's work. His mechanistic approaches failed, but it would be ridiculous to affirm that as a result, Maxwell did not succeed in explaining anything!

It is interesting to note that in 1780 A. L. Lavoisier and P. S. Laplace in a joint paper affirmed that for the validity of their results it was immaterial whether one accepted the caloric theory or the theory that heat was kinetic energy. Yet it was some time before scientists saw that many important and firm results could be derived even prior to solving the riddle of the *nature* of heat. Ironically, this idea became widely accepted only when it was too late, that is, when the kinetic–statistic theory was beginning to assert itself and to show the "true" nature of heat.

An interesting mediation between Fourier's position and that of model builders was set forth by S. D. Poisson in 1835. According to Poisson, one has to start from a model, derive its results by means of mathematical analysis, and eventually compare them with the results of experience. In this way, one can choose among different models. This represents the birth of the *hypothetic–deductive* method, largely followed even today.

As a matter of fact, the decisive step of thermodynamics did not come from the theorists arguing for or against models, but from S. Carnot who was merely concerned with the technique of steam engines, and who was addressing engineers. Carnot's results, however, remained virtually unknown to scientists for some time. E. Clapeyron, ten years later, resumed and developed them in the form in which they were to be known later.

The debate so far had taken place almost exclusively among the scholars of the French *Ecole Polytechnique*, the heirs of the illuministic tradition and of Cartesian rationalism. New voices were heard from Great Britian (J. P. Joule, Lord Kelvin, W. J. M. Rankine, J. C. Maxwell) and from Germany (J. R. von Mayer, H. Helmholtz, R. Clausius).

I do not believe that the validity of answers to questions of science depends on nationality, type of culture, and economical and social conditions of those who ask the questions and give the answers. But these factors certainly influence the quality of the questions and the kind of the answers selected to be compared with experience. If a given T_n is valid (within a certain D_n), it must be valid for everybody; but this does not mean that everyone has the same probability to conceive T_n. Here social and cultural factors may come decisively into play.

In particular, one cannot easily miss the change of direction that occurred in thermodynamics when the French *esprit de géometrie* was replaced by the German romanticism. The stress was shifted to mysterious and hidden properties, such as energy and entropy, to general principles not directly amenable to verification, to an inexorable degradation, until thermal death. Based on this background one has the impression of seeing a parade of well-known characters, opened by Faust and closed by Siegfried!

All this can be thought-provoking and is certainly worth some reflection; however, let us be careful not to conclude the absurdity that once the romantic period was over, the principles of thermodynamics were to be rejected!

Scholars sometimes feel authorized to derive similar conclusions from the fact that the development of thermodynamics has been far from linear. In my opinion, such conclusions are equally unwarranted. When I reach the top of a mountain, I may realize interestingly enough that instead of climbing along the path of steepest ascent, I have made a number of zig-zags and at times have even descended. But neither the scenery I can see from the top nor the pure air I can breathe depend

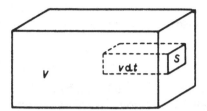

Figure 3.6

on those circumstances. They continue to represent an important conquest to me, even if once on top I realize that there is a still higher top and set out to reach it.

3.9. The kinetic theory

The hypothesis that the particles that constitute material bodies are continuously in motion, that heat is nothing but their kinetic energy, and that the pressure of a gas on the container walls is produced by the countless particles hitting them and recoiling from them, is very old. The first quantitative theory, applied to a gas, was worked out by D. Bernoulli in 1738, but was soon forgotten. Only about the middle of the nineteenth century did ideas start to develop rapidly. The historian would mention the efforts of J. Herapath (1820), J. J. Waterston (1945), and A. Krönig (1856), which preceded the first important contribution by R. Clausius (1857) on the subject.

We shall briefly outline the approach followed in this first stage in order to show, along with its encouraging results, the difficulties encountered. Owing to these difficulties, workers very soon had to appeal to the more refined statistical considerations that will be described later.

Let us consider a container of volume V (Fig. 3.6), holding one mole, or N molecules, of a gas at temperature T. The molecules will be shooting in all directions inside the container. We make the oversimplified hypothesis that all molecules have the same velocity v, that one-third of them move in the x direction, one-third in the y direction and one-third in the z direction. If we denote by $n = N/V$ the number of the molecules contained in a unit volume, the number of those moving parallel to x in the unit volume will be $n/3$, and the number of those moving in the positive direction of x will be $n/6$. Let us now consider a portion S of the wall, perpendicular to x. During time dt, the area S will be hit by all the molecules contained in the volume $Sv\,dt$ and moving in the positive direction of x, that is, by $nSv\,dt/6$ molecules. Each one of these molecules will have momentum mv; because it recoils with velocity $-v$ and momentum $-mv$, conservation of momentum requires that S should "absorb" momentum $2mv$. Altogether, S will absorb

$$Q = \tfrac{1}{6}nSv \, dt \, 2mv \tag{3.44}$$

Let us now assume that the portion S of the wall has mass μ. If for the very small time dt, S is left free to move, it will move away from the container and acquire momentum Q. Much the same effect can be caused by a pressure P, which we now evaluate.

The pressure P gives rise to a force PS acting on S. By the second law of dynamics it will confer to it the acceleration PS/μ; in time dt, the velocity will be $(PS/\mu)dt$, and S acquires the momentum

$$Q = \mu \frac{PS}{\mu} \, dt \tag{3.45}$$

Equating both values of Q given by equations (3.44) and (3.45) and simplifying the expressions, we get

$$P = \tfrac{1}{3}nmv^2 \tag{3.46}$$

Multiplying both sides by V and using $nV = N$ (Avogadro's number), we obtain

$$PV = \tfrac{1}{3}Nmv^2 \tag{3.47}$$

If we assume that v depends uniquely on the temperature, we have clearly derived Boyle's law equation (3.1), stating that at constant temperature the product PV is constant.

On the righthand side of equation (3.47) there appears an expression proportional to the translational kinetic energy of one molecule $e_c = mv^2/2$. Introducing e_c and comparing equation (3.47) with the gas equation (3.7), we derive

$$e_c = \tfrac{3}{2}kT \tag{3.48}$$

where k (*Boltzmann's constant*)[18] is defined by

$$k = \frac{R}{N} \tag{3.49}$$

In CGS we have $k = 1.38 \times 10^{-16}$. Taking into account this value and introducing the mass of the hydrogen atom 1.6×10^{-24} g, one readily finds from equation (3.48) that at room temperature the molecules have velocities of hundreds of m/s.

Thus we have found that if we assume that the temperature is proportional to the kinetic energy of the molecules as required by equation (3.48), the state equation of perfect gases can be derived by simple mechanical considerations. This represents a first and momentous success of the kinetic theory.

The hypothesis that the molecules move only in the three directions x, y, z is very rough. But it is not essential, as was shown by Clausius. If the molecules are allowed to move in all directions, and all directions are equally represented, much the same results are obtained.

A second assumption that we have implied is that the molecules should never collide with one another. This is very serious; however, we might say in our defense that such a condition is met in the limit, when the gas is extremely rarefied and the proper volume of the molecules is negligible with respect to the total volume at their disposal. We shall discuss this subject presently.

Let us now heat the gas at constant volume. We assume that the whole internal energy U is represented by the kinetic energy of the molecules. By equation (3.48) the increase due to one molecule is $3k \, dt/2$, hence we have $3R \, dT/2$ for the whole gas. On the other hand, equation (3.13), which refers to 1g and should therefore be multiplied by the molecular weight M, yields $dU = Mc_V \, dT$. On comparing, we get

$$c_V = \frac{3}{2}\frac{R}{M} \qquad (3.50)$$

Then, from equation (3.18) we have

$$c_P = \frac{5}{2}\frac{R}{M} \qquad (3.51)$$

Thus the ratio between the specific heats turns out to be

$$\gamma = \frac{c_P}{c_V} = \frac{5}{3} \qquad (3.52)$$

This relation is in good agreement with the result of experience ($\gamma = 1.67$) for *monoatomic gases*, whose molecules consist each of one single atom, such as He, Ne, and so forth. For other gases, however, the situation is different and the theory often clashes with experience. An explanation of the discrepancy found for *polyatomic gases* can be sought in the fact that their kinetic energy is not only of translation but also of rotation.

To begin with, let us consider a point mass, that is, a body whose size can be neglected. Its mechanical state is perfectly specified by giving its three coordinates, x, y, z and their time derivates dx/dt, dy/dt, dz/dt, or the three components v_x, v_y, v_z of the velocity. The point mass is said to have three *degrees of freedom*, precisely, the three *translational* degrees of freedom. Free atoms (or molecules of mono-

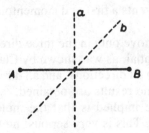

Figure 3.7

atomic gases) behave, with respect to the kinetic theory, as point masses. From equation (3.48) we have in this case the kinetic energy $kT/2$ per degree of freedom per molecule.

When the body has a nonnegligible size, besides the translational degrees of freedom there are also *rotational* degrees of freedom. For example, a system formed by two point masses A, B (Fig. 3.7), rigidly connected a fixed distance apart, will also have, in addition to the three translational degrees of freedom, two rotational degrees of freedom–that is, because the system can turn around either one of two axes a, b perpendicular to both each other and AB. A diatomic molecule (such as H_2 or HCl) behaves approximately in this way. Hence it will have altogether five degrees of freedom.

Let us now make the hypothesis that the energy available is shared in equal proportion by all the degrees of freedom. This is the *equipartition of energy*, a law supported by much better arguments than simply using analogy, even though it is difficult to justify exactly. The overall kinetic energy of a diatonic molecule will then turn out to be $5kT/2$. Proceeding in the same way, as for a monoatomic gas, we find

$$c_V = \frac{5}{2}\frac{R}{M}, \qquad c_P = \frac{7}{2}\frac{R}{M}, \qquad \gamma = \frac{7}{5} \qquad (3.53)$$

Therefore we should have $\gamma = 1.40$. This result is fairly correct for a few diatomic gases (such as O_2 and HI) but other diatomic gases show conspicuous discrepancies. The situation for polyatomic gases is even worse (three rotational degrees of freedom); in most cases the theory is in flagrant disagreement with experience.

On the other hand, even if the accord between this kind of theory and experience were perfect, one might ask: "Why should an atom, consisting of the nucleus and electrons, be treated as a pointlike body?" Furthermore, why should the connections between different atoms in a molecule be considered as rigid? Only quantum mechanics has been able to answer these questions adequately. Let us set this aside for

Figure 3.8

now until we pass from classical to quantum physics. However, even within the scope of classical physics, the previous discussion can be shown to be oversimplified and to lack something essential.

First, we should ask: "Are we really justified in disregarding collisions between the molecules?" It is readily seen that this is not so, for by neglecting collisions, we are ignoring an important aspect of the real process.

Let us put into contact two different gases occupying two adjacent portions of the same container. We observe that the two gases mix because each one diffuses into the portion occupied by the other. However, the diffusion turns out to be fairly slow, whereas in the case with no collisions it should be virtually instantaneous[19] because the speed of the molecules is hundreds of meters per second. Diffusion is slow, for each molecule, far from moving freely, continuously collides with the other molecules, hence traveling in a very complicated zigzag path. One has to evaluate the mean free path, that is, the average distance covered by a molecule between two collisions.

As soon as we know the essential role played by the collisions, we must also conclude that the hypothesis that all molecules have the same speed is far from real. Even if such a condition were realized in a particular case, it would be destroyed in no time at all.

For simplicity, let us assume that the molecules are perfect and rigid spheres, such as billiard balls, and that their surfaces are perfectly smooth (frictionless). The last condition has the result that during a collision one molecule acts on another with a force perpendicular to both surfaces, or parallel to the line joining the centers of both spheres. Let us then consider two molecules having equal velocities before collision (Fig. 3.8). Let the velocity v_1 of the first molecule be parallel to the line joining the centers, whereas the velocity v_2 of the second molecule is at 45° with that line. We select the x axis parallel to v_2; hence the modulus v_2 will coincide with the x component of v_2. A moment of reflection will convince the reader that the force acting during the collision can only increase the x component of v_2. As a result, the overall velocity of the second molecule can only increase, and because of energy conservation, the velocity of the first molecule can only decrease. Thus the equal velocity condition cannot be conserved.

If the speeds of the molecules are different from one another, we must start to reason statistically and to consider averages. For example, we can assume that the temperature is proportional to the mean kinetic energy of the molecules, according to equation (3.48). But how are the velocities distributed among the molecules? How can one derive the results of the kinetic theory exactly?

Now we can no longer talk about the precise velocity of a molecule; we must instead give the probability that the molecule has a given velocity. From the mechanics of precise and well-defined events we pass to statistical mechanics[20] one of the most fascinating and difficult chapters of twentieth century physics. The main concept of this subject is that of *probability*. It is necessary to discuss probability at length, as well as information theory, before reverting to statistical mechanics.

3.10. Probability

The notion of probability is very commonly used, and the concept is believed to be clear to everyone. Yet explication of the concept is fairly difficult and can be grasped only at an advanced stage of intellectual development.

Children under seven do not know the "*probable*," because they do not admit the "*fortuitous*" (see Piaget, 1947, §5.3, and §6.2). To children everything is connected with everything in a deterministic way, and their questions presuppose that adults are omniscient. Piaget says: "A child can say 'I don't know' to get rid of you, but he will say only very late (at about 11 or 12) 'one cannot know.'" At the early stage of development only true and false exists, not probable.

But it is not only individuals who acquire the notion of probability at a fairly late stage of their development. All humankind has had to wait for science to come of age before attempting to clarify and make that concept precise.

In 1654 Chevalier De Méré posed to B. Pascal a question about casting dice. Pascal correctly solved the problem, and thus the first step in the modern theory of probability was taken (see, e.g., Costantini, 1970; Jeffrey, 1961; Popper, 1959, Chap. 8). Later contributions to this theory were made by P. Fermat, C. Huygens, and J. Bernoulli, but it was not until 1812 that P. S. Laplace gave us a systematic foundation of the concept.

According to Laplace's classical definition, the probability of a given event is the ratio between the number of cases favorable to the event and the total number of possible cases. Therefore we write for the classical probability p, the expression

$$p = \frac{\text{number of favorable cases}}{\text{number of possible cases}} \tag{3.54}$$

Thus when we throw a die, there are six possible cases. If the expected event is, say, the drawing of number 5, there is only one favorable case. Accordingly, the probability is 1/6. Let us play heads and tails and toss the coin twice in succession. The possible results are: (1) first toss heads, second toss heads; (2) first toss heads, second toss, tails; (3) first toss tails, second toss heads; (4) first toss tails, second toss tails. Thus we have four possible cases. What is the probability that the game ends in a draw? The cases favorable to this event are clearly (2) and (3), or two cases over four. Hence the probability is 2/4 = 1/2.

Counting cases is not always so easy as this, and may involve a large amount of work. Fortunately, the attainment of the end result can be considerably facilitated through the application of some simple rules derived from equation (3.54), which embody the essential principles of probability theory. We shall mention some of these rules, leaving their easy proofs to the reader.

If two or more events are *mutually exclusive*, the probability that one or the other of them will be verified equals the sum of the respective probabilities. If the events are A and B and their probabilities are denoted by $p(A)$ and $p(B)$, we have

$$p = p(A) + p(B) \tag{3.55}$$

for the probability that either A or B occurs. For instance, in the preceding heads-and-tails game the outcome will be a draw if either (2) or (3) occurs. These cases are mutually exclusive and each has probability 1/4; so the overall probability will be 1/4 + 1/4 = 1/2.

If N events A_i, besides excluding one another, are also *exhaustive*, in the sense that one of them must perforce occur, the sum of their probabilities is 1. Thus we have

$$\sum_{i=1}^{N} p(A_i) = 1 \tag{3.56}$$

If two or more events are *independent* of one another, the probability that all of them should occur, equals the product of their probabilities. In the particular case of two independent events A and B we have

$$p(A, B) = p(A)p(B) \tag{3.57}$$

where $p(A, B)$ denotes the probability that both A and B occur. In our example, the probability of obtaining twice heads, which is 1/4, equals the probability of having heads in the first toss times the probability of having heads in the second toss (i.e., $1/2 \times 1/2$), because the result of the second toss does not in the least depend on the result of the first.

The situation is different when the events are not independent, in the sense that the probability of at least one event is conditioned by

the occurrence or nonoccurrence of some of the others. In such a case
we can make use of a *conditional probability*. Let $p(A)$ represent the
probability of occurrence of A and $p_A(B)$, the probability of occurrence
of B once A has occurred. The product

$$p(A, B) = p(A)p_A(B) \tag{3.58}$$

will now represent the probability that both events occur. For example,
put two black balls and a white ball inside a box. What is the probability
of two black balls coming out in the first two drawings? We have to
decide beforehand whether or not the first ball drawn is returned to
the box. Let us begin with the case where the ball is returned. The
probability of the first ball being black is $p(A) = 2/3$; the probability
of the second ball being black is the same, hence $p(B) = 2/3$. The
probability of drawing two black balls will therefore be given by $(2/3)$
$\times (2/3) = 4/9$. On the contrary, if the first ball is not returned to the
box, the probability of the second drawing will depend on the outcome
of the first. If the first ball was black, the probability $p_A(B)$ of drawing
a second black ball is no longer 2/3, but 1/2. Hence the probability
sought is now $(2/3) \times (1/2) = 1/3$. Finally, it is evident that an *impossible* event has probability 0, whereas a *certain* event has probability
1.

With these rules, we can build the whole of probability calculus.
Trouble ensues when we apply the results of this calculus to experience
in a somewhat critical way. Why should we obtain a greater probability
for an event that occurs frequently and a smaller probability for an
event that occurs rarely? In other words, what has the mere counting
of possible and favorable cases to do with the frequency of occurrence
of a given event in the physical world?

The most naïve answer one can give (which actually has sometimes
been given) is that one must count as possible cases the equiprobable
cases. It is evident that this circular definition says nothing.

The appeal to the principle of *insufficient reason* or of *indifference*,[21]
as was done by classical authors is not much better. According to this
principle, if there is no sufficient reason for one of the possible cases
to occur in preference to the others, all cases must be considered as
equally probable. For example, when we throw a perfect die, there is
no reason that number "2" or number "5" should be preferred to the
other numbers; as a result, all results from 1 to 6 should be equally
probable.

Really, this principle is deceptive and cannot stand an elementary
analysis. The main difficulty is to know how to select the cases to
which the principle of indifference is to be applied. We shall see in due
course that in microphysics the problem is serious, even when one is

dealing with discrete cases. In classical physics the solution is hopeless or ambiguous, especially when the cases form a continuous set.[22] For instance, if we select colors at random in a spectrum, what are the equiprobable colors?

But even if we assume to be able to enumerate unequivocally the various possible cases, how can we decide a priori that there is insufficient reason to prefer one case to the others? How can we know that carrying engraved number "6" is not a sufficient reason for a facet of a die to be preferred and to occur more frequently than the others? Only experience can teach us that that is so. The principle of insufficient reason therefore represents one of those cases in which, having long learned a rule of the empirical world, we deceive ourselves to be able to affirm it a priori, as a result of rational analysis. In other words, we say that there is insufficient reason to prefer one case to another, when experience itself has taught us (either directly or in analogous cases) not to have preferences. As stated by E. Borel, the equiprobability of *heads* and *tails* is an experimental fact, or else the downright definition that the coin is good.[23]

Once the need to appeal to experience has been recognized, we arrive at the *frequency* interpretation of probability, according to which probability of an event coincides with the relative frequency of its occurrence. The idea, introduced mainly by J. Venn about the middle of the last century, has asserted itself in modern physics more and more.

One must deal with an experiment that can be repeated at will under identical circumstances. If the experiment is repeated n times and the event of interest e is seen to occur n_e times, the relative frequency f of the event is defined as

$$f = \frac{n_e}{n} \tag{3.59}$$

In the past there has been a certain conceptual difficulty in identifying f with the probability p of e. For f depends on n and varies with the number of trials, whereas the concept of probability seems necessarily to imply independence from n. The question was thoroughly discussed after World War I, mainly by R. von Mises and H. Reichenbach. Both believed it necessary, although with different approaches, to postulate the existence of a limiting value for f when n tends to infinity, and to identify probability p with that limit.

This view has often been criticized by noting that the limit $n \to \infty$ can only make sense for a mathematical sequence, not for a sequence of physical experiments.[24] In my opinion, this kind of criticism is right. This is due to confusion between mathematics and physics, which leads people to believe that the quantities of physics should be measured by real numbers, without considering the necessarily finite accuracy of any procedure of measurement.

The difficulty disappears as soon as one realizes that probability can be conceived of as a physical quantity, operationally defined by means of repeated experiments and by counting how many times event e occurs. What we called the frequency f is simply a measure of probability p, obtained with a given precision ϵ, in agreement with equation (1.2). Accordingly, we write

$$p = f \pm \epsilon \tag{3.60}$$

As the number n of identical experiments becomes greater, the quantity ϵ becomes smaller, that is, the measurement is more precise.[25]

Obviously, we have to extend the postulate of *spatiotemporal invariance* of physical laws (§1.4) to include probability as well. In other words, if we carry out two sets of identical experiments (except for their spatiotemporal locations, which may be different), we find for the event e the same probability (up to the respective precisions). Experience has never falsified this postulate.

We may now ask: "Because probability can be thought of as an experimental quantity in this concept, how is it possible that Laplace's classical method of counting cases a priori may have been so successful?" The answer is that the exclusion of any appeal to experience, in a way, was the result of self-deception. Indeed, as already noted, it is not difficult to show that the "reasons" implied in the principle of insufficient reason are largely represented by our previous knowledge of the physical world, learned from experience! The fact that the number marked on a face of a die does not influence its frequency of appearence can only be suggested by what we know beforehand about empirical reality. One might notice that, strictly speaking, the principle of insufficient reason holds for a die with *identical* faces, all marked with same number. But in that case they would be indistinguishable and we would not be able to verify the equiprobability (for further discussion on this point see §4.7). As soon as the faces bear marks that enable us to distinguish them from one another, only experience or previous knowledge can tell us if they remain equiprobable. Suppose that the numbers are marked by little lead balls. In that case we do not hesitate to say that the faces are not equiprobable. Why? The answer is because we know the laws of gravity and thereby can predict some easy qualitative consequences.

As a conclusion, in the application of the classical theory of probability, we meet once again with one of those circularities that are inherent in the method of physics. Such circularities should be known rather than rejected. The fact is that the application of classical probability to physics has yielded splendid results.

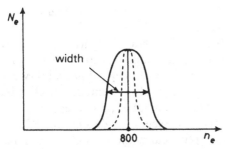

Figure 3.9

Once we have admitted that the possible cases are equiprobable, or occur with equal frequency, we can derive from classical theory a very important law, known as the *law of great numbers*. In its simplest form, it was proved in a celebrated theorem by J. Bernoulli.

Suppose we make n trials for an event of constant probability p (perhaps known from counting cases), and find the relative frequency f. We are interested in knowing the difference $|f - p|$, sometimes called the *error*. A naïve idea of probability might suggest that such a difference should tend to vanish when n increases beyond any limit. But that is not exactly so. Bernoulli's theorem states that for any small ϵ that we may choose, the *probability* that $|f - p|$ exceeds ϵ tends to vanish when n increases. Beyond a certain value of n, such probability can therefore be neglected.[26]

Let us clarify these notions with an example. Suppose that we have 8 white and 2 black marbles in an urn and that every time we draw a marble we replace it in the urn. The probability of drawing a white marble is then the constant $p = 0.8$. In a series of $n = 1000$ drawings we must expect that a white marble is drawn $n_e = pn = 800$ times. But nobody will expect that this number be *exactly* 800. There will be fluctuations. Let us repeat the series of one thousand drawings a great number N of times and make a graph (Fig. 3.9). We plot the number N_e of the series in which the white marble is drawn n_e times, against n_e. By joining all the points thus obtained by a continuous curve (solid line), a characteristic bell-shaped curve is obtained (*Gaussian distribution*). The value $n_e = 800$ appears more frequently than any other value; however, the values not much different from 800 have nonnegligible probabilities. The curve is characterized by its width, that is, by the range of values about 800 that have nonnegligible probability; it gives an idea of the departure from 800 which is to be expected. The width is found to be proportional to $\sqrt{n}$, hence increases with increasing n. But because the number of times the white marble is drawn is about pn, we see that the relative departure $\sqrt{n}/pn$ tends to vanish like $k/\sqrt{n}$ with constant k. By scaling down the curve obtained with a larger value n (so as to have the same value of the maximum), we would obtain a narrower bell (broken line). This should clarify the content of the law of great numbers: The relative departure from the most probable value becomes smaller and smaller with increasing n.

The frequency interpretation of the concept of probability is not the only one possible and is not universally accepted. Another interpretation, which has been adopted by a number of workers (F. P. Ramsey, B. de Finetti) is the subjective one (see, e.g., de Finetti, 1970). In their opinion, probability is not a physical quantity, but a measure of our belief that a certain event will take place. It has to do more with our betting behavior than with the objective development of a process in nature.

Another important interpretation is the *logistic* interpretation (L. Wittgenstein, J. M. Keynes, R. Carnap), in which probability represents a logical relation between two sentences. For instance, the first sentence may express the experimental evidence E which is available, whereas the second sentence expresses an hypothesis H. Probability is identified with the *degree of rational belief* (as distinguished from psychological or subjective belief) that H be true, based on the evidence E. The reader will notice that we are approaching the general problem of inductive inference. In fact, Carnap has developed *inductive logic* on that line (see §4.21).

Thus a few different interpretations of the concept of probability are possible. But now the idea is growing that the arguments that have taken place in the past between supporters of different interpretations do not make much sense.

Probability theory is a mathematical doctrine, placed on axiomatic and rigorous bases by A. Kolmogorov in 1933, which does not need an interpretation. Giving it an interpretation means merely applying it to some field of science. Hence there can be a number of different, but all equally valid, interpretations.[27]

The interpretation of probability that is useful in physics is the one based on the frequency of occurrence of a given event in a series of repeated (or repeatable) experiments.[28] For this frequency the postulate of the spatiotemporal invariance is assumed to exist in the same way as for all physical phenomena.

3.11. Information

All living creatures, human or otherwise, are naturally endowed with the means by which to *react* to the environment. The reaction may consist in flying from danger or tackling a source of food or approaching a sexual partner, and so on. For the reaction to occur in the right way at the right time, the living being has to receive information from the environment. Certain physical phenomena – optical, acoustic, tactile, chemical, thermal – must take place in order to acquaint the individual with some features of its physical environment.

This characteristic of living beings has attained an amazing extension and importance in human species. Intelligence, the faculty that chiefly characterizes humans, needs information as raw material in order to operate. Abundant and precise information must be rapidly received, stored in the memory (or in books) to be recalled and utilized when needed, and then processed and transmitted.

Although the important role of information in human life has always been recognized, it is only in recent times, with the development of science and technology that information has priority with respect to other *goods* at our disposal. During the nineteenth century and the early twentieth, humankind has been mainly concerned with the availability of raw materials, agricultural resources, and manpower. Today instead there is a tendency to learn privileged information – scientific knowledge and technical know-how. In detective stories of yesterday, gold, diamonds, and objets d'art were stolen; in today's scenarios secret documents of political or scientific material are the main targets. These considerations have not remained at a qualitative stage, but have received quantitative and precise treatment of great interest, which is known as *information theory* (see, e.g., Pierce, 1961).

Information theory had first been developed in technical circles.[29] The need for a rigorous and quantitative approach to this theory was felt mainly by workers in the field of telegraph, telephone, and radio communications. Transmitting a message costs money, because equipment and its operation cost money. The technician therefore is interested in studying the means by which to transmit the largest amount of information with maximum reliability, in the shortest time. These requirements have given rise to information theory.

After some pioneer work by H. Nyquist and R. V. L. Hartley, the general foundations of the theory were laid by C. E. Shannon in 1948 and by N. Wiener in 1949. The theory that emerged was of such generality that it has gone considerably beyond the limits of its intended technical applications. It clearly involved language theory, biology, computer technique, philology, law, and, more generally, the theory of scientific knowledge.[30]

This great generality, however, is limited by a certain lack of effectiveness and of concrete results, and these facts are often used to argue against information theory. Sometimes it is maintained that the theory mainly represents a new *language*, which although attractive and precise, does not yield essentially any new solutions to old problems.

This criticism is *partly* correct. However, an adequate and precise language in science can be of paramount importance. It is often responsible for the creation and definition of new concepts. In particular, it is almost inconceivable that one may speak of modern science and technique without using the language of information theory. For this reason and because information theory is closely related to statistical thermodynamics, we shall briefly discuss the main points of this concept.[31]

Clearly, the first problem to be solved is that of *measuring* information, or of characterizing the amount of information in a reasonable way, with a number, as used for any other physical quantity. Surprisingly, this can be done.

Information is transmitted by means of *messages*. We give this word a very broad meaning. A message may be represented by a word, a gesture, a bell ring, a letter of the alphabet, a Morse dot or line, a printed page, a book, and so on.

The person transmitting the message can select it from a set of n different messages. In the case of a *dot* or *line*, $n = 2$. In the case of directions: *right, left, straight on*, $N = 3$. In the case of a typewriter key being tapped, $n = 32$ (assuming that the typewriter has 32 keys, as we always assume for convenience).

A message will convey the greater information, the greater is the number n of the messages out of which it can be selected. For example, a simple *yes* or *no* ($n = 2$) will carry very modest information; a sentence of ten words instead can carry an enormous amount of information, because the number n of the different messages that one can build with ten words is enormous.

Accordingly, it seems convenient to represent the measure of information by a quantity that increases with n. It might be n itself, or n^2 or n^3 or $\ln n$ or e^n, and so on. The number of the functions of n that monotonically increase with n is infinite; in order to make a convenient choice, we have to impose some condition. The most reasonable condition is to require that transmitting two, three, m consecutive messages selected from the same set should convey, respectively, two, three, m times the information of a single message. Two printed pages should contain twice the information of one page; a book of 100 pages should contain 100 times that information.

Suppose now that we transmit a sequence of *elementary* messages, taken from a set of n different messages. If we transmit *two* elementary messages, each one of the n possible messages of the first transmission can combine with one of the n possible messages of the second. Altogether we have $W = n^2$ messages, whereas the information is *twice* that of one elementary message. If we transmit a sequence of N elementary messages, we have altogether $W = n^N$ possible messages and N times the information of one elementary message. The amount of information must therefore be proportional to the *exponent* N, or to the logarithm of W.[32]

We must now choose a base for log W. Natural logarithms, or logarithms to the base e (denoted by ln), are mostly used in pure science, whereas in applied science one often finds the decimal logarithms, or logarithms to the base 10 (denoted by $\log_{10}$). Note that in going from one base to another we simply multiply all logarithms by a constant. Hence in our case, because the amount of information must be proportional to the logarithm, the base can be chosen arbitrarily. Choosing a base amounts to the same as choosing the unit of measurement for a physical quantity.

For the amount of information I, it is customary to choose the base 2. Accordingly, we write

$$I = \log_2 W \tag{3.61}$$

where W indicates the overall number of distinct messages out of which one can select the message to be transmitted. If one single elementary message is transmitted, we have $I = \log_2 n$, if N elementary messages are transmitted, we have $I = N \log_2 n$. For simplicity, we shall from here on omit the base of the logarithm; "log" will therefore stand for the logarithm to the base 2.[33]

Usually we say that the amount of information, as expressed by equation (3.61) is measured in *bits*. Accordingly, when we choose between two messages such as *yes* or *no*, or a *dot* or *line*, we transmit each time 1 *bit* of information. When we choose out of four messages, we transmit 2 *bits*, when we choose from 2^N messages, we transmit N *bits*. If we assume that the letters of the alphabet (including punctuation marks) are $32 = 2^5$, each letter carries 5 *bits* of information. If a printed page consists of 3,000 characters, it will carry $3,000 \times 5 = 15,000$ *bits* of information. A book of one hundred such pages will contain 150,000 *bits*, and so on.

As the logarithm of 1 in any base is 0, there follows that when the message is unique (there is no choice) it carries zero information. In such a case there will be no curiosity; one knows beforehand what the message will say and there is no interest in receiving it.[34]

3.12. Information and probability

Imagine a police inspector asking of a witness who has seen the criminal: "Was the person white or black?"

What has been said here, might lead one to conclude that from the answer to this question the inspector should expect one *bit* of information. Yet there may arise some doubt. One has only to remark that whereas in the United States the inspector would certainly ask that question (probably as the first one), in Italy he would not bother. Why? In Italy there are so few blacks that even uttering the question would most probably be a waste of time. Learning that the criminal was white, would virtually bring no information.

Thus arises the idea that the *probability* of a message should have something to do with the information it can convey.

Evidently, the definition $I = \log n$, given in the previous section, is all right when the n elementary messages are all equiprobable. In that case the probability of each one will be given by $p = 1/n$ and we can write[35]

$$I = \log p \tag{3.62}$$

The information carried by a message whose probability is p, is given by $-\log p$.

Let us now consider the case when the various elementary messages $x_1, x_2, \ldots, x_n$ have different probabilities, given by $p(x_1), p(x_2), \ldots, p(x_n)$. For simplicity, we write P_i instead of $p(x_i)$. Let us assume that all these probabilities are *independent* of one another, that is, that the probability P_i does not depend on the fact that, say, x_n has been transmitted before x_i.

The information carried by message x_i, is $-\log P_i$. If we transmit a very long sequence of N elementary messages, message x_i will appear NP_i times, apart from some departure that becomes comparatively negligible when N is very large (remember §3.10); it will therefore convey the overall information $-NP_i \log P_i$. The total information I_{tot} carried by the sequence of N messages will therefore be represented by the sum

$$I_{\text{tot}} = -\sum_{i=1}^{n} NP_i \log P_i \tag{3.63}$$

The *average* information carried by one message, which we denote by $\bar{I}$, will be obtained by dividing this expression by N, hence we have

$$\bar{I} = -\sum_{i=1}^{n} P_i \log P_i \tag{3.64}$$

This is the fundamental quantity that is of interest in information theory; in the ordinary case of nonequiprobable messages, it must replace the simple definition, equation (3.61).

Note that the probabilities P_i refer to mutually exclusive events (i.e.,

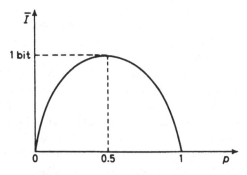

Figure 3.10

when x_i is transmitted, x_n cannot be transmitted at the same time); furthermore, the set is exhaustive (i.e., no other message besides $x_1, \ldots, x_n$ can be transmitted). Hence by virture of equation (3.56), we have

$$\sum_{i=1}^{n} P_i = 1 \qquad (3.65)$$

The probabilities P_i cannot be fixed in a completely arbitrary manner, but must be subject to this condition.

To illustrate equation (3.64) let us start with the case of two possible messages, say, *yes* and *no*. If *yes* has probability p, *no* will have probability $1 - p$, by virtue of equation (3.65). Hence by applying equation (3.64), the average information carried by one message can be given by

$$\bar{I} = -p \log p - (1 - p) \log (1 - p) \qquad (3.66)$$

If we plot $\bar{I}$ against p (which evidently can vary between 0 and 1) in a graph, we obtain the curve of Figure 3.10. The curve has a maximum at $p = 0.5$, that is, when both messages have equal probabilities. In that case each message carries 1 *bit* of information, as we already know. On both sides of the maximum, $\bar{I}$ decreases and vanishes for $p = 0$ and $p = 1$, that is, both when *yes* is certain and when *no* is certain.

In a well-known parlor game one of the participants must guess something, by asking questions to which the others must answer only by yes or no. In order to guess as quickly as possible, one must be so skillful as to select questions to which the answers *yes* and *no* are equiprobable. Only in that case can one obtain maximum information (1 *bit*) from each question. Instead, questions to which the answers are virtually certain must be avoided. For instance, if the first question:

Is it a person? has been answered with *yes*, one should not ask: Does that person have the heart on the right? That would be a blunder, because from the answer to that question one can expect, on the *average*,[36] very little information.

We can generalize the result obtained graphically for two elementary messages, to the case of n messages; what is the maximum value of $\bar{I}$? We have to find the maximum attained by equation (3.56), when the P_i are varied in all possible ways, subject to condition, equation (3.65). This is a problem of conditional maximum, which is easily solved by the methods of calculus. The result is that $\bar{I}$ attains its maximum value when all P_i are equal and have the value $1/n$. In that case we obtain, as we already know, $\bar{I} = \log n$.

As an example, let us consider the case of the 32 characters that we assume for convenience to constitute the alphabet. Does each one of them really carry 5 *bits* of information? As a matter of fact, we must now correct our previous statement and say that each character *could* bring 5 *bits*. But the characters of the alphabet do not occur with the same frequency, hence they do not have the same probability. Their probabilities depend, of course, on the language in which the message is written, on the epoch, the subject, and the author. Nevertheless, one can determine, to a good approximation, the average frequency of each letter in a given written language.[37]

If one takes into account that frequency and applies equation (3.64), one obtains for the average information much less than 5 *bits* per character.[38] Roughly speaking, one can assume that the average information in a Western language is 1 *bit* per character. We have in this case (as in a number of other cases of information recording or transmission), a redundancy. The writing system is redundant with respect to its employment. It could carry 5 *bits* of information per character, but carries instead much less than that.

How many different messages could be carried by a printed page of N characters? If the characters were equiprobable, we would have the number 2^{5N}. But everyone knows that a message taken at random from that set would have no meaning and would be judged *impossible*.[39] Shannon has proved that the number of *possible* messages is virtually equal to $2^{N\bar{I}}$, where $\bar{I}$ denotes the average information per character, as defined by means of the probabilities.[40] Because $\bar{I}$ is much less than 5, the channel of transmission has a much larger capacity of information than we actually exploit. It is like using a tanker to carry a glass of wine.

This shortcoming can be eliminated by properly *encoding* the messages, that is, by transforming them into another set of signals that requires a smaller capacity. The ideal condition could be met by as-

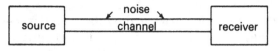

Figure 3.11

signing to each message a number, from 1 to $2^{N\overline{I}}$, and simply writing the number, instead of the message; actually, it could be shown that this would require just the capacity of $N\overline{I}$ *bits*. But in general, this is not a viable approach.[41] In practice, one can resort to a number of different methods of encoding. Sometimes such methods are even obvious. For instance, in the case of the Morse alphabet the letter *e*, which is the most frequent in the English language, is represented by a dot, whereas less frequent letters are represented by longer sequences.

3.13. The transmission of information

In the transmission of information one must distinguish a number of different stages. Let us refer to the simplest kind of system, as shown in Figure 3.11. First of all, we have a *source* of messages represented by a person who speaks or operates a telegraph key, or by a measuring instrument (e.g., placed on board an artificial satellite), and so on.

Ordinarily, the message is transduced and encoded, that is, turned into a *signal* (mechanical, acoustic, electric, etc.) according to well-determined rules, before being let into a *communication channel*. The channel can be a telegraph wire, the atmosphere entertaining acoustical waves, a simple rope activated by pulls, and so on.

Eventually, the signal reaches the *receiver*, where it is *detected, decoded* and utilized.

We must now take into account the *time* needed to generate, or to transmit, or to receive the information. This is an essential factor for the economy and for the efficiency of the transmission. Accordingly, we shall now discuss *bits* per *second* (or *bit/s*) rather than simple *bits*.

The communication channel has the *capacity* of a certain number of *bit/s*. For instance, the capacity of a telegraph channel will be determined by the minimum duration of a pulse and of the subsequent void interval, which correspond to a *dot*. It is impossible for that channel to convey information at a rate faster than that determined by its capacity.

The capacity of the source of messages must be properly *matched* to the capacity of the channel. It is often convenient to reduce the redundancy of the source. A source producing alphabet letters at the rate of N characters per second should require the capacity of 5 *bit/s*.

But we know that if the text transmitted has to have some meaning in some language, the information generated by the source is much less than that, say, 1 *bit/s*. Hence by suitably encoding the message, one can use a channel with five times less capacity, or, what amounts to the same, transmit with the same channel at a rate five times faster. As already mentioned, various coding systems have been worked out, which can reduce, or virtually eliminate, the redundancy.

The receiver, too, will have its well-determined detecting and decoding capacity. If the rate of arrival of the messages exceeds that capacity, some of the messages will be wasted, or, at any rate, temporarily useless. For example, think of a business person who receives every day more letters than he can read. One should, in general, also match the capacity of the channel to the capacity of the receiver.

The communication system hitherto considered is fairly idealized. It is *infallible*, because the received message is either identical to the transmitted message, or at least allows unequivocally to pinpoint the latter.

In reality no communication system attains this degree of perfection. There is always some cause of disturbance or of *noise*, as it is called in the technical jargon. This effect is known to everyone. For instance, in a radio transmission one always perceives some more or less loud noise, which has nothing to do with the actual signal. But the term *noise* is used in a much broader sense; there is noise every time that after message x has been transmitted, a nonvanishing probability exists of receiving a *different* message y. For instance, in the case of a book, a misprint or an error of the reader can alter the message received with respect to the message transmitted by the writer. In the case of a loudspeaker or of a telephone receiver, it is well known that the voice heard is somewhat different from that heard directly, so that sometimes it is difficult to recognize the speaker. Furthermore, there are rustles, creakings, and so on.

If it is a rule that when x is transmitted, a different y can be received, how can one use a communication system with complete confidence? Apparently, one should have to accept some amount of *uncertainty*. Luckily, this is not so; the credit for having clarified this point goes to Shannon.

Suppose that we have at the source a number of possible elementary messages x_i, having probabilities $p(x_i)$. The average information per message, which as is usual, will now be denoted by $H(x)$, is given by equation (3.64)

$$H(x) = - \sum_i p(x_i) \log p(x_i) \qquad (3.67)$$

Note that this quantity also represents our a priori ignorance about the message that will be transmitted each time. Do not be surprised that ignorance can be equal to information! One has only to reflect that ignorance is removed every time we receive the message; losing ignorance is indeed equivalent to gaining information.

Let us now assume that the receiver can receive a set of messages y_j, different in general from the x_i, and having the probabilities $p(y_j)$. If between the x_i and y_j there is no correlation, that is, if the probability of receiving y_j does not depend on the transmitted message x_i, and vice versa, it is evident that no information is transmitted. If the sender transmits the *Iliad* and the addressee receives the *Bible*, no real communication takes place.

But in general, the situation is different. If the addressee receives the word "fact," he knows that there is some probability that the sender has transmitted "pact"; a smaller probability applies to "cat," and so on, until one reaches the virtually zero probability of "teetotaler"!

There exists a *conditional probability* $P_{y_j}(x_i)$, defined as the probability that x_i has been transmitted, once we know that y_j has been received. This probability can be fairly easily evaluated by repeated trials, at least for the simplest communication systems.

As soon as y_i has been received, the ignorance of the addressee turns into $H_{y_i} = - \Sigma_i P_{y_j}(x_i) \log P_{y_j}(x_i)$. Averaging over all the possible y_j, we obtain the average ignorance a posteriori

$$H(y) = - \sum_j p(y_j) \sum_i P_{y_j}(x_i) \log P_{y_j}(x_i) \qquad (3.68)$$

It now is evident that the difference between the a priori and the a posteriori ignorances will represent the information gain. Accordingly, we can define the average information per message, by the expression

$$\bar{I} = H(x) - H(y) \qquad (3.69)$$

In the absence of noise, when the input message is unmistakably related to the output message, we have $P_{y_j}(x_i) = 1$ for a well-determined x_i, and $P_{y_j}(x_i) = 0$ for all the others, hence by virtue of equation (3.68), there follows $H(y) = 0$. On the contrary, if the noise is such that the x_i and the y_j are totally uncorrelated, we simply have $P_{y_j}(x_i) = p(x_i)$, for the probability of x_i is not conditioned by y_j. In this case, because by equation (3.65) we have $\Sigma_j p(y_j) = 1$, there readily follows $H(y) = H(x)$, hence $\bar{I} = 0$. It is the case of the *Iliad* and the *Bible*.

The fundamental result due to Shannon, is that by a proper code, a system such as that described can actually transmit $\bar{I}$ *bit* per elementary message, with any wanted small frequency of errors. In other words, even when noise is present, one can communicate with virtual certainty, although at a smaller rate than would be possible in the absence of noise.

This result may surprise the reader; but a simple and empirical form is known to everybody. For instance, a kind of code that very effectively reduces the probability of error is represented by repetition (think of loudspeakers in railroad stations and airports); of course, repetition doubles the time needed for the transmission. Another well-known type of code is applied when one says "t" for Thomas, "g" for George, and so on. Such codes are very effective, but are not ideal. An ideal code allows $\bar{I}$ *bit* per message to be transmitted with virtual certainty.

These considerations can help to understand why natural languages are redundant. What they lose in speed, they gain in reliability of the communication. This happens naturally where the person receiving the message is perfectly acquainted with the *statistics* of the language and is ready to reconstruct, in a probabilistic way, the original message from the message altered by noise. A foreigner is much less prepared to do the reconstruction; to him, the language is not very redundant.

It is also useful to be familiar with the *long-range* statistics of the source, that is, to know with what probabilities the source can transmit the various messages. The evaluation of these probabilities is the job of the philologist who endeavors to reconstruct a text damaged by time and full of gaps (noise). Woe to him, if he could not count on redundancy! All reconstructions, including the wildest ones, would have the same probability.

Information theory is profitably applied also to the science of vision. The outer world sends us visual messages; each scene we see is a message. How many are the possible and different scenes we can see?

An apparently possible message would consist of a chaotic juxtaposition of countless, more or less bright points, of all colors, without any correlation between them. In this way a single visual scene would be able to transmit an enormous amount of information and it would require an extremely long time for us to perceive and analyze all the details. But the world is different! Such a juxtaposition of points cannot represent a real scene. The visual world is endowed with statistical correlations between its points, which drastically reduce the number of bits transmitted by each single scene.

The adult observer is fairly familiar with the statistics of the visual world and takes advantage of this in order to reconstruct and integrate very rapidly (and mostly subconsciously) information that is not directly acquired. The visual apparatus, from the retina to the visual cortex, is a tremendous reconstructor of information.

The redundancy of the visual world and the ensuing possibility of encoding it was discovered some twenty thousand years ago by our

Figure 3.12

ancestors who discovered drawing. Drawing exempts the artist from reproducing all the points of a visual scene and enables him to represent only contours. But in order for this method to work, the addressee must know the code and be prepared to decode the message. To us, the operation of decoding a drawing appears only natural; but in reality, it is the result of long training, to which we are subjected as children by first learning the statistics of the visual world, and then by learning from adults how one can take advantage of the redundancy. As is well known, no animal, except humans, is able to read a drawing.

Obviously, difficulty arises when we do not know the statistics of what we are seeing. A sensational case was that of Galileo, when he turned his telescope to Saturn. He perceived a scene, which, owing to the limited resolving power of the telescope, appeared as that shown in Figure 3.12. He announced his discovery to Kepler by an anagram, whose solution was: *altissimum planetam tergeminum observavi*. Thus he had observed that Saturn consisted of three bodies! This was due to the fact that his "statistics" of the sky included point objects and circular objects, but did not include bodies with rings. If planets with rings had been known to him, he would have observed that Saturn has rings.

Analogous observations could be made for the sense organs other than sight (see, e.g., Lindsay and Norman, 1972). From the preceding discussion, we derive the conclusion that the amount of information provided by the sense organs is actually considerably less than what we usually believe. The additional information comes from what we know a priori about the visual world, the acoustical world, and so forth. This integration of prior information with that directly received, takes place very quickly, and mostly by a subconscious process.

Once again, we must draw attention to the fallacy of the ordinary concepts of direct observation, of immediate data, and so on. It is dangerous to forget that any observation implies some data processing, based on the statistics of the world with which we are familiar.

3.14. Microstates and macrostates

P. Laplace had written in a celebrated passage of 1814:

An Intelligence capable of knowing, at a given instant, all the forces and the disposition of all the entities present in nature, and sufficiently profound to submit all these data to analysis, would grasp with the same formula the motions of the largest bodies of the universe and of the lightest atoms; to her, nothing would be uncertain and her eyes would perceive both the future and the past.

This passage is very often quoted as an illustration of the strictly deterministic world view. For the time being, we are not concerned with that view; we only wish to emphasize the unmeasurable amount of information that would be required in order to get that kind of knowledge of the world. One should have to know, at a given time, both the location and the velocity of every single atom (or subatomic particle) of the universe, along with the interactions between them.

Now this type of knowledge, besides requiring a frightening number of *bits* of information, would be absolutely useless for almost all of our purposes!

For example, let us think of a gas enclosed in a container. Who is interested in knowing its *microstate*, that is, the position and the velocity of every single molecule? What we are interested in and what is sufficient to predict the future behavior of the gas is the *macrostate*, that is, the distribution of the temperature and of the pressure, along with the shape and size of the container. Fortunately, the number of the *bits* needed to describe the situation in this case is much more reasonable, so that the problem can be dealt with by human means.

The molecules are considerably smaller (a few 10^{-8} cm) than the physical regions with which we are usually concerned. Accordingly, we are interested only in the mean or statistical properties of the molecules contained in each region. Those properties constitute the macrostate. It is evident that a well-determined macrostate can correspond to an enormous number of different microstates. We are not interested in knowing which one in particular is realized.

Statistical mechanics is based upon a postulate, which at first, appeared evident, whereas only later revealed its profound, and by no means obvious, significance. The postulate essentially asserts that *all microstates are equiprobable.*[42] It is clear that a postulate of this kind can be accepted only if equiprobability, once assumed at a given instant, is conserved at later times.

Let our macroscopic system be a gas formed by N identical molecules. Its microstate will be fixed by assigning 6 numbers per molecule, that is, the three coordinates x, y, z, and the three velocity components v_x, v_y, v_z. Altogether, we have $6N$ numbers, which, following J. W. Gibbs, we can consider as the coordinates of a point in a $6N$-dimensional abstract space, called the phase-space.[43] As time passes, the point will move in phase-space and will describe a line.

Let us now consider an ensemble of systems, that is, a large number of systems of N molecules, all macroscopically identical with the one of interest. Each system will be represented by a point in phase-space, and the ensemble will occupy a volume (or *hypervolume*) of phase-space. A celebrated theorem of analytical mechanics, due to J. Liouville, assures us that this volume does not vary with time, while each representative point describes its path. As a result, if the systems of our ensemble are initially distributed with uniform density in phase-space, the density will remain uniform. If the probability of a microstate is assumed to be proportional to the volume of phase-space in which its representative point can be found, and if initially the microstates are all equiprobable, they will remain equiprobable at all times.

The probability of a given macrostate will be assumed proportional to the volume occupied by all the microstates that can realize it, or to the number of such microstates.

In Gibbs's approach the value of a physical quantity of the macroscopic system is computed as an *ensemble average* over all the possible microstates. According to Boltzmann's *ergodic* hypothesis,[44] the ensemble average should be identical with the *time average* of the values taken by the quantity during the evolution of a single microsystem. This time average is what we actually measure when we do a measurement on a macroscopic system. It is difficult to justify precisely the ergodic hypothesis, even subject to appropriate restrictions (in its most general form it is probably false). An enormous amount of literature exists on the subject. After a number of contributions of both the nineteenth and the twentieth centuries (in particular, G. D. Birkhoff's contribution of 1930), the problem has been clarified progressively better, but it is still not exhausted (see, e.g., Caldirola, 1974 Chap. III).

Regarding the postulate of the equiprobability of the microstates, apparently, it arises from an application of the principle of insufficient reason or of indifference. But as already mentioned, it is erroneous to believe that the principle can be applied a priori without any experimental confirmation. One has only to think of the absolutely arbitrary way in which the distinct microstates are counted. The way chosen in classical statistics was particularly fortunate in that its results were confirmed by experience. But later, quantum physics has taught us that the microstates must be counted in different ways, depending on the kind of particles constituting the system (see §4.7 and §4.12). As a result, the principle of indifference cannot be applied a priori without having resource to some knowledge that only experience can furnish.

Let us consider N identical molecules of an ideal gas, enclosed in volume V. To simplify, let us assume for a moment that all the molecules have the same velocity, hence the same kinetic energy given by equation (3.48), and are shooting uniformly in all directions of space.

In an ideal gas the molecules are point masses (i.e., occupy a volume
≪V), not acting on one another at a distance. We ask how the molecules
are distributed in the volume.

To begin with, we divide V into n equal subvolumes or elementary
cells V_1, V_2, ..., V_n, each having the smallest size measurable with
ordinary, macroscopic means. Let us denote by N_1, N_2, . . ., N_n the
numbers of the molecules contained in the various elementary cells.
We evidently have

$$\sum_{i=1}^{n} N_i = N \qquad (3.70)$$

The macrostate is perfectly determined by assigning the numbers N_i.
But there are many different microstates corresponding to this same
macrostate.

Every time we carry out a *permutation* between the N molecules,
or interchange the places of at least some of them, we change the
microstate but not the macrostate. A not too difficult reasoning shows
that the number of all possible permutations between N objects is given
by the product $1 \times 2 \times 3 \times \ldots \times N$. This is called the *factorial* of
N and is denoted by $N!$.

However, the distinct microstates are not exactly $N!$. If we permute
the molecules contained in one and the same cell V_i, we obtain the
same microstate. Indeed, that microstate is characterized by the fact
that those molecules are placed in V_i and this circumstance is not
changed. The number of permutations within V_i is evidently $N_i!$. Hence
for each one of the $N!$ possible microstates, there are $N_i!$ microstates
equivalent to it. The overall number of possible microstates thus reduces to $N!/N_i!$. By repeating the same argument for all cells V_i, we
obtain

$$W = \frac{N!}{N_1!, N_2!, \ldots, N_n!} \qquad (3.71)$$

where W represents the total number of distinct microstates corresponding to the same macrostate.

As we have assumed that all microstates are equiprobable, the most
probable macrostate will be that corresponding to the greatest number
of microstates. In order to find this, we have to find what distribution
N_1, N_2, . . ., N_n of the molecules renders W a maximum.

Note that when W is a maximum, its logarithm also has maximum
value. By the well-known properties of logarithms, we have from
equation (3.71)

$$\ln W = \ln N! - \ln N_1! - \ln N_2! - \cdots - \ln N_n! \qquad (3.72)$$

There is a formula (Stirling's formula) that gives an approximate value for the logarithm of the factorial of an integer M; the relative approximation is very good when M is very large. Precisely, one can write

$$\ln M! = M \ln M - M \qquad (3.73)$$

Applying this formula to each term of equation (3.72), and remembering equation (3.70), we have

$$\ln W = N \ln N - \sum_{i=1}^{n} N_i \ln N_i \qquad (3.74)$$

the term $N \ln N$ does not depend on the N_i and is constant; hence it will suffice to render maximum the sum following it.

Let us assume that a given microscopic state has been realized, characterized by the distribution $N_1, N_2, \ldots, N_n$ of the molecules. What is the probability P_i of finding a given molecule in cell V_i? There are evidently N_i favorable cases, whereas the possible cases are N in number, thus we have $P_i = N_i/N$. Taking this into account, the expression to be rendered maximum can be written as

$$-\sum_{i=1}^{n} N_i \ln N_i = -\sum_{i=1}^{n} NP_i \ln NP_i$$

$$= -N \sum_{i=1}^{n} P_i (\ln N + \ln P_i) \qquad (3.75)$$

$$= -N(\ln N + \sum_{i=1}^{n} P_i \ln P_i)$$

having made use of the relation

$$\sum_{i=1}^{n} P_i = 1 \qquad (3.76)$$

which is readily derived from equation (3.70).

As the additive constant $N \ln N$ and the constant factor N in the last equation (3.75) are immaterial for our purpose, we have to maximize the expression

$$H = -\sum_{i=1}^{n} P_i \ln P_i \qquad (3.77)$$

with the condition, equation (3.76).

But we already know this problem and its solution.[45] The maximum is attained when all the P_i are equal. We can conclude, then, that the most probable macrostate is that in which the gas has uniform density; in our case, it is the state of *thermodynamic equilibrium*.

Encouraged by this result, let us remove the unrealistic restriction that all molecules should have equal velocities. We characterize the state of a molecule by means of its Cartesian coordinates x, y, z and its velocity components v_x, v_y, v_z. One can also imagine that these six numbers represent the coodinates of a point in a six-dimensional space, sometimes called the μ-space. Thus each molecule occupies a point in μ-space. All these points define a volume (hypervolume) S of μ-space.

We apply a procedure similar to the previous one. Let us divide volume S into many elementary cells S_1, S_2, . . ., S_n and denote by P_1, P_2, . . ., P_n the respective probabilities of occupation by a molecule. Repeating the previous argument, we should maximize equation (3.77), subject to condition, equation (3.76). But in this case there will also be a second condition. Precisely, let us assume that the total energy U of the gas is known. If the gas does not exchange energy with the environment, U must be a constant.[46] If we denote by ϵ_i the energy of each molecule located in S_i, we should write

$$\sum_{i=1}^{n} N_i\epsilon_i = N \sum_{i=1}^{n} P_i\epsilon_i = U \qquad (3.78)$$

Thus we are confronted with a problem of conditional maximum, where the fixed conditions are equations (3.76) and (3.78). As already mentioned, there are simple methods of differential calculus that allow us to solve these problems. As we do not assume the reader to be familiar with such methods, we are providing the solution here which in our case is

$$P_i = \alpha \exp(-\beta\epsilon_i) \qquad (3.79)$$

where the notation exp x stands for e^x, and α and β are constants to be determined.

Even before determining the constants, we can realize a few interesting facts. First, let us recall that in our case the energy of the molecules is purely kinetical (ideal gas). Therefore ϵ_i will depend on v_x, v_y, v_z, but not on x, y, z. Hence P_i does not depend on the coordinates of the molecule, and the most probable distribution has again uniform density. As to the velocities, they do not have uniform probability; the probability decreases exponentially with increasing energy. This important result has a general validity and does not depend on the particular case treated. To see this, one has to remember only that the circumstance that the ϵ_i are purely kinetic has not been used in the calculation; one could deal even with more general kinds of energy.

Of course, if ϵ_i depended on x, y, z (potential energy), the density would cease to be uniform.

Let us rewrite equations (3.76) and (3.78), utilizing equation (3.79). We obtain

$$\sum_{i=1}^{n} \alpha \exp(-\beta\epsilon_i) = 1, \qquad N \sum_{i=1}^{n} \alpha\epsilon_i \exp(-\beta\epsilon_i) = U \quad (3.80)$$

These two equations are sufficient to determine both constants α and β. However, this cannot be easily done in an elementary way.

It is useful to transform both sums into integrals; this is possible because each one is the sum of many terms whose values vary slowly with the index. In place of the discrete probability P_i, one has to consider a continuous function $f(x, y, z, v_x, v_y, v_z)$, introduced by Boltzmann,[47] called the *distribution function*. Precisely, $f(x, y, z, v_x, v_y, v_z)$ $dx\, dy\, dz\, dv_x\, dv_y\, dv_z$ represents the number of molecules found in the small volume $dx\, dy\, dz\, dv_x\, dv_y\, dv_z$ of μ-space centered on the point x, y, z, v_x, v_y, v_z. The probability that a molecule be inside the volume is obtained by dividing that number by N. In our case we already know that the distribution does not depend on x, y, z, and will simply write $f(v_x, v_y, v_z)$.

Inserting in place of ϵ_i the kinetic energy of a molecule

$$\epsilon_i = \tfrac{1}{2} m v^2 = \tfrac{1}{2} m(v_x^2 + v_y^2 + v_z^2) \tag{3.81}$$

we make the substitution

$$P_i \rightarrow \frac{1}{N} f(v_x, v_y, v_z) \, dx\, dy\, dz\, dv_x\, dv_y\, dv_z$$

$$= \alpha \exp[-\beta\tfrac{1}{2}m(v_x^2 + v_y^2 + v_z^2)] \, dx\, dy\, dz\, dv_x\, dv_y\, dv_z \tag{3.82}$$

Now notice that the argument leading to equation (3.48) in the case of equal velocities could be repeated, with the same result, provided that by e_c we understand the *mean* kinetic energy of the molecules. The overall energy U will be N times e_c and $U = 3NkT/2$ will result. Hence equations (3.80) become

$$\int_{-\infty}^{+\infty} \alpha \exp[-\tfrac{1}{2}m(v_x^2 + v_y^2 + v_z^2)]$$

$$dx\, dy\, dz\, dv_x\, dv_y\, dv_z = 1 \quad (3.83)$$

$$N \int_{-\infty}^{+\infty} \alpha\tfrac{1}{2}m(v_x^2 + v_y^2 + v_z^2) \exp[-\tfrac{1}{2}m(v_x^2 + v_y^2 + v_z^2)]$$

$$dx\, dy\, dz\, dv_x\, v_y\, dv_z = \tfrac{3}{2}NkT \quad (3.84)$$

Both these integrals are sixfold integrals. One has to carry out an integration over x, then over y . . . , and finally over v_z.

The integrations over x, y, z are readily done, because the functions under the integral sign do not depend on these coordinates; the sum of all the elementary volumes $dx\,dy\,dz$ is simply equal to the volume V of ordinary space, occupied by the gas. As to the integrals over v_x, v_y, v_z, they are of a kind well known in the calculus and are readily evaluated. Having done this, one can derive α and β from equations (3.83) and (3.84), and finally arrive at

$$f(v_x, v_y, v_z) = \frac{N}{V} \left(\frac{m}{2\pi kT} \right)^{3/2} \exp\left(-\frac{mv^2}{2kT} \right) \qquad (3.85)$$

This is the famous formula first derived by J. Maxwell in 1859. It shows that in the most probable state the velocities are not all equal, but have an exponential distribution. The most probable velocity of a molecule is zero, but all velocities, however large, are represented, with a probability tending rapidly to zero. By means of Maxwell's distribution, one can derive all the properties of interest of a rarefied gas, such as the mean free path of the molecules, the viscosity, the thermal conductivity, and so on.

3.15. Statistical irreversibility

We have found the most probable state for a gas and have assumed that it represents the equilibrium state. Is this assumption justified?

In reality, we have only proved what follows. Let us select at random a gas enclosed in a volume V, and having energy U. We shall hit on a well-determined microstate. If all microstates are equiprobable, we most probably shall have selected the macrostate M which corresponds to the maximum number of microstates. We also add, without proof, that due to the enormous value of Avogadro's number, the maximum is very sharp. In other words, the probability of finding the gas in a state far from M is extremely small, virtually zero.

What happens if instead of choosing at random, we purposely set up a very improbable case? Let us suppose, for example, that we have realized the case of Figure 3.3, and consider the instant when we remove the dividing wall between A and B. How can we be sure that the gas will rush also to occupy B, so as to give rise to the most probable state?

This is one of the most difficult and fascinating problems of statistical mechanics, which is still giving rise to perplexities and discussions. However, by and large, it is sufficiently clarified. The decisive start was given in 1872 by Boltzmann, who proved a celebrated theorem, known as the *H-theorem*.

Having introduced the distribution function $f(x, y, z, v_x, v_y, v_z)$, let us consider the quantity H, defined by equation (3.77). Apart from a coefficient and an additive constant, which need not concern us here, we can write[48]

$$H = -\int_{-\infty}^{+\infty} f \ln f \, dx \, dy \, dz \, dv_x \, dv_y \, dv_z \qquad (3.86)$$

where, for simplicity, we have omitted the indication of the variables within f; the integral must be extended to the whole of μ-space.

Due to the collisions between the molecules, the value of f will vary with time and, consequently, H will also be a variable. Boltzmann succeeded in proving that in all cases

$$\frac{dH}{dt} \geq 0 \qquad (3.87)$$

In other words, H can never decrease; it can either increase or remain constant. Precisely, it will have its maximum value and will remain constant if and only if f coincides with Maxwell's distribution equation (3.85).

Thus it seems plausible that if one starts from a distribution t different from that of the most probable state M, the system will evolve until it has reached M. This conclusion is not mathematically precise, but we shall be satisfied with its plausibility. Hence M represents the equilibrium state.

But what physical entity is represented by H? If we introduce into equation (3.86) Maxwell's distribution, equation (3.85), we find integrals of the kind, such as equations (3.83) and (3.84), which are readily evaluated, and thus obtain

$$H = N(\tfrac{3}{2} \ln T + \ln V) + \text{constant} \qquad (3.88)$$

Let us now recall equation (3.43), expressing the entropy S of the gas. If we replace c_V with its value, equation (3.50), we find

$$S = R(\tfrac{3}{2} \ln T + \ln V) + \text{constant} \qquad (3.89)$$

Finally, knowing that $R/N = k$ represents the Boltzmann constant, we obtain

$$S = kH \qquad (3.90)$$

apart from an additive constant.[49]

We have defined the entropy S at equilibrium, or in the case when the system can at least be divided into a finite number of parts, each one at equilibrium. In such conditions, S coincides with kH. But then we can assume kH as the definition of entropy, even in the most general case when the system is far from equilibrium. Boltzmann's theorem assures us that even in these general conditions, the entropy can only increase until the equilibrium state of the entire system has been attained. This is the statistical interpretation of the second law of thermodynamics. It says that a macroscopic system goes naturally from the less probable to the more probable situation. The most probable one is the equilibrium condition.

In addition, recall that H, apart from two unessential constants, represents the sum, equation (3.77), which in turn coincides with the sum, equation (3.67), encountered in the information theory. From what we have said in §3.13, we can conclude that the *entropy* represents our *ignorance* regarding the system. The greater the entropy, the greater the number of *bits* we need to determine the microstate.

Sometimes instead of ignorance, entropy is called the *disorder*. Take all the objects contained in a room and arrange them at random. The greatest probability by far is that a situation of disorder will result. Sure enough, a situation of order corresponds to a possible case, but its probability is so small that only by a miracle could it be obtained from a random distribution. Entropy is a very reasonable measure of the amount of disorder. The disorder of an isolated system can only increase. Thermal death, for example, represents maximum disorder.

Boltzmann's H-theorem may seem to have furnished a mechanical explanation of irreversibility and of entropy increase. If we consider a sequence of instants $t_1 < t_2 < \cdots < t_n$, we have for the entropy

$$H(t_1) \leq H(t_2) \leq \cdots \leq H(t_n) \tag{3.91}$$

whereas the reverse sequence cannot occur.

But at this point, serious doubts can arise, and have actually arisen very soon. Is it not absurd to justify an irreversible process by means of reversible processes, like those of mechanics? It was argued by J. Loschmidt; let us consider the microstate corresponding to $H(t_n)$ in equation (3.91) and build an identical microstate, but with all velocities reversed. This is a possible microstate, yet it would follow the sequence, equation (3.91), in reverse order!

Furthermore, there is a famous theorem proved by J. Poincaré in 1880, which states that starting from any microstate and waiting sufficiently long, the system gets back as close as we wish to the original state. This embodies a property of *quasi-periodicity*, but its practical consequences are the same as those of true periodicity. Any state is virtually repeated. E. Zermelo argued: How can a process be irreversible, if all its states are repeated?

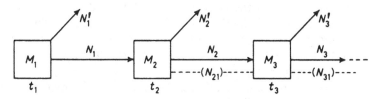

Figure 3.13

The situation is very curious: Everyone is right, including Boltzmann, who based irreversibility on statistics! The paradoxes arise mainly from the bewildering order of magnitude of Avogadro's number, of which we cannot succeed in forming an intuitive idea.

An important contribution to the clarification of these problems was given by Paul and Tatiana Ehrenfest. We shall try to give an elementary and extremely simplified idea of the state of affairs.

Let us consider at time t_1 a macrostate M_1 (Fig. 3.13), from which at later times $t_2, t_3, \ldots$, the system proceeds to the macrostates M_2, $M_3, \ldots$, predicted by classical thermodynamics. We have, of course, $H(t_1) \leq H(t_2) \leq H(t_3) \leq \cdots$.

Only N_1 microstates, out of all those possible for M_1, go to M_2, whereas N_1' microstates go to other macrostates, possibly with $H(t_2) < H(t_1)$. However, the ratio N_1'/N_1 is so small that choosing at random one microstate of M_1, we are virtually *certain* to go to M_2. If we went on and on, always choosing microstates at random from M_1, not only human life, but even the duration of the universe would not be sufficient for one of the microstates that do not go to M_2 to be fished out!

Of all microstates possible for M_2, N_2 go to M_3, while N_2' go elsewhere. Note that since M_2 is more probable than M_1, its microstates are larger in number than those of M_1. Only a negligible fraction of the microstates of M_2 comes from M_1. Accordingly, not only N_2, but also N_2' is a number much larger than N_1. Let us denote by N_{21} the number of those microstates that go to M_3, coming from M_1. From M_3, N_3 microstates will go to M_4 (N_{31} of them coming from M_1), whereas N_3' will go elsewhere. From M_4, N_4 microstates will go to M_5, and so on.

For any microstate of M_1, there exists a sufficiently long time during which it has taken one of the channels $N_1', N_2', N_3', \ldots$, and it has come back to the initial conditions. However, it would take about $10^{10^{19}}$ years for 1 cm^3 of a gas to return to the initial conditions (Boltzmann, 1896). No one would be so crazy as to start waiting!

Let us now consider the N_1 microstates that go from M_1 to M_2. As already mentioned, N_2' is a number much larger than N_1. It is therefore not paradoxical that the N_1 Loschmidt states with reversed velocities are all comprised within the N_2' states that leave the regular sequence. Such states do belong to M_2, and go back to M_1. However, if we select at random one microstate of M_2 we are virtually certain not to hit on one of those microstates.

This is the reason underlying the apparent contrast between the microscopic reversibility and the macroscopic irreversibility. The explanation is subtle but not difficult to grasp. Let us recall, however, that we made an important assumption regarding the equiprobability of all microstates, which is neither provable, nor evident a priori.

3.16. Does time have an arrow?

The considerations about irreversibility developed in the preceding section must have some bearing on the problem of the direction of time, or of the time arrow, as is commonly used, after A. S. Eddington.

If time is just a coordinate, like the spatial coordinates, why is it not symmetric? Why do past and future exist? Why are we always forced to go toward the future, whereas the x axis can be traveled forward or backward indifferently? Why does a "now" exist, a privileged point on the time axis, although all points are equivalent on the space axes?

It goes without saying that these problems, in one way or another, have puzzled scientists and philosophers for centuries on end (see Whitrow, 1972, 1973). These kinds of problems are never *solved* or *exhausted*. However, modern physics can undoubtedly shed much light on them (see Grünbaum, 1973, Chaps. VIII and X). Of course, a completely adequate treatment should be based on *quantum statistics*. However, even the results of classical statistical mechanics, as outlined in the previous section, can serve to give a good deal of insight in the problem of the time arrow.

In classical antiquity there was a tendency to stress the cyclic rather than the linear aspect of time. Roughly speaking, time did not seem an autonomous concept, but was more or less identified with the repetition of some astronomical or terrestrial phenomena. Sometimes it was even maintained that the world itself has a cyclic nature and that all events will be identically repeated an infinite number of times. The real world was viewed more as a sequence of *discrete* events than as a *continuous* flow.

The linear and irreversible nature of time asserts itself in the Christian Middle Ages. The incarnation of God's son takes place only once and divides a past of ignorance and immaturity from a future of awareness of revealed truth. There is not only a genesis, but also an eschatology.[50]

With the intensification of commerce, and later with the birth of industry, time is more and more reified; it becomes money. It is measured independently of the astronomical events, with more and more perfect clocks.

The *linearity* and *continuity* of time become fundamental tenets of Galileo's doctrine and of his school. Newton completes the process of objectivation, speaking of an absolute time that flows uniformly, without regard to anything external.

But is it really true that time flows? With what speed does it flow? Perhaps at a second per second? (see Smart, 1954; Black, 1959; Park, 1972; Davies, 1974).

A view favored by many physicists is that expressed by H. Weyl (1949) "the objective world simply *is*, it does not *happen*." Things are in space–time, that continuous manifold described by relativity. Each one of us describes a world line in this continuum and sees a varying scenery. In other words, the flow of time would be a psychological or subjective, rather than objective phenomenon. Much can be said to support this conception; but it is impossible to remove all doubts.

Nevertheless, the study of statistical irreversibility can suggest the following considerations. According to the ergodic hypothesis, an isolated macroscopic system goes through virtually all the microstates compatible with its constraints. But for most of its time, it remains in the equilibrium state M, in which its entropy has the maximum value S_M. This is due to the fact that the number of the possible microstates corresponding to M is much larger than that of the remaining microstates. Hence if we observe the system at a randomly chosen instant, we are virtually certain to find it in the equilibrium state, with entropy S_M. Major fluctuations away from M, even if very rare, can sometimes occur. In such cases, entropy is less than S_M. The more entropy differs from S_M, the rarer is the fluctuation.

If one plots entropy as a function of time, one finds a curve almost always identical with a horizontal straight line, at the height S_M over the time axis. Every now and then, there is a fluctuation and the curve goes down for a very brief interval. But there is *no anisotropy* in the shape of the curve, the positive direction of t being absolutely equivalent to the negative direction. Nevertheless, there is no paradox, and the second law of thermodynamics is not contradicted. In order to verify this statement, we would find it useless to select a generic instant, when the entropy is S_M and will remain S_M. We must select one of the few instants t_1, when the entropy has a value $S_1 < S_M$. It is a very rare fluctuation; but a larger fluctuation is still rarer; we can therefore be virtually certain that at a later time t_2, we shall have the entropy $S_2 > S_1$ and not the other way around!

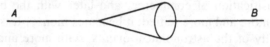

Figure 3.14

In this way, the irreversibility is amenable to a question of initial conditions. The irreversibility of an isolated macroscopic system is de facto and not nomologic or due to a law (see Mehlberg, 1961; Costa de Beauregard, 1971; Davies, 1974). Let us clarify this point.

In the overwhelming majority of cases we find an isolated system at maximum entropy S_M. In this situation we are virtually certain that the entropy will not vary. In a limited number N_1 (virtually zero) of cases the entropy, starting from S_M, will decrease. In a number N_2 of cases the system will be found with entropy less than S_M, and we are virtually certain that it will increase. Now the de facto condition is that N_2 is considerably greater than N_1, for in the real world we do find systems with $S < S_M$ (hence $N_2 \neq 0$), whereas N_1 is virtually zero. Hence if we take an isolated system at random, we may be certain that it is either in equilibrium or evolving toward equilibrium. But no law prescribes irreversibility (at least at this classical stage).

The fact that irreversibility is not nomologic, but is due to the initial conditions, does not prevent us from finding a physical basis for time asymmetry. There are in the physical world such things as "before" and "after," definable in physical terms.[51] Precisely, considering two instants t_1, t_2, we can say that t_2 is after t_1 if for the overwhelming majority (virtually all) of the isolated systems $S(t_2) > S(t_1)$. Never mind if the before–after relation can only be defined de facto in this universe, and not by a law existing in all possible worlds. We live in this world!

But there is a subtler question, which has caused many mistakes. As was emphasized, for example, by Grünbaum, the possibility to define before and after gives only a physical basis to the asymmetry of time, that is, to the fact that both directions in time are not equivalent. It does not tell us, however, which is *the direction* of time, the one in which time is flowing. Nothing in this argument implies that time flows from before to after.

As a rough analogy to help understand the question, let us consider the cone of Figure 3.14. The direction of the axis from A to B can be distinguished from the opposite direction by the increasing cross section of the cone. But there is nothing in the cone to tell us whether A is left or right of B (that depends on the observer!), or if something is flowing from A to B.

We shall return to this kind of question. Now let us instead discuss the problem of the initial conditions. Consider a glass of water and an ink drop, which at the initial time is concentrated in a small volume near the surface. At time t_0 the entropy S_0 of the system is less than the maximum entropy S_M, which is reached only when the ink has uniformly diffused through the water. If the system is isolated, we are sure that at later times $t_1 > t_0$, $t_2 > t_1$, ..., the system will evolve toward the state S_M. But because the curve of the entropy as a function of time does not present any systematic asymmetry, we should also be certain that by reversing the direction of time, that is, by going to earlier times $t_{-1} < t_0$, $t_{-2} < t_{-1}$, ..., the entropy should go closer and closer to S_M. Apparently, observing the system at t_0, we should derive the conclusion that there was previously a uniform mixture of water and ink and that at t_0 a very rare fluctuation has occurred, so that the ink has gone to a very small volume near the surface. After t_0, the system will start returning to the uniform situation.

All this may seem absurd. The reason is that we have tacitly assumed that our system was isolated, even before t_0. If we do not have any information about the previous state of the system, we can make two different hypotheses:

1. The system was isolated prior to t_0, and there has been a fluctuation; it is the apparently absurd conclusion just mentioned.

2. The system was not isolated prior to t_0, and a person or a mechanism has dropped some ink into the water.

We wish to evaluate the probability of each one of these hypotheses. Let us indicate by $p(D)$ the probability that observing the system at t_0, we see the ink drop at the surface, by $p(I)$ the probability that the system was isolated before t_0, and by $p(N)$ the probability that it was not isolated. The corresponding *conditional* probabilities will be denoted by $p_I(D)$, $p_N(D)$, $p_D(N)$, $p_D(I)$: for instance, $p_I(D)$ is the probability that at the instant t_0 an ink drop should appear at the surface of the system, under the assumption that the system was previously isolated, and so on. The joint probabilities will be denoted by $p(I, D)$, $p(N, D)$: for instance, $p(I, D)$ is the probability both that the system *is* isolated prior to t_0 *and* that the ink drop appears. By virtue of equation (3.58), we have

$$p(I, D) = p(I)p_I(D) = p(D)p_D(I) \tag{3.92}$$

$$p(N, D) = p(N)p_N(D) = p(D)p_D(N)$$

From the first equation we obtain

$$p_D(I) = \frac{p(I)p_I(D)}{p(D)} \tag{3.93}$$

and from the second

$$p_D(N) = \frac{p(N)p_N(D)}{p(D)} \tag{3.94}$$

These are the probabilities of the hypotheses (1) and (2), respectively. Indeed, we know that D has occurred; in that case, equation (3.93) gives us the probability that the system was previously isolated, and equation (3.94) gives us the probability that the system was not isolated.[52]

In order to compare equation (3.93) with equation (3.94), the value of $p(D)$ is immaterial, being a common denominator to both expressions. As to the other probabilities, it is reasonable to assume that $p(I)$, $p(N)$, $p_N(D)$ are possibly very small, but never negligible; $p_I(D)$ is instead virtually zero. As a result, equation (3.93) shows that $p_D(I)$ will vanish, too, and we shall be forced to choose hypothesis (2).

We can now form the following image of the world (see Grünbaum, 1973). The universe, for some reason, is found de facto in a low entropy state. As already mentioned, it is dangerous to speak too freely of the entropy of the universe. But for our purpose, we can restrict the argument to a very large portion of it, which may be considered nearly isolated (e.g., our galaxy). The different parts that we can distinguish in the world often interact with each other and temporarily build larger and more complex systems. After the interaction, the individual systems may branch off from the total system (Reichenbach, 1956), remain isolated for some time, and tend to reach equilibrium. Then there occurs a new interaction with a large system which alters the equilibrium, a new branching off, and so on. Naturally, the overall entropy of the portion of the world considered is steadily increasing. But the time required for it to reach a maximum is of cosmic order (e.g., billions of years), whereas the small branch systems reach maximum entropy in a time that we can observe.

A good example of a branch system that periodically interacts with a bigger system is the surface of the earth, which is illuminated by the sun during the day and is isolated from it during the night. A frequent case – an occurrence in everyday life – is represented by a small system, which interacts with a human being (or an animal) and is removed from the equilibrium state, as in the case of the water with the ink drop, previously described. We have already explained why on encountering a system far from equilibrium, we think of an interaction rather than of a fluctuation.

These considerations help us to understand why there may be traces or memories of the past, but not of the future. Let us refer to a famous example used by M. Schlick (Schlick, 1948, p. 106). If we look at a beach with uniform surface (i.e., at equilibrium in the gravitational field) and discern some footmarks on it, we infer that a person *has passed,* not that a person *will pass* on the beach. This should not be surprising. We have defined before and after by means of the entropy. It is therefore natural that on finding a quasi-isolated system with entropy less than S_M, we should think of an interaction that has occurred in the past. In the future the entropy will increase and tend to S_M; the footprints will gradually vanish.

Now let us digress somewhat and turn to a still rather controversial subject – converging and diverging waves. Let us drop a stone in a pond. A system of circular waves arises that diverge from the point of impact. Now the wave equation also gives as a solution a system of circular waves that converge onto the point of impact. However, the latter type of waves are never observed or used.[53] Someone has said that in this case we are concerned with an irreversible and *non-entropic* process.

In my opinion, one should approach the subject in this way. Let us replace the stone by a vibrator consisting of a small sphere, which is made to oscillate up and down by a motor, located at point A of the surface of the pond. If we assume that a wave motion of maximum entropy is present on the surface, it will consist of a chaotic distribution of peaks and troughs, in which no definite pattern can be discerned. But mathematical analysis shows that the distribution can always be considered as a superposition of many well-ordered wave systems of different frequencies, with random amplitudes and phases. For instance, one can use linear wave-trains (the analog of plane waves in space) or circular wave-trains of different kinds[54] diverging from A or converging onto A. Each one of these ordered wave-trains represents one degree of freedom of the surface and on the average has the same energy (equipartition of energy). If the pond has been perfectly isolated for a long time, the only energy available is thermal energy. Each degree of freedom will have the mean energy kT (a harmonic motion has $kT/2$ of kinetic and $kT/2$ of potential energy). At ordinary temperature this energy is so small that the tiny ripples of the surface will defy observation (even neglecting surface tension). If the vibrator also has the small energy kT, it will on the average neither give nor receive energy. It will sometimes yield energy to a system of diverging waves or take energy from a system of converging waves, depending on the relative phase. No temporal asymmetry will arise. If, on the contrary, the vibrator has a macroscopic energy, it will be so much greater than kT that the system vibrator, plus pond, will be very far from equilibrium. In order to go toward equilibrium, the vibrator will have to pour much more energy into the divergent waves than it absorbs from the convergent waves. In this case we are authorized to neglect the converging waves (which, however weak, will still subsist) and to consider only a very intense diverging wave-train generated by the vibrator. But we are not authorized to conclude that we have an irreversible, non-entropic process.

If for the sake of argument we assume that the pond will continue to be completely isolated and the energy will be scattered into all of the degrees of freedom (perhaps by the irregular shoreline), each degree of freedom will eventually acquire the same mean energy as the vibrator (of course, the water will boil and evaporate long before that condition is reached). Equilibrium between converging and diverging waves will be reestablished and the asymmetry will disappear.

Before leaving this subject, we should state that some irreversible and nonentropic processes might be possible in microphysics (see §4.27).

Let us return to the fascinating problem of the traces of the past. A particularly efficient system that can record and store these traces for a long time is human memory. Of course, humans are not an isolated system. But this fact does not seem to be essential for understanding memory. If one stays for a few seconds in the dark without eating or breathing, and so on, memory is perfectly conserved. The preceding discussion helps us to understand why we can remember the past and not the future. If we consider the brain as a physical system, it is not surprising that it can make a distinction between before and after, in accord with what happens in the world outside. Event A precedes event B, if at the time of occurrence of B, one can happen to remember A, and not the other way around.

As a conclusion, we can say that as far as the *asymmetry* of time and the distinction of before and after are concerned, there seems to be in the (macroscopic) physical world, everything that is needed for an adequate definition.

The same is not true for the problem of the *flow* of time and for the definition of *now*. Is there anything in the physical world that can distinguish my *now* from Dante's *now* or from the *now* of my descendants? Why is my *now* running along my world-line?

In my opinion, all those who have attempted to give *explicata* in physical terms for these *explicanda* have incurred vicious circles and tautologies. I agree with those authors who tend to view such problems as intimately tied up with the problem of the ego. They are largely psychological problems (see e.g., Bergmann, 1929; Grünbaum, 1973). In the same way, the problem posed at the beginning of this section, concerning the speed of the time flow seems to be purely psychological. To a child a year is close to infinity, but an aged man exclaims with Horace: "*Eheu, fugaces, Postume, Postume, labuntur anni!*"

3.17. Fluctuations

It has already been observed that the microstates corresponding to the state of equilibrium are greater in number than all of the other states put together. For this reason, if we take a system that has been isolated for a sufficiently long time, we are sure that it is in equilibrium. As noted, this is due to the enormous value of Avogadro's number, that is, to the enormous number of particles that are ordinarily found in a macroscopic volume. What happens when that number is sufficiently reduced, for instance, by rarefying drastically the gas, or simply by isolating a very small volume of it? In that case one can note some statistical fluctuations or departures from the mean values of the macroscopic theory.

Let us consider a gas of N molecules contained in volume V. At equilibrium, all the molecules will be uniformly distributed in the volume. A portion V_1 of V will contain $N_1 = N(V_1/V)$ molecules. We can also assert that each molecule has probability $p = V_1/V$ of being found in V_1, and that by virtue of the law of great numbers, we have $N_1 = pN$. But we know that, in general, there will be a departure from that value, on the order of $\sqrt{N_1}$.

If V_1 is a few cm^3, it will contain, say, $N_1 = 10^{20}$ molecules, hence $\sqrt{N_1} = 10^{10}$. We have in this case $\sqrt{N_1}/N_1 = 10^{-10}$, which means that the density of matter in V_1 differs by a few ten billionths from its mean value. There is no means to show such a difference experimentally.

If V_1 is instead 10^{-15} cm^3 (10^{-5} cm linear size), we have $N_1 = 10^5$, hence $\sqrt{N_1}/N_1 \simeq 3 \times 10^{-3}$. The density V_1 will differ by only a few thousandths from the mean value, and this difference is no longer negligible.

On the basis of these considerations, Lord Rayleigh could explain quantitatively the characteristics of the skylight. An electromagnetic wave is only slowed down by a transparent and homogeneous medium (refractive index), but preserves an ordered propagation. If the medium presents instead a number of small volumes of greater or smaller density, diffraction will take place, and the light will be partially scattered in all directions. Scattering is particularly efficient when the inhomogeneities are on the order of the wavelength λ.

Radiowaves, down to microwaves, have wavelengths that are too long to be appreciably scattered by this process. But light waves have λ on the order of 10^{-5} cm and some scattering occurs.

The light from the sun goes through the atmosphere and part of it is undisturbed, so that we can see the sun disk. But a substantial part is scattered by the inhomogeneities of the atmosphere, and this process is responsible for the brightness of the sky. A more detailed study would show that in the range of wavelengths considered, smaller wavelengths are more efficiently scattered. The blue–violet portion of the spectrum is scattered more intensely than the red portion. This is why the sky is blue. Conversely, the light coming directly from the sun, being stripped of the blue rays, appears shifted to the red. This is why the sun (or the moon) near the horizon, when the rays traverse a longer path in the atmosphere, appears red.

Another interesting case of fluctuation is represented by the *Brownian motion*. Consider a solid sphere immersed in a liquid. If the sphere has the same density as the liquid, it can remain stationary at a certain point. Due to thermal motion, the molecules of the liquid will hit the walls of the sphere, but on the average there will be as many shocks on one side as on the other. The very small difference, due to fluctuation, is negligible, in the case of a macroscopic sphere.

But the situation is different if the sphere is also very small. When one observes a sphere of about one μm diameter immersed in a liquid with a microscope, one notices that the sphere performs a curious dance, jumping in different directions, following a zigzag path, which eventually brings it far from its original position. In 1905, A. Einstein correctly interpreted this random-walk process as being due to fluctuations of the molecule motions. On the basis of this theory and the experiments made by J. Perrin, it was possible to evaluate Avogadro's number. A similar evaluation can be made starting from the measurement of the light scattered by the sky. The existence of the fluctuations can stimulate some intriguing thoughts, or even arouse some exceedingly optimistic hopes.

Clearly, the equilibrium state represents thermal death, only because we look at it with an insufficiently sharp eye. Below a certain linear size there is a sort of life in perpetual motion (the Brownian motion was for some time believed to be of biological origin). One may then wonder if it will be possible to obtain work from fluctuations by exploiting the favorable ones and discarding the others. Will it be possible to circumvent Lord Kelvin's postulate and obtain work from the Brownian motion that takes place in a liquid at uniform temperature?

These dreams were given a very clear expression by Maxwell, who in 1871 imagined "a being whose faculties are so sharpened that he can follow every molecule in its course." This being, who later became known as Maxwell's demon, can do surprising things:

For we have seen that the molecules in a vessel full of air at uniform temperature are moving with velocities by no means uniform, though the mean velocity of any great number of them, arbitrarily selected, is almost exactly uniform. Now let us suppose that such a vessel is divided into two portions, *A* and *B*, by a division in which there is a small hole, and that a being, who can see the individual molecules, opens and closes this hole, so as to allow only the swifter molecules to pass from *A* to *B*, and only the slower ones to pass from *B* to *A*. He will thus, without expenditure of work, raise the temperature of *B* and lower that of *A*, in contradiction to the second law of thermodynamics.

Many different solutions to this paradox have been proposed. Some workers believed that the important point is the intervention of an intelligent being, for whom one could postulate that some of the ordinary laws of physics are not valid. But as a matter of fact, this hypothesis is not of great help, because Maxwell's process could be carried out by an automatic device with the same result.

The correct solution of the paradox was approached by L. Szilard in 1929 and completed by L. Brillouin in 1950 (see Brillouin, 1956, p. 162). The demon, no matter how sharp his faculties are, must receive information about the molecules to be able to open or close the hole at the right moment. For instance, he will be able to see the molecules and to this end he must illuminate them. The light source must necessarily be at a temperature T_1 greater than the temperature T of the gas. For if it were $T_1 = T$, the gas would emit light of the same intensity as that emitted by the source, and the demon would see a uniformly bright background and could not distinguish the molecules. If the source emits the energy Q, its entropy will decrease by Q/T_1. The gas (or the eye of the demon) will absorb the energy Q and its entropy will increase by Q/T. As a result, the entropy of the overall system increases. Furthermore, one can show that the increase is at least as large as the decrease of entropy brought about by the demon's opening or closing the hole. As a net result, the second law of thermodynamics is not violated.

A quantitative relation, demonstrated by Brillouin, is suggested by equation (3.90). Each bit of information is to be paid for by an entropy increase of at least $k \ln 2$.[55]

In order to enunciate a more general principle, let us consider an isolated system, not in equilibrium. Let S represent its entropy and S_M the maximum entropy, which will be reached at equilibrium. The quantity

$$N = S_M - S \tag{3.95}$$

will be called the *negentropy* (i.e., the negative entropy).

Negentropy is like money, and can be spent to obtain information. Information, in turn, can be reconverted into negentropy (as would be done by Maxwell's demon). By measuring both negentropy N and information I with the same units (e.g., with thermodynamical units, which implies multiplying the number of bits by $k \ln 2$), we can express the second law of thermodynamics by

$$\Delta(N + I) \leq 0 \tag{3.96}$$

This means that the sum $N + I$ can at most remain constant (reversible process), whereas in general it decreases. When thermodynamic equilibrium is reached, the sum vanishes. One cannot extract any information from a system at equilibrium.

This complete formulation of the second law of thermodynamics had to wait until the middle of the twentieth century because Boltzmann's constant $k = 1.38 \times 10^{-16}$ erg/K is extremely small. Information is so inexpensive in terms of energy that one gets the impression that it is free!

Our great source of negentropy is, of course, represented by the sun, which steadily sends to the earth the radiant energy $W = 1.38 \times 10^6$ erg/(cm^2 s). This energy comes from the sun's surface, whose temperature T_0 is about 6000 K, and carries to the earth's surface entropy $S = W/T_0 = 2.3 \times 10^2$ per cm^2 per s. The energy is then degraded until it reaches the temperature T of about 300 K of the earth's surface. This corresponds to an entropy $S_M = W/T = 4.6 \times 10^3$ (which is later radiated into space and lost). We have therefore at our disposal an amount of negentropy $N = S_M - S = 4.4 \times 10^3$ per cm^3 per s. Dividing by $k \ln 2$, we find that with the negentropy coming from the sun on 1 cm^2 we could produce 4.6×10^{19} bits of information per second. This is an amount of information greater than that transmitted by all the television systems of the world!

If we have a gas enclosed in a container, we can make measurements to derive the volume, temperature, and pressure. All this does not represent a large number of bits of information, and we are ordinarily allowed to neglect the increase of entropy brought about by the measurements. But the situation is very different when we want to know the state of each molecule, as Maxwell's demon would wish to know. In that case one has to use Avogadro's number, that is, such an enormous factor that the increase of entropy is anything but negligible.

We can now return to the precision of a measurement (§1.9), with some additional information to help clarify its significance.

Suppose we want to measure the temperature of a certain amount of water. If the water neither boils nor is frozen, we can conclude that the temperature is (by definition) between 0°C and 100°C. We assume that we have no other prior information, so that all the temperatures between these values are equiprobable. If our thermometer has the precision $\epsilon = 0.01$°C, we have $100/\epsilon = 10^4$ possible results for the measurement. Hence the probability of each result will be $p = 10^{-4}$. Consequently, the amount of information we derive from the measurement is $I = -\log_2 10^{-4} = 13.2$ bits. In order to obtain this information, we have to increase the entropy of the environment by at least $13.2 \times k \ln 2 = 1.33 \times 10^{-16}$ thermodynamic units. This is very little indeed. However, if we make ϵ tend to zero, I and the entropy tend to infinity. Hence we derive a shatterproof reason that it is impossible to make $\epsilon = 0$.

Thus thermodynamics warns us that exact measurements represent only a dream and cannot be reached, for theoretical reasons, not only for practical difficulties. Further reasons for the impossibility of exact measurements will be offered by quantum mechanics.

4 Microphysics

4.1. The objects of physics

The world appears to us to be made of *things* and *objects*. We conduct our investigations on and establish relationships between them.

The term *thing* (*res*)[1] historically precedes the term *object* (which was introduced by the scholastics) and is accompanied by a different conceptual meaning. The concept of object, unlike that of thing, implies an intentionality and activity of the conscious subject who, according to the circumstances, isolates and distinguishes, or links and fuses together, the elements of the real world.

Furthermore, objects are not necessarily purely physical, but can also be psychological, mathematical, philosophical, and so on. For example, thirst, the number five, and the concept of cause can be objects of our thought.

In my opinion, "objectuation"[2] is a primitive activity; that is, it logically (and chronologically) precedes all other activities of thought. It is for this reason that all attempts to define the concept of object are hopelessly destined to circularity. The pattern of the definitions given by several philosophers goes more or less like this: object is that which, and so on. As is evident, that "that" is already an object.[3]

Concerning the objects that are commonly called *material things*, a well-known and interesting approach is that set forth by Bertrand Russell. Let us consider a material thing that is seen in different ways (or in different perspectives) according to the point at which the observer finds himself. Russell says:

> We can now define the momentary common-sense "thing" as opposed to its momentary appearances. By the similarity of neighbouring perspectives, many objects in the one can be correlated with objects in the other, namely with similar objects. Given an object in one perspective, form the system of all the objects correlated with it in all the perspectives; that system may be identified with the momentary common-sense "thing." Thus an aspect of a "thing" is a member òf the system of aspects which is the "thing" at that moment (Russell, 1926, p. 96).

But what do all these perspectives have in common? They can all be derived one from another by means of operations belonging to the group of translations and rotations. This property is an invariant, as we go from one perspective to another. This suggests that something is preserved notwithstanding our movements or those of the object. Piaget writes,

The conservation of the object is therefore the result of the co-ordination of the plans carried out by the sensor-motor intelligence whose preliminary work is the continuation of habitual movements. The object consists first of an extension of these co-ordinations of habit and is finally constructed by the intelligence, of which it constitutes the first invariant. This invariant is essential for the elaboration of space, of spatialized causality, and generally for all forms of assimilation which go beyond the actual field of perception (Piaget, 1943).

So the idea of the permanence or conservation of the object is born, which also extends beyond the contingent fact of whether or not the object is seen from some perspective.

But from the simple vision of the things of common sense, we easily go (and in my estimation, necessarily) to quite different generalizations. Again, Russell says:

Now physics has found it empirically possible to collect sense-data into series, each series being regarded as belonging to one "thing," and behaving, with regard to the laws of physics, in a way in which series not belonging to one thing would in general not behave. If it is to be unambiguous whether two appearances belong to the same thing or not, there must be only one way of grouping appearances so that the resulting things obey the laws of physics. It would be very difficult to prove that this is the case, but for our present purposes we may let this point pass, and assume that there is only one way. We must include in our definition of a "thing" those of its aspects, if any, which are not observed. Thus we may lay down the following definition: *Things are those series of aspects which obey the laws of physics.* That such series exist is an empirical fact, which constitutes the verifiability of physics (Russell, 1926, p. 115).

Clearly, this generalization is very bold. From the things of common sense we have arrived at a mere series of sense data that obey the laws of physics.[4] One may object greatly (and objections have been raised) to this conception. Among other things, its circularity meets the eye immediately: Knowledge of the laws of physics should show us which series of sense data are things, although it is impossible to arrive at the laws of physics except by experimenting on things. Yet we can note that the developments in microphysics, which came after Russell set forth these ideas (1914), have done more to confirm than to belie them. Above all, it is quite clear today that an object of physics, in order to be such, need not necessarily have the same properties as the macroscopic things of common sense. The laws it obeys can be notably different from those of classical physics, but still be physical laws. Let us not get too far ahead, and proceed in an orderly way.

Objectuation constitutes a preliminary stage through which we must necessarily pass if we want to investigate, or even just think, reality.

This is the role of objectuation in classical physics. To enunciate the laws of mechanics, we must consider material objects or bodies; to enunciate the laws of electromagnetism, we must consider electric charges or magnetic dipoles. But no law tells us how these bodies are necessarily made; what masses, what charges, what dipolar moments they have. In classical physics, objects are only de facto and their various characteristics are not nomologically determined.

In the last century we have witnessed an event of exceptional importance. Objects have entered, almost by force, into the nomological sphere of physics. Nomological objects,[5] that is, those objects that have fixed and prescribed characteristics, escaped the attention of scholars for many centuries, simply because they are too small to be observed with ordinary means. Molecules, atoms, nuclei, and particles in general, belong to *microphysics*. Today the existence of nomological objects is beyond discussion, and constitutes one of the fundamental points of empirical science.

If one wished to synthesize the meaning of this event in one expression, one could say that it is characterized by the recovery of *whole numbers* or of *natural numbers*. It seems very likely to me that the concept of natural numbers should be in some way interdependent with the process of objectuation, which separates or links together the elements of the real world. To begin with, as regards *unity*, one may recall a particularly expressive affirmation of Plotinus, which was picked up by several philosophers: "Separated from the one, beings are no more. The army, the chorus, the flock, would not exist were there not *an* army, *a* chorus, *a* flock." Passing from unity to the following numbers, and coming to our own time, it is possible to note that Russell's abstract definition of number, like a class of classes (of objects) (Russell, 1948), is intimately tied to "objectuation."[6]

From the point of view of psychogenesis, J. Piaget maintains that the birth of the concept of number is tied to the scheme of the conservation of the object, and that "arithmetical concepts dispose themselves progressively as a function of the demands of conservation" (Piaget and Szeminska, 1941; see also Čapek, 1971, p. 446).

The result of having discovered the nomological objects of physics is that today an object can be measured by a whole number of atoms and particles, by a whole number of elementary electric charges, and so on. The problem of the natural unit of measurement turns out equal to the problem of converting all measurements into whole numbers.

But why has the recovery of whole numbers, of those numbers from which humans certainly set out at the beginning of civilization, been necessary? Why had they been lost since and the need felt to go on to rational numbers and then to real numbers?

The incredible smallness of the nomological objects of physics has generated an erroneous opinion since antiquity. It was a matter of empirical observation that any material body could be divided into parts. Thus naturally, rational numbers were born. Once the initial body was assumed to be *one*, its parts had to be assigned numbers that could not be found among the whole numbers. In this way, fractional numbers were brought into use. But the parts into which the body was divided were also *bodies* and therefore they too could be divided into parts. At this point it was possible to be tempted to make an *extrapolation*, and to conclude that the procedure could be followed indefinitely. So the idea of the *continuum* was born, which Aristotle defines as "that which is divisible into parts which are always divisible."[7] That matter should be continuous in this way easily ends up appearing as a commonsense assertion and may become evident a priori.

However, the critical sense of the Greeks immediately realized the serious dangers hidden in this mode of reasoning. That extrapolations of this kind were not permissible, must have come to mind when one considered the classical example of the sandpile. If we remove a grain of sand from a sandpile it remains a sandpile. Yet it is absurd to conclude that one can repeat the operation an infinite number of times and always have a sandpile left over! The credit for calling attention to the difficulty of the continuum goes largely to Zeno. I am not referring here to the paradox of Achilles and the tortoise (which stems mainly from the absolutely gratuitous admission that the sum of an infinity of terms must necessarily be infinite) but, rather, to Zeno's observation, reported by Aristotle:

If that which is, were not one and indivisible, but could be divided into a multiplicity of entities, nothing would be truly one, for once it is admitted that the continuum can be divided, the divisibility should go on to infinity; on the other hand, if nothing is truly one, neither is it multiple, if it is true that multiplicity is constituted by many unities. It follows then that nothing would exist, unless something which is one and indivisible existed.

It was no accident that Zeno had Leucippus as a student, and Leucippus had Democritus as a student, and that the latter was the founder of the atomic theory of antiquity. According to this theory, atoms, as the etymology indicates, are indivisible and, together with the vacuum in which they move, constitute the only reality. All the sensible qualities of bodies are mere appearances and derive from different arrangements of the atoms.[8]

For more than two millennia the atomic hypothesis and the hypothesis of continuous matter confronted one another in various forms, each strengthened by the conceptual difficulties of the other. Kant even sanctioned the inevitability of the conflict, making it the second antinomy of reason. The *thesis* is: "Every substance in the world consists of simple parts, and there does not exist in any place anything which is not either simple or composed of simple parts." The *antithesis* is: "No compound thing in the world consists of simple parts and there exists nothing simple in the world, in no place whatsoever." According to Kant the antinomy rests on the fact that both of these assertions are demonstrable.

But by the early nineteenth century precise experimental results began to strengthen the atomic hypothesis. The discovery of multiple proportions in chemistry, which reintroduced whole numbers, brought J. Dalton in 1808 to propose an explanation based precisely on the existence of atoms. Then in 1811 came the studies of A. Avogadro and, from the second half of the century onward, the series of successes of the kinetic theory and of statistical mechanics. Yet all this was not enough to alleviate all mistrust and opposition. Poor Boltzmann had a hard life, while scholars such as W. Ostwald and E. Mach remained tenaciously attached to their antiatomic choice.[9] What people probably required was to *see* the isolated atoms before believing in their existence. But this was impossible with the means of that time. Only many decades later the electron microscope enabled people to "see" atoms. However, not everyone demanded such trivial proof. Poincaré, for example, said that atoms exist because they can be counted (again, whole numbers!).

In any case, after the account of Brownian motion given by Einstein and the experiments of Perrin, practically all the scientific community began to be convinced of the existence of atoms and molecules.[10]

But was Kant wrong, then? Was the thesis of the antinomy true and the antithesis false? I believe one must be very cautious in affirming so. In reality, the poison of the antinomy continued to work, so much so that once the existence of the atom was accepted, scientists began to ask how it was made. And once the constituents of the atom were discovered, one began to ask how they in turn were made!

If one accepts classical physics, there is not much hope of pulling out of this regression into infinity. But the world of microphysics has shown itself to be rebellious toward every attempt to enclose it in the laws that are valid in the macroscopic world. It has been necessary to work out a new theory, *quantum mechanics*, whose principles we shall put forth shortly. It is very different from classical physics and is much less intuitive. But it has shown itself extraordinarily adequate in dealing with the phenomena of microphysics.

It would be hazardous to affirm that quantum mechanics has resolved all the difficulties inherent to the nomological objects of physics. But it has shifted them on to other planes, into more profound zones. In regard to the continuum–discontinuum opposition, one could perhaps say that quantum mechanics has accomplished a *dialectic* operation, realizing a synthesis of opposites. But these are things that we shall see later on.

4.2. Spectral lines

We shall now discuss the structure of the atom. The history of this problem, from the last decade of the nineteenth century to the first decades of the twentieth, is rich and complicated and, of course, it is not linear. The characters and ideas that comprise the scene are many. The historian's job is vast.

But those, like us, whose first aim is not historical reconstruction, may choose from two approaches. We could simply limit ourselves to describing the structure of the atom as we know it today without mentioning the gradual development of the ideas and the experimental data that have conditioned them (this is not a dishonest method, although it lacks cultural content). Or we could briefly trace the history of the subject, necessarily accepting many simplifications and linearizations. In my opinion, the second alternative is the best, provided we pay attention to its hidden dangers. To be exact, we risk propagating and enlarging mistakes, as the authors of many treatises and elementary texts have done. The first writer simplifies and linearizes history, the second simplifies the account of the first, and so on, until we arrive at involuntary, but not harmless, falsifications. It is therefore advisable, whenever possible, to refer to the original texts or to works of history (see, e.g., Bellone, 1973, part III). With this advice in mind, we shall recall just a few conceptually important points.

As soon as scientists were sure of the existence of atoms, they began to wonder about their structure. This attitude corresponded, as we have already seen, to a kind of intellectual need by which one did not accept the continuum or the discrete model as definitive. But this was not all. Many signs deriving from the empirical world suggested that atoms were not objects that corresponded completely to their etymology, but that they were compound systems.

First, the periodic system[11] had led many scientists to surmise that the atoms of the various elements were all composed of the same raw material in different proportions. Second, there were the *spectral lines* of the light emitted by the atoms. Gases, when excited by a flame or by an electric discharge, emit characteristic electromagnetic radiation. The emission of monoatomic gases, which consists of distinct systems of monochromatic radiations, is especially simple. Each monochromatic radiation has a well-determined wavelength (or frequency), and appears in the spectroscope as a line. Each element has a characteristic system of lines that helps us to recognize it. This was determined in 1859 by G. R. Kirchhoff and R. W. Bunsen, who inaugurated *spectroscopic analysis*, and thereby discovered new elements. It was very difficult to understand how an atom, not divisible or composite, could produce such a phenomenon.

More impressive yet is Balmer's discovery, in 1882, that the frequencies of the lines emitted by the hydrogen atom obey a very simple formula, which today is usually written in general form as

$$v = R \left(\frac{1}{n_1{}^2} - \frac{1}{n_2{}^2} \right) \tag{4.1}$$

where R is a constant (Rydberg's constant) and n_1 and n_2 are whole numbers. Thus reappears the whole number, in a phenomenon that presumably occurs inside the atom. More generally, W. Ritz announced in 1908 that for each element there exists an (infinite) characteristic succession of terms $T_1, T_2, T_3, \ldots$, such that the frequency of each spectral line is determined by the difference $T_m - T_n$ of two terms (the *combination principle* of spectral lines).

4.3. Electrons

Meanwhile, some exceptionally valuable information concerning a nomological object emerged from the study of the electric conductivity of gases.

The electric conductivity of solids had been studied in the early nineteenth century and had led to Ohm's law in 1825. Faraday in 1834 had discovered some fundamental laws that govern the conductivity of liquids and which were to lead in 1887 to the theory of electrolytic dissociation of S. Arrhenius. Whole numbers reappeared here, too, and everything led scientists to think that the carriers of electric charges were discrete.

Complex phenomena occur in gases. When one passes a current through a glass tube containing a rarefied gas, vivid colors appear, which depend on the type of gas and on the pressure (gas lamps). If one constantly diminishes the pressure, the tube gives off a luminescence (fluorescence) clearly produced by special rays which originate from the negative electrode or *cathode*, and excite the opposite wall of the tube. The discovery of these *cathode rays* was made by J. Plücker in 1859. W. Crookes showed in 1879 that the cathode rays were deflected by a magnetic field, as negatively charged electric particles would be. But the particle hypothesis of cathode rays was not accepted

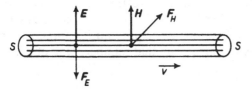

Figure 4.1

immediately, and found opposition particularly in Germany where scientists maintained that cathode rays were not rays but waves.

The matter was complicated by the fact that W. C. Röentgen discovered X rays, in 1895. These are generated when sufficiently energetic cathode rays strike a target, which today is called the *anticathode*. Because X rays are not deflected by a magnet, they lead one to think of the propagation of electromagnetic waves. However, the question remained controversial for a long time, until, in 1912, M. von Laue proposed that diffraction phenomena could be obtained by passing a beam of X rays through a crystal. The experiment yielded positive results, demonstrating that cathode rays were in fact waves of very small wavelengths, on the order of interatomic distances.

The nature of cathode rays was clarified by J. J. Thomson[12] in 1897. He submitted the rays to the combined action of an electric field E and a magnetic field H. Let us suppose that E and H are parallel (e.g., vertical) and perpendicular to the direction of propagation of the rays (Fig. 4.1). If the rays consist of (all equal) particles of electric charge e and mass m, each particle, in the presence of the electric field, will be subjected to a force

$$F_E = eE \tag{4.2}$$

parallel to E and, consequently, to an acceleration

$$a_E = \frac{eE}{m} \tag{4.3}$$

parallel to E. The direction of a_E is opposite to that of E because e is negative. On the other hand, if S is the beam cross section and v the velocity of the particle, the particles that pass in a second occupy a cylinder with base S and height v, therefore volume $V = Sv$. If n is the number of particles per unit volume contained in the beam, the number of particles that pass in a second will be nSv and their overall charge $enSv$. This is the charge that passes in a second, that is, the intensity of current I. Therefore we write $I = enSv$. Now, because of equation (2.31), a beam element with length l will be subjected to a force perpendicular to H and to the direction of the beam, given by $E = enSvHl/c$. On the other hand, the number of particles contained in that beam element will evidently be nSl and the force F_H on each particle is obtained by dividing F by that number. One obtains

$$F_H = \frac{1}{c} evH \qquad (4.4)$$

and, correspondingly, an acceleration

$$a_H = \frac{1}{c} \frac{evH}{m} \qquad (4.5)$$

Measuring the ratio between the horizontal and vertical deflections of the beam, or between the vertical and horizontal displacements of the fluorescent spot on the opposite wall, one can obtain the value of the ratio a_H/a_E. On the other hand, because of the preceding equations, this ratio is equal to vH/cE and because c, H, and E are known, one obtains the velocity v of the particle. These velocities are generally quite high (appreciable fractions of the speed of light).

Knowing v, one can calculate the time t that the particles take to get to the fluorescent wall of the tube. Knowing t, and measuring the vertical displacement, one can derive the vertical acceleration a_E that the particles underwent in the zone in which they were subjected to the electric field. Equation (4.3) then allows one to determine the ratio e/m. These elements were in fact deduced by J. J. Thomson. Thomson's bodies were called *electrons*.[13]

The last act of this fascinating story was represented by a famous experiment of R. A. Millikan, who in 1912 was able to directly measure the charge e. Then from e/m the value of m could also be determined. Today we have the values $e = (4.80325 \pm 0.00002) \times 10^{-10}$ e.s.u. and $m = (9.10956 \pm 0.00005) \times 10^{-28}$ g. The result is that the electron is about 1837 times lighter than the hydrogen atom.

4.4. Classical models of the atom

What role do electrons have in the composition of matter? It was immediately evident that the role was an important one. H. A. Lorentz developed an *electronic theory* of matter that, by assuming that such particles existed in every material body and that they interacted with electromagnetic fields, widened the scope of Maxwell's theory. In this way, one could deal with the conductivity of metals (in which electrons are relatively free to move and constitute the current), the polarization of dielectrics (in which each electron is elastically bound to a position of equilibrium), refraction, dispersion, the emission of light (due to the oscillation of electrons), absorption, and so on.

In 1896 P. Zeeman had discovered that spectral lines were influenced by a magnetic field acting on the source. To be exact, there is a division of every line in a number of lines with different polarizations. The simplest form of this phenomenon (the *normal* Zeeman effect) is in perfect agreement with Lorentz's theory. It even yields a value for the ratio e/m, equal to that found by Thomson.

It was becoming clearer and clearer that electrons play an essential role in the constitution of the atom. But how?

We should mention another important character that had come into play: *radioactivity*. This phenomenon was discovered rather by accident by H. Becquerel, who noticed that uranium salts were capable of making an impression on a photographic plate, even through black paper. This was a new type of highly penetrating radiation that was emitted by certain substances, apparently in a way that was absolutely independent of external physical conditions. The principal substance of this type, *radium*, was discovered and isolated by Marie and Pierre Curie.

The properties of the new radiation were soon clarified, largely through the efforts of E. Rutherford. There are three kinds of radiation: β rays, identical to cathode rays, that is, composed of electrons; α rays, composed of particles thousands of times heavier than electrons and positively charged; and finally, γ rays, analogous to X rays, that is, composed of electromagnetic waves of extremely short wavelengths ($\lambda \simeq 10^{-11}$ cm). In 1908 Rutherford demonstrated that α particles are none other than helium atoms that have lost two electrons (and are therefore doubly *ionized*). Rutherford and F. Soddy are credited for having recognized and clearly described the fact that when a radioactive atom emits an α or β particle, it disintegrates and decays into an atom of another element. To be exact, when it emits an electron it jumps one place ahead in the periodic system; its atomic Z increases by one unit, whereas its atomic weight remains approximately the same. When it emits an α particle, it jumps two places back in the periodic system; its Z diminishes by two units, whereas its atomic weight diminishes by four units. It is clear from these laws that *isotopes* must exist, that is, atoms that have the same Z (and therefore occupy the same place in the periodic system and have identical chemical properties), although they have different values for A.

We shall return to the interpretation of these facts when we discuss the atomic nucleus. Here we want to stress the fact that the radioactive decay of atoms occurs with a constant but absolutely *casual* rhythm.

Given a certain quantity of radioactive substance, a certain number of atoms (always the same fraction of the total number) decay in every unit of time. But there is no way to predict which atoms will decay and which will remain unchanged.

Everyone was convinced that atoms contained electrons and that they had a role of fundamental importance in atomic structure. But, in general, because atoms are electrically neutral, there must exist some positive charge that balances the electrons. Here the imagination was free to wander. For example, it was hypothesized that positive electrons existed and that the atom was made up of an equal number (several thousands) of positive and negative electrons. But this idea encountered several difficulties. For instance, it can be demonstrated that a system of discrete charges obeying Coulomb's law cannot be stable; either it dissipates or it collapses into one point (Earnshaw's theorem).

J. J. Thomson imagined that the atom was made up of a diffuse cloud of positive charge, in which the electrons were immersed (the system was irreverently called the "plum pudding"). The electrons order themselves on various rings and circulate inside the positive cloud. With this model, Thomson was able to obtain several partial successes. Moreover, by interpreting the results of X ray diffusion experiments, he showed that the number of electrons must be on the order of A (and therefore not on the order of thousands). Actually, this number is equal to Z, as was definitely established by N. Bohr.

A different conception, in which the atom is analogous to a solar system, the positive charge representing the sun and the electrons the planets, had been proposed several times (e.g., by H. Nagaoka). But as a matter of fact, this model did not present fewer difficulties than Thomson's and there seemed to be no reason to favor it.

However, this conception was taken up again successfully by E. Rutherford in 1911. The reasons that led him to adopt it derived mainly from the results of experiments of α particle scattering from atoms. Important experiments had been conducted by Geiger and Marsden in 1908. The α particles emitted by a radioactive substance were shot against a material target and their deflection was studied. Thus it was observed that a certain number of particles (about one in 8,000) were scattered at very large angles, and that some were even reflected back. Considering the mass of α particles, many times greater than that of electrons, this result could be interpreted only with great difficulty, using Thomson's model. It led instead to the supposition that a large part of the mass was concentrated in a small volume of the atom.

To understand this, we can make an analogy. Let us consider a bale of uncompressed hay in which the mass of say 10 kg is distributed in

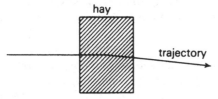

Figure 4.2

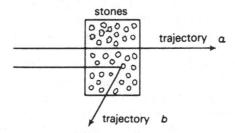

Figure 4.3

a large volume (Fig. 4.2). If one shoots a gun at it, the bullet will slow down somewhat and will generally deviate slightly from its path, because of the dishomogeneity it encounters in the hay. If we now remove the hay and introduce the same total mass, but concentrated in a few small volumes, like stones (Fig. 4.3), things will occur differently. Many bullets will be practically undisturbed and will follow a path like trajectory *a*. But some will strike a stone and follow a highly deviant trajectory, like *b*.

After making the precise calculations, Rutherford established that the results were consistent with the nuclear model of the atom. And this nuclear model is still valid today.

The mass of the atom is almost all concentrated in a *nucleus*, which is very small with respect to the atom as a whole. Its size is on the order of 10^{-13} cm, whereas that of the atom is on the order of 10^{-8} cm, that is, one hundred thousand times greater. The hydrogen nucleus is the smallest. Its mass today is known to be $(1.672614 \pm 0.000012) \times 10^{-24}$ g, which is 1,836 times greater than that of the electron. A nucleus of atomic number Z has a positive electric charge equal, in absolute value, to that of Z electrons. Around the nucleus, distributed throughout the volume of the atom, is a cloud of Z electrons, so that the whole turns out to be electrically neutral. Because the size of the electron is on the order of that of the nucleus, one is led to conclude that the atom is largely empty (emptier, in fact than the solar system!).

Radioactive properties originate in the nucleus, which disintegrates and transmutes from one species to another, whereas chemical and spectroscopic properties depend on the Z electrons that circle the nucleus.

Naturally, an atom made in this way cannot be static, for it would be unstable. So one must think of a dynamic equilibrium. The electrons circle the nucleus, and centrifugal force, which (together with the mutually repulsive force) tends to make the electrons escape toward the outside, is balanced by the Coulombian attractive force of the nucleus. Because Coulomb's force, like Newton's force of gravitation, is inversely proportional to the square of the distance, it appears that as a first approximation, electrons should move like the planets around the sun.[14]

Unfortunately, even this model comes up against difficulties which cannot be overcome in the sphere of classical physics. An electron circling in an orbit is equivalent to an oscillating current,[15] that is, to an antenna that radiates an electromagnetic wave, and therefore emits energy. In this way, the electron must end up losing all its energy and falling into the nucleus. It is true that in the case of many electrons that move in a suitable way, the total energy emitted can be extremely small, for the effect of one can neutralize the effect of the other, as was shown by J. J. Larmor. But in the case of hydrogen, which has only one electron, this way out does not work. One must think of something else, as N. Bohr did.

4.5. Planck's quanta

At this point we must go back for a moment to thermodynamics. As we have seen, this science, which prescribes very general laws for the physical world, had been satisfactorily translated into statistical mechanics, at least as far as material bodies are concerned. But for electromagnetic radiation, which also had to enter into the general picture, it was insufficient.

Every material body, when struck by electromagnetic radiation, absorbs more or less of this radiation, in proportion to what is called its *absorption coefficient a*. This coefficient (between 0 and 1) indicates what fraction of the incident energy is absorbed by the surface of the body. Furthermore, every material body, when heated, emits electromagnetic radiation (thermal radiation) in proportion to its *emissive power e*. G. R. Kirchhoff had established in 1859 that the ratio e/a does not depend on the nature of the body, but is a universal function of the temperature T and of the frequency v of the radiation. This is fairly easily shown, because if it were not the case, the second law of thermodynamics would be violated. Therefore we have

$$\frac{e}{a} = u(v, T) \tag{4.6}$$

where $u(v, T)$ indicates a universal function.

In simple words, we can express Kirchhoff's law by saying that a body is capable of absorbing the same kind of radiation as it is capable of giving off. If, for example, it has a conspicuous emissive power for a given frequency, as in the case of a spectral line, the body will also be highly absorbent for that radiation. In this way we can explain the dark lines in the spectrum of the sun, which were discovered by Fraunhofer in 1811. The radiations generated in the hottest and innermost layers of the solar surface are partially absorbed by the colder, more external layers. The elements found in these layers absorb particularly those lines that they are capable of emitting. This is why the spectrum exhibits dark lines, exactly in the positions in which these elements normally emit bright lines. In this way we can pinpoint the elements contained in the sun and in the stars.

Following the nomenclature introduced by G. Kirchhoff, a *blackbody* is a body capable of totally absorbing *all* the radiation that strikes it, that is, a body for which $a = 1$, at all temperatures and for all frequencies. From equation (4.6) there follows

$$e = u(v, T) \tag{4.7}$$

Thus the universal function u represents none other than the emissive power of a blackbody.[16]

We can measure $u(v, T)$ experimentally. To do so, we must heat a blackbody. A rather rough approximation to a blackbody can be a body whose surface is covered with lampblack. But we can do much better with a material body M (Fig. 4.4), inside of which is a cavity C, communicating with the outside only through a small hole H. A generic ray r that penetrates through H will bounce many times from one wall of the cavity to another and will virtually be absorbed before finding its way out through H. Therefore we can say that every ray that strikes the surface of the hole is absorbed. But then, by definition, the surface of the hole behaves like the surface of a blackbody.[17] Then to obtain $u(v, T)$ experimentally, one has only to heat the body M to a temperature T (uniform for the whole body) and measure the radiation that emerges from H, at different frequencies. One obtains a diagram such as that of Figure 4.5, in which u is represented[18] as a function of frequency for a few increasing values of temperature $T_1, T_2, T_3, \ldots$. In any case u starts from zero at frequency $v = 0$, then grows, reaches

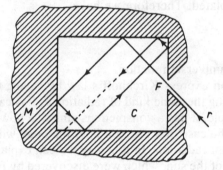

Figure 4.4

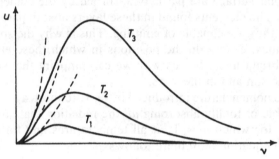

Figure 4.5

a maximum, and decreases rapidly, tending toward zero for high frequencies. The maximum emission shifts gradually toward high frequencies, as temperature increases.[19] This corresponds to the known phenomenon that a cold body emits almost no visible radiation (it emits radiation at lower frequencies); then when heated it begins to glow a dark red, and then grows lighter and lighter in color (white heat).

The curves of Figure 4.5 represented a true enigma for those who wanted to interpret them according to classical physics.

For example, one can reason as follows. The emission from the hole at each frequency is proportional to the density of the electromagnetic energy contained in the cavity C at that frequency. The electromagnetic radiation inside C is in thermodynamic equilibrium with the walls, and we can say that it is at the temperature T of the walls. Naturally, the radiation inside the cavity is chaotic and has the maximum entropy possible at the temperature T. The case is perfectly analogous (except that there are now three dimensions instead of two) to that of the water in the pond discussed in the preceding chapter (§3.16). One may think of the chaotic radiation as a superimposition of many ordered wave trains, of different amplitudes, frequencies, and directions of propagation and polarization. Each of these wave trains may be thought of as a degree of freedom of the radiation. Mathematical analysis shows that a discrete series of these waves is enough to reproduce whatever distribution of electromagnetic radiation is found inside the cavity. To be exact, if we use V to indicate the volume of the cavity, the number g_v of the degrees of freedom whose frequency is between v and $v + 1$ turns out to be[20]

$$g_v = \frac{8\pi}{c^3} V v^2 \tag{4.8}$$

If we assume that the classical equipartition of energy (in which each degree of freedom has on the average evergy kT) is valid, we obtain a total energy given by $g_v kT$ and a density of energy (i.e., energy per unit volume) given by

$$u = \frac{8\pi}{c^3} kT v^2 \tag{4.9}$$

As mentioned, the emission of radiation from the hole H is proportional to this density of energy. Equation (4.9) represents the classical Rayleigh–Jeans formula. It gives an emission that increases as a function of v^2, as is represented by the broken lines in Figure 4.5, for various temperatures. The law works very well at low frequencies but is definitely incorrect at high frequencies. Besides, it would lead to the absurdity that the emission tends toward infinity as v increases. P. Ehrenfest called this the *ultraviolet catastrophe*.

In 1900 M. Planck made a fundamental discovery.[21] If one assumes that exchanges of electromagnetic energy can take place only for finite *quanta* whose values are proportional to the frequency, one obtains a formula that perfectly explains the experimental results of Figure 4.5.

These quanta, or granules of energy, are represented by

$$E = hv \tag{4.10}$$

where h is a universal constant called *Planck's constant*. Today we know its value to be $h = (6.626196 \pm 0.000050) \times 10^{-27}$ when E is measured in *ergs* and v in vibrations per second, or *hertz*. We shall soon see a derivation of Planck's formula.

4.6. Photons

Now let us go ahead to 1905, when A. Einstein interpreted the *photoelectric effect* by means of quanta. The photoelectric effect, which had been discovered some years before, reveals that when a beam of light strikes the surface of some metals (e.g., the alkaline metals), some electrons are knocked off, leaving the surface at a certain velocity. Experiments had shown that contrary to one's expectations, the speed of the electrons did not depend on the intensity of the light, but on its frequency. To be exact, a *photoelectric threshold*, or a frequency v_0 below which the effect did not occur, existed for each metal. Above v_0, the speed of the electrons emitted increased with frequency.

Taking a heuristic point of view, Einstein hypothesized that quanta not only appeared in energy exchange phenomena (emission and absorption), but represented actual granules of which light is made, as though they were minute particles. These granules were later called *photons*.

In order to explain the law of photoelectric effect, let us suppose that each electron is tied to the structure of metal in such a way that an energy A must be expended in order to free it. Therefore a photon of energy hv arriving on the surface of the metal, will be able to set free an electron only if $hv \geq A$. If its frequency is less than $v_0 = A/h$, the effect will not occur. For greater frequencies, an electron will be freed, and the energy $hv - A$ will still be available. This energy reappears as the kinetic energy E_k of the electron emitted. And so we write

$$E_k = hv - A \qquad (4.11)$$

This equation interprets the data of the photoelectric effect. Although Einstein had spoken solely of a "heuristic point of view," it was difficult to ignore the idea that photons really existed.

Photons are very peculiar particles. First of all, they travel at the speed of light. Furthermore, each photon has a fixed and nonvanishing energy, given by equation (4.10). Now if we recall equation (2.68), which gives the energy of a particle of mass m in the form $E = mc^2 / \sqrt{1 - v^2/c^2}$, we can also write

$$E \sqrt{1 - \frac{v^2}{c^2}} = mc^2 \qquad (4.12)$$

When v approaches c, the lefthand expression tends to zero. Therefore the right side must also tend to zero, hence we obtain $m = 0$. Thus photons have zero mass.

But particles of zero mass can have a nonzero momentum. This is given by equation (2.73), so that

$$p = \frac{hv}{c} \qquad (4.13)$$

will represent the momentum of a photon of frequency v.

If photons have momentum, they must exert a pressure when they strike a specular surface and bounce back, as molecules do in kinetic theory. Let us return to Figure 3.6, and imagine that a beam of light carrying N photons per second falls perpendicularly on S and that the light is perfectly reflected. The momentum transfered to S in time dt is given by

$$Q = N \, dt \, 2 \frac{h\nu}{c} \tag{4.14}$$

The pressure P exerted on S is tied to Q by equation (3.45), which we can write as

$$Q = PS \, dt \tag{4.15}$$

Comparing the last two equations, we get

$$P = 2 \frac{Nh\nu}{cS} \tag{4.16}$$

But $Nh\nu/S$ represents the light energy that arrives on a unit surface per unit time. We can call it the light *intensity*. Indicating it with I, we can write

$$P = 2 \frac{I}{c} \tag{4.17}$$

The fact that this pressure is actually exerted on the surface was verified by P. Lebedev in 1900.[22]

The final "proof" of the particle nature of electromagnetic radiation was given by A. H. Compton in 1923, by means of the scattering of X rays by electrons. According to the classical theory, an electromagnetic wave meeting a free electron at rest makes the electron oscillate at the wave frequency. The electron, behaving like the oscillating current in a small antenna, in turn radiates a wave of the same frequency. The *Compton effect* consists of the fact that the radiation scattered is of a lower frequency than the incident radiation.

This fact is easily explained, if we consider the process as a collision between two particles, the photon and the electron. Before the collision the photon has the energy $E = h\nu$ and the momentum $p = E/c$, whereas the electron at rest has zero energy and momentum. After the collision the electron has been put in motion and possesses a certain momentum and a certain energy, which necessarily have been taken from the photon. So the energy of the photon is decreased and, consequently, its frequency is also decreased. If one makes precise calculations, applying conservation of energy and of momentum, one obtains results that perfectly agree with the experiment.

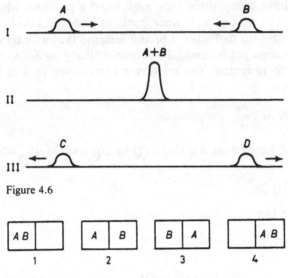

Figure 4.6

Figure 4.7

4.7. The Bose–Einstein statistics

Once we accept the existence of photons, it is natural to want to consider them from a statistical point of view, as in kinetic theory. However, we find ourselves confronted by a new fact.

In Boltzmann's kinetic theory, which we examined in Chapter 3, it is essential to be able to single out (at least conceptually) each molecule and therefore to be able to give a precise operational meaning to the exchange of two molecules.[23] But for photons it seems that the same cannot be said.

Photons are in some way associated with electromagnetic waves. This association is still quite mysterious, but it can lead us to make the following considerations. To begin with, let us suppose that we have a "solitary" wave A (Fig. 4.6) which propagates toward the right and an identical wave B which propagates toward the left (I). When they meet (II), they give rise to a superimposition A + B. Afterward they separate again and give rise to wave C that moves toward the left and wave D that moves toward the right. Which is which of these C and D waves?

Two different interpretations are possible:

1. C is wave B, which continues along its path after the superimposition and, correspondingly, D is wave A.

2. The superimposition represents a collision, and the waves bounce back, therefore C is A and D is B.

Wave theory allows each of these interpretations, and there is no way to decide between them. The reason is that waves A and B do not have an individuality that permits one to distinguish between them continuously.

The same is true for photons. There is no way to distinguish between them continuously, consequently, there is no sense in saying that they interchange.[24] Recalling our reasoning concerning classical statistical mechanics, it is clear that for a gas of photons, we must proceed differently. We shall find what is usually called the *Bose–Einstein* statistics.

In order to explain in a simple way what all this is about, let us proceed as follows. Consider two cells in which two particles A and B have been placed. Classical statistics distinguishes the four cases illustrated in Fig. 4.7. We allot the probability 1/4 to each of them. The new statistics distinguishes only three cases, for cases 2 and 3, obtained by exchanging the two particles, must be considered as one single case. Therefore the probability of each case will now be 1/3.

Let us go back and consider the cavity C of Figure 4.4.[25] We shall think of every plane wave (or degree of freedom) as a cell that can contain a certain number of photons, all of the same energy $h\nu$. Now equation (4.8) provides the number g_ν of the cells that correspond to a frequency between ν and $\nu + 1$ (i.e., that have virtually the same frequency, because 1 is negligible with respect to ν). Among these cells, there will be a certain number g_ν^0 that contain zero photons, a certain number g_ν^1 that contain one photon, a certain number g_ν^2 that contain two photons, and so on. Evidently, we have[26]

$$\sum_{n=0}^{\infty} g_\nu^n = g_\nu \qquad (4.18)$$

The energy contained in the g_ν^0 cells is evidently zero, whereas that contained in the g_ν^1 cells is $g_\nu^1 h\nu$, that contained in the g_ν^2 cells is $g_\nu^2 2h\nu$, and so on. Therefore the energy E_ν contained in the g_ν cells will be

$$E_\nu = \sum_{n=0}^{\infty} n g_\nu^n h\nu \qquad (4.19)$$

If E is the total energy contained in the cavity, we then have[27]

$$E = \sum_{\nu=0}^{\infty} E_\nu \qquad (4.20)$$

We consider this value as fixed (and in effect, it is, except for negligible fluctuations).

A microstate of the photon gas is assigned once we establish how many photons are found in each cell. Therefore we can form the same macrostate in several different ways. It is senseless to rearrange the photons, as we have noted, but we can interchange the cells, or their roles. Let us clarify this concept.

Let us suppose that among the cells of frequency v, cell A contains four photons and cell B contains seven. After exchange of the roles of the cells, cell A contains seven photons and cell B contains four.

The g_v cells will produce $g_v!$ permutations. But among these cells, g_v^0 have the same role (that of containing zero photons), g_v^1 have the same role (that of containing 1 photon), and so on. Thus as far as the frequency v is concerned, the same macrostate will be obtained in

$$N_v = \frac{g_v!}{g_v^0! g_v^1! g_v^2!, \ldots} \tag{4.21}$$

distinct ways.

Let us now suppose that all the microstates are equally probable. Then in order to find the most probable macrostate, it is necessary to render maximum the product

$$N = N_0 N_1 N_2 \ldots \tag{4.22}$$

of the numbers N_v given by equation (4.21), under the fixed conditions, equations (4.18) and (4.20).

This is a problem of conditional maximum, analogous to that encountered in regard to equation (3.71). With the methods of differential calculus already mentioned, one finds a result analogous to that of equation (3.79); precisely,

$$g_v^n = \alpha_v \exp(-\beta n h v) \tag{4.23}$$

α_v and β being constants.

To agree with classical statistics, in the limit in which it can be applied, we shall see that we must make $\beta = 1/kT$. So using equation (4.18), we have

$$\alpha_v \sum_{n=0}^{\infty} \exp\left(-n \frac{hv}{kT}\right) = g_v \tag{4.24}$$

The sum represents a *geometric* series,[28] and one immediately obtains

$$\alpha_v \frac{1}{1 - \exp(-hv/kT)} = g_v \tag{4.25}$$

from which, because g_v is given by equation (4.8), one obtains the constant α_ν,

$$\alpha_\nu = \frac{8\pi}{c^3} V\nu^2[1 - \exp(-h\nu/kT)] \tag{4.26}$$

With these results, equation (4.19) becomes

$$E_\nu = \frac{8\pi}{c^3} V\nu^2\left[1 - \exp\left(-\frac{h\nu}{kT}\right)\right]$$
$$\times h\nu \sum_{n=0}^{\infty} n \exp\left(-n\frac{h\nu}{kT}\right) \tag{4.27}$$

The sum is a series analogous to the geometric one, whose result we know.[29] One obtains

$$E_\nu = \frac{8\pi}{c^3} V\nu^2\left[1 - \exp\left(-\frac{h\nu}{kT}\right)\right]$$
$$\times h\nu \frac{\exp(-h\nu/kT)}{[1 - \exp(-h\nu/kT)]^2} \tag{4.28}$$

Making some easy simplifications and dividing by V, we finally obtain the density of energy u,

$$u(\nu, T) = \frac{8\pi}{c^3} \frac{h\nu^3}{\exp(h\nu/kT) - 1} \tag{4.29}$$

This is Planck's formula. It gives, for the distribution of energy in the spectrum of the blackbody, a behavior that agrees perfectly with experience, that is, with the curves of Figure 4.5. We can also demonstrate that when ν is very small, it gives results practically equal to those of the Rayleigh–Jeans equation (4.9). Naturally, the same result is obtained making h tend to zero, that is, going to the classical limit.

Dividing equation (4.29) by $h\nu$, we obtain a quantity that we call $f(E)$,

$$f(E) = \frac{8\pi}{c^3} \frac{\nu^2}{\exp(h\nu/kT) - 1} \tag{4.30}$$

This is evidently the number of photons per unit volume that have energy $E = h\nu$ and, in some way is the analog to the distribution function that appears in equation (3.85).

When we are sufficiently near to $\nu = 0$, the granularity is so fine that the radiation has practically the same properties of classical waves. The statistics is that of Rayleigh–Jeans. Instead, when ν increases and particle properties prevail, the exponential $\exp(h\nu/kT)$ becomes much larger than unity. Neglecting the latter in the denominator of equation (4.30), it results that $f(E)$ is proportional to[30] $\exp(-E/kT)$, in agreement with what we found with equation (3.85), for the classical statistics of molecules.

The discovery of Bose–Einstein statistics recalls the considerations we made in our discussion of the principle of indifference in probability. We stated that only experimental evidence can decide which are the equiprobable cases. In other words, any effort to choose a criterion a priori to enumerate equiprobable cases is only tentative, for the final choice is up to experience. The failure of classical attempts to interpret the blackbody spectrum can be intuitively explained by the ignorance that scientists had then of the existence of photons. But more precisely and conceptually, we can say that they did not know how to count the equiprobable cases in the right way. Experience proved that the usual way of enumerating was incorrect, and compelled scientists to choose another one.

4.8. Bohr's atom

We have seen how attempts to discover the composition of the atom, applying the ordinary laws of physics, had failed one after the other. The situation was cleared up by N. Bohr in 1913.

Bohr, who had spent some time in Thomson's laboratory, and then had moved to that of Rutherford, formed the conviction that the properties of the atom were not explainable by means of the ordinary laws of classical physics. Although daring, the idea was not fantastic, for Planck's quanta, which were absolutely extraneous to classical physics, already existed. Bohr actually had the idea to use quanta[31] and to apply them to Rutherford's atom.

In Rutherford's atom, electrons move around the nucleus as the planets move around the sun. But whereas the possible orbits of the planets form a continuous infinity, Bohr postulated that only some well-determined and discrete orbits were possible. Moreover, he postulated that as long as it remains in a fixed orbit, the electron does not emit electromagnetic energy.

The electron possesses a well-determined energy in each orbit. Consequently, a discrete succession of permitted energies, or of energy levels $E_1, E_2, E_2, \ldots$, exists for the atom. When the atom is on level E_m, it can jump to level E_n, emitting or absorbing the difference of energy (according to whether $E_m > E_n$ or $E_m < E_n$) in the form of a photon. Thus, for example, when the atom jumps from a higher level E_m to a lower one E_n, it emits a photon of frequency v, such that

$$E_m - E_n = hv \qquad (4.31)$$

This explains why spectral lines are well defined and the relative frequencies can be found as differences of terms, according to the principle of combination.

In order to be able to determine the several discrete orbits one must assign a condition of quantization. The quantization of the angular momentum of the electron in its movement of revolution around the nucleus can be postulated. Precisely, according to Bohr, the angular momentum K of the electron is always a multiple of $h/2\pi$, so we write

$$K = n\frac{h}{2\pi} \qquad (4.32)$$

n being an integer. This condition turns out to be of very general significance, as we shall see.

Let us now suppose that an electron of charge e moves in uniform circular motion around a nucleus of charge Ze and that the influence of the other electrons on it is negligible. If the electron is at a distance r from the nucleus, the mutual attraction, according to equation (2.26) will be

$$F = \frac{Ze^2}{r^2} \qquad (4.33)$$

On the other hand, if v is the velocity of the electron, the centripetal acceleration, equation (2.8), is v^2/r, and to it corresponds the centripetal force

$$F = \frac{mv^2}{r} \qquad (4.34)$$

Equations (4.33) and (4.34) must represent the same force. Thus we have

$$\frac{Ze^2}{r^2} = \frac{mv^2}{r} \qquad (4.35)$$

By definition, the angular momentum of the electron is given by mvr; therefore from equation (4.32) we obtain

$$mvr = n\frac{h}{2\pi} \qquad (4.36)$$

Squaring this, we have $m^2v^2r^2 = n^2h^2/4\pi^2$, from which we deduce

$$mv^2 = \frac{n^2h^2}{4\pi^2mr^2} \qquad (4.37)$$

Substituting this expression on the right side of equation (4.35), we obtain

$$\frac{Ze^2}{r^2} = \frac{n^2h^2}{4\pi^2mr^3} \tag{4.38}$$

from which we can immediately determine the value of r

$$r = \frac{n^2h^2}{4\pi^2Ze^2m} \tag{4.39}$$

We notice that this result gives us r in terms of known constants. Making $Z = 1$ (hydrogen) and $n = 1$, we obtain the Bohr radius, which equals 0.528×10^{-8} cm. The order of magnitude is correct; that is, it is what different experimental methods give for the radius of an atom.

Now we come to the energy. The electron possesses a kinetic energy $E_k = mv^2/2$, which, substituting the value, equation (4.39), for r in equation (4.37), yields

$$E_k = \frac{2\pi^2Z^2e^4m}{n^2h^2} \tag{4.40}$$

But the electron, besides kinetic energy, possesses a potential energy E_p, which, as we shall see presently, is given by

$$E_p = -\frac{Ze^2}{r} \tag{4.41}$$

In fact we recall that the force that acts on the electron is the Coulombian force, equation (4.33). If the electron passes from distance r to distance $r - dr$ (with dr very small) this force performs work $dL = F\,dr = (Ze^2/r^2)\,dr$. This work must be equal to the decrease in potential energy, $E_p(r) - E_p(r - dr)$, which according to equation (4.41) is given by $-Ze^2/r - [-Ze^2/(r - dr)]$. Therefore we must have

$$\frac{Ze^2}{r^2}\,dr = \frac{Ze^2}{r - dr} - \frac{Ze^2}{r} \tag{4.42}$$

If on the right side of the equation, we multiply the numerator and denominator of the first fraction by r and those of the second fraction by $r - dr$, we obtain two fractions with the same denominator, which we can add immediately to get

$$\frac{Ze^2}{r^2}\,dr = \frac{Ze^2dr}{r(r - dr)} \tag{4.43}$$

Neglecting dr in respect to r in the denominator (a relative error that we can make as small as we like), we obtain an identity. Thus the choice, equation (4.41), is the correct one.[32] Substituting the value of

r from equation (4.39) in equation (4.41) yields

$$E_p = - \frac{4\pi^2 Z^2 e^4 m}{n^2 h^2} \tag{4.44}$$

Therefore the total energy $E_k + E_p$ on the energy level characterized by the quantum number n turns out to be

$$E_n = - \frac{2\pi^2 Z^2 e^4 m}{n^2 h^2} \tag{4.45}$$

Applying equation (4.31) with this value of E_n, one finds that the frequency v emitted when the atom passes from level E_{n_1} to level E_{n_2} is

$$r = R\left(\frac{1}{n_1^2} - \frac{1}{n_2^2}\right) \tag{4.46}$$

the constant R being given by

$$R = \frac{2\pi^2 Z^2 e^4 m}{h^3} \tag{4.47}$$

But equation (4.46) is none other than Balmer's formula, equation (4.1). If we make $Z = 1$ and calculate the constant R using equation (4.47), one obtains frequencies that agree very well with those found experimentally for the series of lines of hydrogen.

But things work well also if Z is greater than 1, that is, for atoms heavier than hydrogen, provided they are ionized and are left with only one electron to rotate around the nucleus. Things work out even better if we bear in mind that the nucleus, instead of being fixed as we have supposed for simplicity, also moves around the center of gravity of the system.

From this and other successes[33] of Bohr's treatment one knew that something basically correct had been found. But a complete and coherent theory that would precisely account for all atomic phenomena was still a long way off.

4.9. Waves and particles

Electromagnetic radiation presents some very strange properties. On one hand, it behaves as though it were made up of waves, because it produces interference and diffraction phenomena. On the other hand, it behaves as though it were a swarm of photons, that is, of particles. Therefore we ask: Are we dealing with waves or with particles?

There is no doubt that we must accept both aspects as substantially indispensable. Thus there is a kind of *duality* in these processes (see

§4.16). The duality manifests itself in those physical entities that are classically described as waves. It can be natural, then, to wonder if, by symmetry, an analogous duality also exists for those physical entities that are classically described as particles. L. de Broglie did this in 1924,[34] starting from ingenious relativistic considerations. We shall limit ourselves here to a simple analogy which suggests the plausibility of the conception.

Let us recall equation (2.37) $\lambda = c/\nu$, which ties wavelength to the speed of light and to frequency. Substituting in equation (4.13), we see that the momentum of a photom is tied to its wavelength by the relation $p = h/\lambda$. The same relation was postulated by de Broglie for material particles, assuming that a wavelength given by

$$\lambda = \frac{h}{p} \tag{4.48}$$

is associated with each particle of momentum p.

To verify if material particles really have wave properties, one must be able to make them produce effects that are characteristic of waves – interference or diffraction. This is not easy because in general, the wavelength given by equation (4.48) is very small. For example, one can calculate that for an electron that moves at a speed on the order of $c/100$, λ becomes on the order of 10^{-8} cm, that is, on the order of the size of an atom or of the interatomic distances in a solid body.[35] One cannot hope to obtain conspicuous diffraction phenomena with ordinary diaphragms and slits such as those used to produce diffraction of light.

But by recalling Laue's X ray diffraction experiment, one can think of diffracting a beam of electrons with a crystal. Such experiments had already been executed by C. J. Davisson, but had not been interpreted correctly. Later, J. Franck and W. Elsasser showed that these experiments confirmed the wave nature of electrons. Repeated in various forms with greater precision, the experiments confirmed perfectly de Broglie's conception and formula, equation (4.48).

Starting from de Broglie's discovery, one can arrive at a very general formulation of the behavior of material particles. Elastic and electromagnetic waves obey a very general differential equation called the *wave equation*. Naturally one wonders if it is possible to write a differential equation governing the propagation of microscopic particles in the same way as the differential wave equation governs true waves.

This program was in fact carried out by E. Schrödinger (1926), who established a differential equation of fundamental importance for elementary particles, which bears his name. This is substantially a wave equation. Thus *wave mechanics* was born.

To understand the meaning of Schrödinger's equation, we note that if we consider a particle fired into free space at a certain initial speed, according to the law of inertia, its momentum will remain constant during the movement. Thus there is very little to add to the result established by de Broglie. If the particle is instead subjected to force (e.g., an electron in the atom), the momentum will generally vary and, consequently, the associated wavelength will, too. To obtain the value of the wavelength at each point, one can proceed as follows.

The total energy E of the particle is composed of the sum of the kinetic energy E_k and the potential energy E_p, and it remains constant during the movement (it is a constant of the movement, as we say). Then we have $E - E_p = E_k = mv^2/2 = p^2/2m$,[36] from which $p = \sqrt{2m(E - E_p)}$. Finally, from equation (4.48) there results

$$\lambda = \frac{h}{\sqrt{2m(E - E_p)}} \qquad (4.49)$$

Generally, the potential energy E_p is given as a function of the position of the particle. For example, in the case of the electron in the hydrogen atom, we have from equation (4.41) $E_p = -e^2/r$. Then for each r we have the corresponding value of p, and therefore of λ. Having this knowledge of the wavelength at the points at which the particle can propagate, we can now write a wave equation. In this way de Broglie's relation is able to enter into an equation in which total energy appears as a constant, and in which potential energy is determined by the kind of forces to which the particle is subjected.

As Schrödinger himself observed, the ordinary mechanics of the material point is to wave mechanics as ray optics, or as geometrical optics is to wave optics. The classical trajectories of particles must be assimilated to the trajectories of rays. These have a very well-defined physical meaning, as long as the "structure" of the trajectories themselves (radius of curvature) does not become so small as to be comparable to the wavelength. In this case, which actually corresponds to the situation holding for the electron paths inside the atom, the wave properties prevail, as in the case of diffraction.

Schrödinger's equation, as soon as it was applied to the hydrogen atom, gave very satisfactory results. Physicists immediately had the impression that Schrödinger had put his hands on something very important. This turned out to be so true that Schrödinger's equation has lasted through the years, and represents even now the perfectly reliable basis for many spectroscopic calculations.

Let us now see how Schrödinger's equation accounts for Bohr's postulations. To begin with, let us recall what happens when one makes a stretched rope vibrate so that a wave propagates along it (Fig. 4.8).

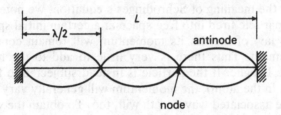

Figure 4.8

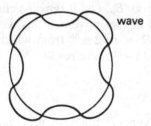

Figure 4.9

When the wave reaches a fixed end which cannot oscillate, a reflection occurs and the wave returns. The phase of the reflected wave
is always in opposition to that of the incident wave. Therefore the
pattern resulting from the presence of the two waves traveling in opposite directions along the rope is that shown in Figure 4.8 (*stationary
wave*). The distance between two *nodes* or two maxima is $\lambda/2$. It follows
that between the fixed ends of the rope there must be a whole number
of $\lambda/2$, because the ends must always be nodes. Not all wavelengths
can yield stationary waves, only those for which $L = n(\lambda/2)$ (n being
an integer), where L is the length of the rope.

More generally, when one confines waves in a finite, closed system,
they can only assume a few special configurations with well-determined
wavelengths. One immediately thinks of Bohr's quantum conditions
as due to this phenomenon. Let us consider, for example, an electron
in a circular orbit. If we think of it as a wave, we must imagine a
progressive wave on the circle like that shown in Fig. 4.9. If the length
of the circle is not a whole multiple of λ, the phase with which the
wave reappears at a given point after one revolution, varies with each
revolution. Thus to obtain a uniquely defined wave that propagates
along the circle, one must have the wave reappear every time in the
same way, or for the length of the circle to be a whole multiple of λ.

In this case we have $2\pi r = n\lambda$ or $2\pi r = n(h/p)$ and therefore $mvr
= nh/2\pi$, which is Bohr's condition, equation (4.36), for stationary

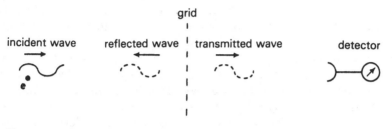

Figure 4.10

orbits. Thus we have a physical meaning for the quantum condition postulated by Bohr. The electron, because of its wave nature, when confined in a finite region of space, can be localized only in certain particular configurations and can oscillate only with certain discrete frequencies. These configurations are called *orbitals*. Orbitals substitute what Bohr thought of as orbit in the classical sense. The discrete frequencies, by way of the relation $E = h\nu$, become discrete energies, that is, a succession of energy levels.

Schrödinger's equation can be successfully applied to atoms more complex than H and to many other problems of microphysics. We are left with the ever more dramatic question: waves or particles?

4.10. The probabilistic interpretation

After having seen the successes of Schrödinger's equation in atomic mechanics, let us now ask: What is the quantity that vibrates in the case of Schrödinger waves, or of the waves associated with the wave aspect of material particles?

In his celebrated thesis of 1924, L. de Broglie had purposely limited himself to discussing "the periodic phenomenon" generically, realizing that the theory had to be considered as "a form whose physical content is not entirely specified." E. Schrödinger had indicated the oscillating quantity with the letter ψ, and this symbol is universally accepted. As for its physical nature, Schrödinger thought it to be a material wave or, more precisely, the distribution of an electron charge. But this position turned out to be completely untenable. In fact, it led to strange conclusions, disproved by the evidence.

Let us think, for example, of an electron meeting a potential barrier in the form of a grid (Fig. 4.10) to which a negative tension is applied. If its kinetic energy is sufficiently high, the electron can overcome the repulsive force of the barrier, pass through the grid and be detected by an instrument. If instead the speed of the electron is low, it will bounce back and never reach the detector. If ψ is calculated for this particular problem, one finds that part of the wave is reflected by the barrier, whereas another part is transmitted. Thus the instrument should reveal the fraction of electron that is associated with the wave transmitted. But experiment shows that the detector records either an entire electron or does not record it at all. Fractions of electrons have never been detected. A diffused wave distribution cannot be found.

Photons had already presented an analogous case. There are no particular problems of interpretation, as long as one considers an electromagnetic wave that vibrates with a certain distribution of field. But when one realizes that energy must travel in clusters, and that these clusters can be absorbed or emitted only as wholes and *not* in fractions, it becomes apparent that attributing a possible physical significance to the process is very difficult.

Schrödinger's ψ thus constitutes a counterpart to the problem encountered with photons. In the case of photons one had an easily understandable wave that was "ruined" by the particle conception introduced with the photon. In the case of electrons one had a particle that is "ruined" by the wave conception introduced with ψ. Could these be two reconcilable aspects of the same reality?

To overcome this situation we must resort to a daring hypothesis, according to which ψ is a *distribution of probability*. This hypothesis is due to M. Born, who writes:

> Once more an idea of Einstein's gave the lead. He had sought to make the duality of particles (light quanta or photons) and waves comprehensible by interpreting the square of the optical wave amplitudes as probability density for the occurrence of photons. This idea could at once be extended to the ψ-function: $|\psi|^2$ must represent the probability density for electrons (or other particles). To assert this was easy; but how was it to be proved? (Born, 1956, p. 183).

According to this famous interpretation, the value of ψ^2 at a point or, better yet, of $|\psi|^2$, because ψ is generally complex,[37] represents the probability of finding the particle at that point. Schrödinger's equation thus provides a distribution of probability of localization of the particle.

The probabilistic significance of ψ must be understood in the relative frequency sense. If in solving Schrödinger's equation for an electron subjected to a certain force field, one finds, for example, that at a certain point $|\psi|^2 = 0.8$, the probability of finding the electron at that point is 0.8, that is, in 100 measurements performed, preparing the system in exactly the same way every time, about 80 times the electron will be actually localized at that point.

We must be even more precise. The probability of finding the electron at a mathematical point does not make much sense and we must instead proceed as follows. The solution of Schrödinger's equation, in a certain case of interest, gives ψ as a function of x, y, z, and t; thus sometimes we write $\psi(x, y, z, t)$. This quantity is called the *wave function*. The probability of finding the electron in a small volume dV around the point x, y, z, at the instant t is given by $|\psi(x, y, z, t)|^2 \, dV$.

Of course, if we know that there is an electron, the probability of finding it at any point in space is 1. We shall therefore impose on ψ for every t the condition of *normalization*

$$\int |\psi(x, y, z, t)|^2 \, dV = 1 \qquad (4.50)$$

which translates equation (3.56), with the integral extended to all space. This is always possible, because it can be demonstrated that Schrödinger's equation provides ψ only apart from a constant factor and that furthermore, if ψ is normalized for a certain t, it will remain normalized for every t.

Probability is not a new tool in physics. We have also seen it in statistical mechanics. But in that context it was introduced only because it is *practically* impossible for us to know and, subsequently, process mathematically, an enormous amount of data, such as those that determine a microstate. But no one had ever said that this operation was *theoretically* impossible. In the new conception instead it does not matter if we are ignorant of initial conditions. The probabilistic aspect is inherent to the nature of things.

The probabilistic interpretation of M. Born became a major point of contention among physicists. For those who totally agreed with this conception, physics was definitely losing its deterministic character, and the physicist could no longer hope to predict future events with certainty once he knew present events. He was permitted only to assess with what probability these events would occur.[38]

On the other side were the physicists who were more devoted to the deterministic conception, among them was Albert Einstein (with de Broglie, Planck and Schrödinger, just to name the most famous), who since the beginning, never wanted to accept Born's formulation, nor believe that "God plays with dice."

We shall return to this question. But it is important to note here that the probabilistic conception has never been "refuted" by experimental evidence and even today is the basis of the interpretation of all microphysics.

The probability used in microphysics is a very strange probability, with properties that are surprising, from the classical point of view. For example, let us consider again the device with the two holes in Figure 2.50, with which Young demonstrated the interference of light. This device works in exactly the same way if instead of the monochromatic wave of light, the source S emits a beam of electrons, all with the same speed and the same de Broglie wavelength. On the screen *HK* (which in this case we consider fluorescent such as that of a TV set) interference fringes are formed. Many electrons arrive on the bright fringes, whereas no electrons arrive on the dark ones.

According to Born, each electron has a great probability of going on to a bright fringe and practically no probability of going on to a dark fringe. Let us call ψ_1 the wave function of an electron that passes through the hole F_1 and ψ_2 the wave function of an electron that passes through the hole F_2. The probabilities that an electron will arrive at point P on the screen passing through F_1 or F_2 are respectively $p_1 = |\psi_1(P)|^2$ and $p_2 = |\psi_2(P)|^2$. Classically, the two events are mutually exclusive, so that for equation (3.55) the total probability p that an electron will arrive at P (passing through either F_1 or F_2) is given by $p = p_1 + p_2$. In this case brightness would vary very slowly across the screen, and we would not be able to see any fringe.

But the fringes are seen because the probabilities are not added directly. Rather, the ψ's are added, as[39] Schrödinger's equation demands. If one has $\psi = \psi_1 + \psi_2$, the probability sought is, as can be verified by simple calculation,

$$p = |\psi_1 + \psi_2|^2 = |\psi_1|^2 + |\psi_2|^2 + \psi_1\psi_2{}^* + \psi_2\psi_1{}^* \qquad (4.51)$$

This probability is different from $p_1 + p_2$. The difference is represented by the last two terms on the right, which are responsible for the formation of the interference fringes.

The violation of equation (3.55) is of utmost importance in characterizing the behavior of the microphysical world. Once more we must reflect on the rashness of calculating probability by enumerating a priori the possible cases and the favorable cases, without letting the experiment first decide what they really are.[40]

Returning to Figure 4.10, another fact of paramount importance must be emphasized. The wave function ψ derived from Schrödinger's equation does not vanish on the right of the grid, even when the negative potential applied is so high that all the electrons should be thrown back according to classical mechanics. This is the *tunnel effect* of quantum mechanics. A particle, encountering a potential barrier or potential step too high to be overcome by classical mechanics has a nonvanishing probability to get through. There is some analogy to the case of a vehicle going across a mountain through a tunnel.

4.11. Spin, atoms, and molecules

Once Schrödinger's equation is established, the investigation of the structure of the atom should become purely a mathematical problem;

that is, it should be a question of finding solutions to Schrödinger's equation that represent the orbitals of the different electrons. In reality, one can perfectly explain in this way only the simplest atomic structure, that of hydrogen.

For more complex atoms, Schrödinger's equation alone is not sufficient. Furthermore, one cannot explain why all the electrons do not occupy the orbital nearest the nucleus. This is a problem that concerned even Bohr for a long time. To overcome these difficulties, scientists were compelled to assume a couple of hypotheses ad hoc.

First, one must assume that the electron possesses an intrinsic angular momentum, called *spin*. This is distinct from the orbital angular momentum, equation (4.32), and is added to it.

By the term spin one thinks that the electron is revolving by itself like a top; but actually, this intuitive idea is too naïvely realistic (or macroscopic). Spin has very strange properties. The spin of an electron is always equal to $h/4\pi$; comparing equation (4.32), one sees that it is *half* a quantum of orbital angular momentum. In microphysics, angular momentum is usually measured in units of $h/2\pi$. Thus we say that orbital angular momentum is an integer, whereas the spin of the electron is 1/2.

Furthermore, if the component[41] of the intrinsic angular momentum of the electron is measured in respect to any preestablished direction, one finds it always equals 1/2 or $-1/2$! It is as though the axis of rotation of the top were always parallel or antiparallel to the established direction. We shall see that this singular fact fits in the general rules of quantum mechanics.

The second ad hoc hypothesis[42] is represented by the *principle of exclusion* proposed by W. Pauli in 1925. It consists of assuming that, given a configuration of ψ as a solution to Schrödinger's equation, not more than two electrons can be found in it. These electrons must have antiparallel spins, or spins opposed to one another.

We can then synthetically describe the structure of atoms and of the periodic system in the following way. Given an atomic nucleus and an electron bonded to it, one solves Schrödinger's equation and finds several distinct configurations of ψ. The electron can be found in any one of these. Each configuration has a well-defined energy (energy level). The lowest possible energy E_0 corresponds to a configuration that is conventionally indicated by 1s. The electron of the hydrogen atom is normally found in it, and in this case one says that it is in *ground state*. The energy immediately above, E_1, corresponds to the configuration indicated by 2s. Greater still is energy E_2, which is common to the three configurations that follow, indicated by 2p, and so on.

When the hydrogen atom is excited, the electron can pass from E_0

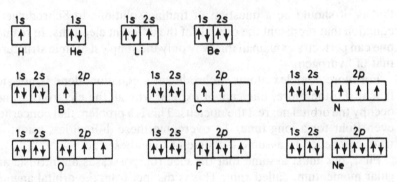

Figure 4.11

to one of the higher energy levels E_1, E_2, . . ., absorbing the corresponding energy. It can return to the ground state by reemitting the energy in the form of a photon. The results obtained agree very well with equation (4.1).

If we pass from hydrogen to helium (He), an electron is added. It can also occupy the 1s orbital, in the ground state, as long as its spin is antiparallel to that of the preceding electron. This is symbolically represented in Figure 4.11. The various squares represent the various configurations, whereas the arrows represent the spin of the electrons. With lithium (Li) we add an electron. This cannot occupy the 1s orbital, which is already occupied by two electrons. Thus it will have to go in 2s. The farther electron of beryllium (Be) also goes in 2s. The electrons of boron (B), carbon (C), nitrogen (N), oxygen (O), fluorine (F), and neon (Ne) go on to occupy successively the 2p orbitals. Then a sodium (Na) electron occupies the 3s orbital, which makes sodium analogous to lithium. Magnesium (Mg) follows, which is analogous to beryllium. Continuing in this way, we construct the periodic system.

When the 2s and 2p orbitals have been filled, we say that an electron shell has been completed. A shell is characterized by distributions of ψ with practically the same energy and with maxima almost equidistant from the nucleus. This configuration gives a particular stability to the atom, together with a typical insensitivity to external influences. The atoms that have a complete shell are called *noble gases* (helium with the 1s completed; neon, 2s + 2p, etc.), so called because they do not easily combine with one another or with other atoms to form molecules.

Two other interesting things occur when the number of electrons is one less than that necessary to complete an electron shell (as fluorine); or one more than that number (as sodium, Na = 11). In the first case the atom shows a strong electronegative character, or a tendency to capture an electron to complete its shell (becoming a negative ion). In the second case instead there is a strong tendency to give up the extra electron in order to go back to the complete configuration (becoming a positive ion). In this case the external electron is strongly screened with respect to the nucleus by the electrons of the complete shell, and thus can be easily captured by another atom. The case of sodium and chlorine is typical (chlorine lacks the electron necessary to reach the configuration of the noble gas argon). Chlorine captures an electron from sodium, which readily gives it up, forming a molecule of NaCl, which stays together because of the force of attraction between the two ions.

The periodic properties of the system of elements are thus explained by the progressive filling of the different electron shells. Every time an electron is added to a noble gas configuration, one obtains an alkaline metal (Li, Na, K, Rb, Cs, Fr), characterized by a hydrogenlike spectrum, low ionization potential, and so on. Every time a shell is completed, there is a noble gas (He, Ne, Ar, Kr, Xe, Rn). Every time one electron is lacking for the completion of the shell, we find the halogens (F, Cl, Br, I, At), all strongly electronegative. In the intermediate cases we find elements of different properties which repeat almost periodically.

We thus understand how, with the ψ, the spin, and the principle of exclusion, it is possible to explain the atom and its properties. One can also explain molecular bonds and practically all of chemistry. But we cannot go further into these subjects.

The existence of discrete energy levels for atoms and molecules solves the difficulty that we encountered in kinetic theory with regard to specific heats. At temperatures that are not too high the number of the degrees of freedom of atomic and molecular systems appears smaller than it should be. For example, atoms behave as rigid spheres without internal movements. The degrees of freedom of these movements are as though *frozen* and do not participate in the equipartition of energy. This depends on the fact that the energy necessary to bring the atom from the fundamental state E_0 to the first excited state E_1 is usually much greater than the available thermal energy kT.

Therefore in the thermal bath, atoms have very little probability of being excited and behave as though they did not have degrees of internal freedom. Analogous considerations can be made for the vibrational movements of polyatomic molecules, for which the freezing of the various degrees of freedom may not occur, according to the ratio of the binding energies to the thermal energy.

A famous discrepancy with respect to classical physics, which appears in the specific heats of solids at low temperature, had already been explained by Einstein on the basis of Planck's theory, in 1906. The physical reasons are analogous to those we have set forth.

4.12. Bosons, fermions, antimatter

Pauli's principle of exclusion is intimately connected to the inability to distinguish equal particles in microphysics. To explain this, we must introduce a generalization concerning the wave function.

An electron located at the point x, y, z at time t and having the value s_z for the z component of spin;[43] is said to have the coordinates x, y, z, t, s_z.[44] Consequently, we write the wave function in the form, $\psi(x, y, z, t, s_z)$ and say that $|\psi|^2 \, dV$ represents the probability of finding the electron in the volume dV at time t with spin s_z with respect to the z axis. The distribution of this probability gives us the *quantum state* of the electron (e.g., state $1s$ with upward spin).

Let us now suppose that we have two electrons. The first one, whose coordinates we symbolically indicate with 1, is in quantum state m and the second, whose coordinates we indicate with 2, is in quantum state n. We indicate the wave functions respectively by $\psi_m(1)$ and $\psi_n(2)$. If we neglect the interaction between the two particles, the two probabilities $|\psi_m(1)|^2$ and $|\psi_n(2)|^2$ are independent of one another and, consequently, if the classical formula, equation (3.57), is valid, the total probability $P(1, 2)$ or the probability of finding certain coordinates for one particle and certain coordinates for the other, is given by $P(1, 2) = |\psi_m(1)|^2 \, |\psi_n(2)|^2 = |\psi_m(1) \times \psi_n(2)|^2$.

One is led to think that an overall wave function $\psi(1, 2) = \psi_m(1)\psi_n(2)$ should exist. In this way, however, one generally obtains $P(1, 2) \neq P(2, 1)$, that is, a probability that changes when the particles are interchanged. But these are indistinguishable particles and this idea is absurd.

There are only two reasonable possibilities of obtaining a P that does not change when the particles are interchanged. Either $\psi(1, 2)$ remains unvaried exchanging 1 with 2 or only its sign changes. In both cases $|\psi|^2$ remains unvaried. These two possibilities can be realized respectively by the two linear combinations[45]

$$\psi_s(1, 2) = A\psi_m(1)\psi_n(2) + \psi_m(2)\psi_n(1) \tag{4.52}$$

$$\psi_a(1, 2) = A\psi_m(1)\psi_n(2) - \psi_m(2)\psi_n(1)$$

where A is a suitable factor of normalization.[46] Wave function ψ_s is *symmetrical*, whereas ψ_a is *antisymmetrical* with respect to the exchange of particles. Let us now observe that in the case of ψ_s, nothing prevents $m = n$; that is, both particles can be in the same quantum state. But this cannot happen in the case of ψ_a, for otherwise the wave function simply vanishes.

One can generalize this result to the case of more than two equal particles, as P. Dirac and J. Slater did. The wave functions must be either symmetrical or antisymmetrical by exchanging any two particles. The symmetrical wave functions correspond to particles of which two or more can stay in the same quantum state, whereas the antisymmetrical ones correspond to particles of which not more than one can be in the same quantum state. We find, by comparing the results of the theory with experiment, that the particles that have integral spin $(0, 1, 2, \ldots)$ behave in the former way and that the particles that have semiintegral spin $(1/2, 3/2, 5/2, \ldots)$ behave in the latter way. Photons (which have spin 1)[47] are of the first kind, whereas electrons (which have spin 1/2) are of the second kind. The particles of the first kind are called *bosons*, because they obviously obey Bose–Einstein statistics, whereas the particles of the second kind are called *fermions*, because they obey another statistics, called Fermi–Dirac statistics.

The latter, conceived by E. Fermi and, independently, by P. Dirac, accounts for the principle of exclusion. Let us take the various quantum states in which equal particles can be found, as elementary cells in which the particles must be distributed. It is immediately evident that Fermi–Dirac statistics leads to results different from those of Bose–Einstein statistics. It is sufficient, for example, to consider Figure 4.7 again. Bose–Einstein statistics cannot distinguish between cases 2 and 3. Therefore there are three different cases. Fermi–Dirac statistics considers cases 1 and 4 impossible and now we are left with a single case!

We shall not go further into the details of Fermi–Dirac statistics, but only say that it can be approached exactly like that of Bose–Einstein, however, keeping in mind that for each cell it allows only two possible cases: either zero or one particle.

The new statistics permits one to interpret successfully the phenomena in which a large number of electrons participate, for example, the conductivity of metals and the properties of semiconductors.

In regard to the Fermi–Dirac statistics, we must comment again on probability and the principle of indifference as was suggested by the Bose–Einstein statistics. It is absurd to establish a priori which cases are equiprobable. Only experiment can tell us. In the case of fermions, experiment tells us that some cases, which a priori could be thought of as equiprobable to the others, have no probability whatsoever; that is, they are impossible!

We still have to point out an important fact. Schrödinger's equation is not relativistic; that is, it is not invariant under Lorentz's transform. We are indebted to P. A. M. Dirac (1928) for having written a generalization of Schrödinger's equation for electrons that accounts perfectly for experimental data and eliminates some ad hoc hypotheses. First of all, it predicts that the electron must have a spin ($= h/4\pi$); second, it follows that the electron must also have a magnetic dipolar moment ($= 1\ Bohr\ magneton$). Thus spin and magnetic moment derive necessarily from Dirac's equation as a result of relativity and are not ad hoc hypotheses.

Furthermore, a new idea of enormous importance arose as a result of Dirac's equation. Dirac realized that the solutions to his equation fall into two categories: There are solutions both for electrons with positive energy[48] and for electrons with negative energy.

From the classical point of view, the latter solutions could be readily discarded, because a particle with $E > 0$ conserves its $E > 0$ and can never switch to $E < 0$. In quantum mechanics the situation is different, for quantum jumps between levels when $E > 0$ and when $E < 0$ are possible.

Dirac, instead of disregarding the solutions with $E < 0$, kept them and gave them a precise physical meaning. Particles with $E < 0$ do not seem to exist in nature. Therefore if one wants to retain the solutions with $E < 0$, one has to admit that all the states with $E < 0$ are occupied. When we accept this hypothesis, it follows that according to Pauli's principle, no quantum jumps between these states are possible. Therefore a sea of electrons exists with E $<$ 0, but normally we are not aware of them, for the only way to reveal their presence is by a change of state (quantum jump).

Naturally, one can think of making an electron jump from a negative energy state to a positive energy state (where free places exist), but the energy that one must communicate to the electron is, in this case, very great. Let us recall equation (2.68), which gives the energy of a particle of mass m and velocity v as $E = mc^2/\sqrt{1 - v^2/c^2}$. From Dirac's equation it follows that this expression can have either a positive or a negative sign, therefore

$$E = \pm \frac{mc^2}{\sqrt{1 - v^2/c^2}} \qquad (4.53)$$

We can make a graph of the values of the energy E as a function of the velocity v (Fig. 4.12). If we take the positive sign in equation (4.53), we obtain the upper curve that starts from $E = mc^2$ for $v = 0$ and goes to infinity when v tends toward c. If instead one takes the negative sign, one obtains the lower curve, which starts from $E = -mc^2$ and goes to $-\infty$ for $v = c$. As can be seen, the energy needed to pass from

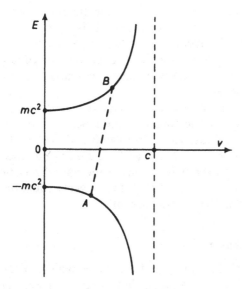

Figure 4.12

point A on the lower curve to point B on the upper curve is always greater or equal to $2mc^2$. Therefore we always need at least twice the energy mc^2 of the electron at rest. This energy is roughly a million times that necessary to make the electron jump from one level to another in the optical range. All this energy is not available in ordinary phenomena.

But it is not impossible with ad hoc experiments to provide a negative energy electron with sufficient energy to make it jump into a state with $E > 0$. In this case the number of electrons with $E > 0$ increases by 1 but, correspondingly, it remains a *hole* in the sea of electrons with $E < 0$. The lack of an electron with negative charge corresponds to the presence of a "positively charged electron." Thus one would have a sea of electrons with $E < 0$, with some holes here and there, which manifest themselves to us as positively charged electrons. In fact, the positive electron or *positron*, was discovered by C. D. Anderson in 1932.

When an electron and a positron meet, they destroy one another. The electron falls into the hole of negative energy and disappears. The corresponding energy, which as we said, equals at least $2mc^2$, is emitted in the form of photons (γ rays). Inversely, a sufficiently energetic photon can (under suitable conditions) disappear, giving rise to an electron–positron pair.

It became clear later that the case of the electron is not exceptional.

Each particle has its antiparticle, which is to it as the positron is to the electron. In other words, antimatter can exist. This is one of the most profound and important discoveries of modern physics.

This is one of the many cases in which theory has preceded experiment in an astonishing way, disproving again the naive idea that the contrary must always occur.

Dirac's equation governs only the behavior of particles with spin = 1/2. But analogous relativistic equations have been written for particles with different spin values. Thus, for example, the Klein–Gordon equation exists for particles with 0 spin and the Proca–Yukawa equation for those of spin 1. All these equations at the nonrelativistic limit of low velocity are essentially equal to Schrödinger's equation, with the addition of the ad hoc hypothesis of spin.

4.13. The uncertainty principle

Just before Schrödinger's formulation of wave mechanics, W. Heisenberg had devised another type of algorithm (*algebra of matrices*) by which the energy levels of atoms can be calculated. M. Born and P. Jordan contributed in a fundamental way to this new formulation, and before long a true and wide-ranging theory was born. Nevertheless, this theory had difficulty in achieving popularity among physicists, probably because of its unfamiliar mathematical apparatus. Apparently, this formulation was more complex than that of wave mechanics and was extraneous to it. But Schrödinger later showed that the two theories were perfectly equivalent.

The idea that the different approaches were no more than two aspects of a more general theory yet to be constructed or at least to be completed, became current. This theory, following Born's example, was called *quantum mechanics*.

We are indebted to Heisenberg for an observation of the general and fundamental character that must be placed before the formulation of quantum mechanics; the *uncertainty principle*.

To illustrate this principle, we start from Schrödinger's wave mechanics. Let us consider a ψ represented by an infinite sinusoidal wave. This wave has a well-determined λ and, according to de Broglie's relation, the particle it represents has a very well-defined $p = h/\lambda$. Thus the infinite wave represents a particle that travels in a fixed direction with a precise velocity. We ask ourselves: Where is the particle? At which point are we more likely to find it? We can easily see that $|\psi|^2$ is constant[49] (i.e., independent of the point). Therefore the probability of localization is uniform all along the axis on which the wave travels. Hence ψ represents a particle of well-defined p (and therefore v), but

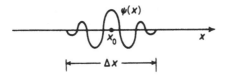

Figure 4.13

with completely indeterminate position (because we cannot establish points of greater probability a priori).

If instead one wants to represent a particle with a more well-defined position x_0, one must use a ψ that is nonzero in a neighborhood of that position, and equal to zero outside. This is a *wave packet* (Fig. 4.13). One finds mathematically that a wave packet such as that shown in the figure is obtained by superimposing a great (infinite) number of infinite sinusoidal waves with suitable phases, amplitudes, and frequencies.[50] Their wavelengths are distributed almost exclusively in an interval that ranges from λ to $\lambda + \Delta\lambda$. From the interference of all these waves, each of which extends with nonzero amplitude from $-\infty$ to $+\infty$, we obtain a wave that has an amplitude different from zero only in a narrow interval Δx. Thus this wave has a position along the x axis fixed within Δx, a wavelength fixed within a band of width $\Delta\lambda$, or a momentum along the x axis fixed within a certain interval Δp_x.

The main result that can be demonstrated mathematically is that the smaller Δx is, the greater must be Δp_x, and vice versa.

To be exact, it follows from Fourier's theorem that by defining Δx and Δp_x in a reasonable way, one necessarily obtains

$$\Delta p_x \times \Delta x \geq h/4\pi \tag{4.54}$$

Following Heisenberg, we interpret this result by saying that it is not possible to determine as precisely as we wish, and at the same time, the momentum and position of the particle. If one determines its position exactly, the momentum remains undetermined, and vice versa. The minimum for the permissible overall uncertainty is given by equation (4.54).

At first sight, it appears logical to think that a particle has well-determined x and p_x and that equation (4.53) concerns the inevitable errors of observation. But this interpretation is difficult to follow through.

According to Heisenberg, Bohr, and Dirac, and according to what later (because of Bohr's fundamental contribution to these concepts) was known as the *Copenhagen interpretation* of quantum mechanics, the particle does not have at the same time a well-determined x and a well-determined p_x. It is senseless to discuss what is not determinable

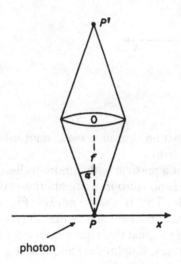

photon

Figure 4.14

because of theoretical reasons, rather than because of practical difficulties.

To clarify the physical significance of the uncertainty principle, let us see how x and p_x[51] can be measured in a specific case. Turn to the Heisenberg microscope (Fig. 4.14), and consider a particle P that travels along the x axis with a well-determined velocity and thus with a well-determined momentum p_x. Let us suppose that we know this p_x from previous measurements. We now want to know the x position at a given moment. To do this we shine light on the particle and observe it with a microscope. The objective O forms the image of P at P'. But P' is not a geometric point. Rather, it is a small disc of diffraction.

The theory of the resolving power of the microscope is analogous to what we saw in the case of the telescope. Equation (2.42) is still valid, and so $r = (\lambda/D)f$. Here r must be considered as the uncertainty of the position of the object P, as observed at the microscope. D represents the diameter of the objective and f the distance from P to the objective. If 2α represents the angular aperture of the objective, that is, the angle under which the diameter D is seen from P, we have approximately[52] $2\alpha = D/f$, therefore by substituting this in equation (2.42) and by making $\Delta x = r$, we obtain

$$\Delta x = \frac{\lambda}{2\alpha} \tag{4.55}$$

So if we make λ sufficiently small, we can find x with an uncertainty as small as we like.

But we must remember that in order for us to see the particle, it has to be illuminated, that is, to scatter photons. And these photons will communicate momentum to the particle. This can destroy the precise knowledge we had of p_x. To disturb the particle as little as possible, we illuminate it with a single photon, of which we know perfectly the direction of origin and wavelength λ, and therefore the momentum h/λ before the diffusion. If we also know the momentum of the photon after the diffusion, we know that the difference will have been given up to P and thus be able to determine the momentum of P with precision.

But the momentum of the photon after the diffusion can be any vector of length h/λ[53] from P to any point on the objective, which is unknown. Its direction is uncertain. Its component along the x axis can vary from $-(h/\lambda)\alpha$ to $+(h/\lambda)\alpha$; that is, it has an uncertainty

$$\Delta p_x = \frac{h}{\lambda} 2\alpha \qquad (4.56)$$

This will also be the uncertainty of the component with respect to x of the momentum of P after the diffusion. By multiplying equation (4.55) by equation (4.56), we obtain

$$\Delta x \, \Delta p_x = h \qquad (4.57)$$

in agreement with the uncertainty principle.[54]

Naturally, analogous relations can be obtained for y, p_y and z, p_z.

With considerations very similar to those made in the case of x and p_x, we can demonstrate a relation

$$\Delta E \, \Delta t \simeq h \qquad (4.58)$$

between the minimum uncertainty ΔE of the energy E of a system and the length Δt of the interval of time available for the determination. Thus we cannot determine the energy of the system very precisely in a very short time interval. According to Heisenberg, the system does not have an energy more definite than ΔE if considered in an interval smaller than Δt.

The uncertainty principle has given rise to infinite discussions about its interpretation and to many errors and misunderstandings. In fact, most authors who approach the subject tend to point out that almost all the other authors are wrong! Nevertheless, Heisenberg's principle has become one of the fundamental pivots for the interpretation of the microscopic world.

One of its most important results is that in dealing with physical systems the role of the observer cannot be neglected. In classical physics one was able to think of observing a physical system without disturbing it, that is, in such a way that the system would develop in the same way it would were it not observed. The Heisenberg principle denies this possibility in microphysics. As soon as one observes a system, the system is disturbed in a way that is not precisely known. The observer becomes inexorably one of the protagonists of the phenomenon.[55]

A very important act on the part of the observer is the choice of the apparatus with which he observes the microsystem. This choice determines which properties of the microsystem will be observed and which will not. According to Bohr, this even leads to the introduction of a new logical category, *complementarity*. For example, the corpuscular and wave nature of a particle are complementary. One can choose to observe the former, but one then gives up observing the latter, and vice versa. These two aspects are not contradictory, because they cannot be observed at the same time.

Much has been written on the principle of complementarity by physicists and epistemologists. In my opinion, they have somewhat exaggerated its importance, which is largely historical. Microscopic particles are not at all waves; they are particles, but particles have properties different from the simple, rough properties that we can observe by experimenting with billiard balls or even by observing the motion of heavenly bodies.[56] However, we shall return to this subject later.

As Heisenberg himself pointed out in a first interpretation,[57] indeterminism can take a wholly unexpected turn on the basis of his principle. The matter is no longer one of being able to predict the results only in a probabilistic way, knowing the initial conditions perfectly. Rather, it is one of not being able to know the initial conditions exactly. The dream of Laplace is absurd, not in its conclusion, but in its premise!

Usually, in a microscopic system we cannot have *complete* information, but we can have *maximal* information, the maximum being reached when we come to the limit imposed by the uncertainty principle for the various quantities that characterize the system.

In order to avoid confusion, we must note an important fact. Heisenberg's principle concerns the future, not the past. A photon with known frequency can pass through two successive holes, however narrow they may be.[58] When it exits from the second hole, we can deduce with great precision its position and momentum at the time of exit from the first hole. But this is not useful in predicting precisely what the photon will do when it leaves the second hole. So the uncertainty principle concerns only the results of measurements that can be interpreted as initial conditions.

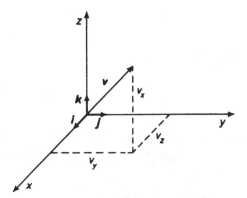

Figure 4.15

4.14. The Hilbert space

The various forms in which quantum mechanics developed made scientists hope from the beginning that a general theory could unify all of them. Such a theory was developed due to the efforts of several people, in particular, P. A. M. Dirac, D. Hilbert, and J. von Neumann. This is an axiomatic theory that at first sight can be frightening because of its formalism and its abstraction. But physicists have found it "convenient" and responsive to their goals, and have adopted it as a standard theory.

In presenting axiomatic quantum mechanics, first we must consider the mathematical instruments that we use,[59] simplifying them as much as possible.

Let us recall some elementary concepts on vectors. A vector v in ordinary three-dimensional space can be characterized by giving its three components v_x, v_y, v_z on three mutually orthogonal axes x, y, z (Fig. 4.15).

The components are (scalar) numbers, whereas v is a vector. But we can also say that v is the sum of three vectors: one of length v_x oriented along the x axis, one of length v_y oriented along the y axis, and one of length v_z oriented along the z axis. To express this, we introduce three vectors with *unitary* length i, j, k, oriented respectively along the x, y, and z axes. We then write

$$v = v_x i + v_y j + v_z k \qquad (4.59)$$

Given two vectors u, v, the product of the length of u (which is a number) and the length of the projection v_u of v on the direction of u

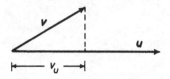

Figure 4.16

is called the *scalar product* and is indicated by $u \cdot v$ (Fig. 4.16). We immediately verify that the scalar product is commutative: $u \cdot v = v \cdot u$. Thus by definition,

$$v_x = v \cdot i, \qquad v_y = v \cdot j, \qquad v_z = v \cdot k \tag{4.60}$$

Furthermore, it is clear that because the unit vectors i, j, k are mutually orthogonal, we obtain

$$i \cdot j = i \cdot k = j \cdot k = 0 \tag{4.61}$$

Expressing u by means of its components, as we did for v in equation (4.59), we write

$$u = u_x i + u_y j + u_z k \tag{4.62}$$

We now multiply u by v scalarly, that is, equation (4.62) by equation (4.59). The scalar product is distributive, thus it can be worked out term by term. Taking into account equation (4.61) and the fact that $i \cdot i = j \cdot j = k \cdot k = 1$, we easily obtain

$$u \cdot v = u_x v_x + u_y v_y + u_z v_z \tag{4.63}$$

Then to determine the scalar product, we can determine the products of the corresponding components of the two vectors and sum the results. A mathematical entity A, which applied to a vector u, transforms it into another vector v whose components are linear combinations of those of u, is called a linear operator.[60] We conventionally write

$$v = Au \tag{4.64}$$

meaning that the following relations are valid

$$v_x = A_{11}u_x + A_{12}u_y + A_{13}u_z$$

$$v_y = A_{21}u_x + A_{22}u_y + A_{23}u_z \tag{4.65}$$

$$v_z = A_{31}u_x + A_{32}u_y + A_{33}u_z$$

$A_{11}, A_{12}, \ldots, A_{33}$ being nine numbers that characterize the operator A.

We have written these relations for a three-dimensional space. But we can easily generalize to a space with any number of dimensions n, except for the rather unimportant fact that one obviously loses visual intuitiveness.

By introducing n mutually orthogonal axes and their unit vectors i_1, $i_2, \ldots, i_n$ (which constitute a *base*), and by calling the components $v_1, v_2, \ldots, v_n$, we generally find that

$$v_l = v \cdot i_l \qquad (4.66)$$

$$v = \sum_{l=1}^{n} v_l i_l \qquad (4.67)$$

$$u \cdot v = \sum_{l=1}^{n} u_l v_l \qquad (4.68)$$

Furthermore, given the linear operator A, the expression $v = Au$ means that

$$v_l = \sum_{m=1}^{n} A_{lm} u_m \qquad (4.69)$$

The operator is called *symmetrical* if $A_{lm} = A_{ml}$. We shall always concern ourselves with symmetrical operators.

In general, the operator A does not transform a vector u into a vector v parallel to u. But for particular directions of u this can occur, and so we can write

$$Au = au \qquad (4.70)$$

a being simply a number by which u is multiplied. In this case we say that u is an *eigenvector* and that a is the corresponding *eigenvalue*. It is evident that if u is an eigenvector of A, any vector proportional to u, that is, any vector having the same direction but different length, is also an eigenvector of A, with the same eigenvalue. We use this property to establish once and for all that eigenvectors must be *normalized*; that is, their lengths must equal unity. Therefore we make

$$u \cdot u = 1 \qquad (4.71)$$

Now let us take the simple case of a two-dimensional space. Equation (4.70), if expanded becomes

$$A_{11} u_1 + A_{12} u_2 = au_1 \qquad (4.72)$$

$$A_{21} u_1 + A_{22} u_2 = au_2$$

or

$$(A_{11} - a)u_1 + A_{12}u_2 = 0 \qquad (4.73)$$
$$A_{21}u_1 + (A_{22} - a)u_2 = 0$$

From the first equation we obtain

$$\frac{u_1}{u_2} = -\frac{A_{12}}{A_{11} - a} \qquad (4.74)$$

and from the second

$$\frac{u_1}{u_2} = \frac{A_{22} - a}{A_{21}} \qquad (4.75)$$

The righthand sides of the latter two equations must then be equal, and we obtain

$$\frac{A_{12}}{A_{11} - a} = \frac{A_{22} - a}{A_{21}} \qquad (4.76)$$

Remembering that $A_{21} = A_{12}$ (symmetrical operator), we can also write

$$a^2 - (A_{11} + A_{22})a + A_{11}A_{22} - A_{12}{}^2 = 0 \qquad (4.77)$$

This is a second degree equation for the unknown a and, as we know, it provides two solutions, which we call a_1 and a_2. It can easily be shown that both solutions are always real. By substituting a_1 in equation (4.74) and by taking into account equation (4.71), we obtain a vector $\boldsymbol{u}_1$ as the determination of $\boldsymbol{u}$.[51] If we repeat the operation with a_2, we obtain a vector $\boldsymbol{u}_2$. We conclude that in a two-dimensional space a symmetrical linear operator A has two eigenvalues a_1, a_2 and two corresponding eigenvectors $\boldsymbol{u}_1$, $\boldsymbol{u}_2$, which we know how to calculate.

One can easily generalize the result finding that in a space with n dimensions, a symmetrical linear operator has n eigenvalues a_1, a_2, . . ., a_n and correspondingly, n eigenvectors $\boldsymbol{u}_1$, $\boldsymbol{u}_2$, . . ., $\boldsymbol{u}_n$. Let us take any two eigenvectors $\boldsymbol{u}_i$, $\boldsymbol{u}_k$ and the corresponding eigenvalues a_i, a_k, and write equation (4.70) for them

$$A\boldsymbol{u}_i = a_i\boldsymbol{u}_i \qquad (4.78)$$
$$A\boldsymbol{u}_k = a_k\boldsymbol{u}_k$$

Using the scalar product, we multiply the former by $\boldsymbol{u}_k$ and the latter by $\boldsymbol{u}_i$, and obtain

$$(A\boldsymbol{u}_i)\cdot\boldsymbol{u}_k = a_i\boldsymbol{u}_i\cdot\boldsymbol{u}_k$$
$$(A\boldsymbol{u}_k)\cdot\boldsymbol{u}_i = a_k\boldsymbol{u}_k\cdot\boldsymbol{u}_i \qquad (4.79)$$

Because A is symmetrical, we are immediately convinced that the left-hand sides of these two equations are equal. By equating the righthand sides and remembering that $u_i \cdot u_k = u_k \cdot u_i$, we obtain

$$(a_i - a_k) \, u_i \cdot u_k = 0 \tag{4.80}$$

If $a_i \neq a_k$ (nondegenerate case),[62] we obtain

$$u_i \cdot u_k = 0 \tag{4.81}$$

This remarkable equation tells us that any two eigenvectors of A are mutually orthogonal.

Again, recalling equation (4.71), we can conclude that the eigenvectors of A form a system of n mutually orthogonal unit vectors and can be considered as a base in the same way as $i_1, i_2, \ldots, i_n$ seen previously. Therefore any vector v of an n-dimensional space can be expressed as

$$v = \sum_{k=1}^{n} v_k u_k \tag{4.82}$$

where the v_k are components with respect to the base of the u_k. It is also clear that

$$v_k = u_k \cdot v \tag{4.83}$$

as we have seen in equation (4.60).

The *Hilbert space* used in quantum mechanics is substantially an extension of the abstract vector space that we have seen up until now. It is more general for two reasons.

First, Hilbert space is a space with an *infinite number of dimensions*.[63] Vectors in it have an infinite number of components. Nevertheless, the vectors we are considering are only those that have finite total length. For this to be possible, the components must diminish suitably in length as their index increases.[64] Furthermore, the components of the vectors in Hilbert space have *complex values*.

A vector in Hilbert space is usually represented by the notation $|v\rangle$ and is called a *ket*. The vector having as components the complex conjugate numbers of the components of $|v\rangle$, is indicated by $\langle v|$ and is called a *bra*. The scalar product in Hilbert space is always obtained by multiplying a bra by a ket. Therefore what in ordinary space is indicated by $u \cdot v$, we indicate by

$$\langle u|v\rangle = \sum_{i=1}^{\infty} u_i^* v_i \tag{4.84}$$

having used $u_i{}^*$ to indicate the components of $\langle u|$, which are complex conjugates of the components u_i of $|u\rangle$.

This strange nomenclature, introduced by P. Dirac, derives from the fact that the symbol $\langle\ \rangle$ is called a *bracket*. A bra and a ket are vectors of Hilbert space. Therefore they are abstract quantities and are not directly measurable or observable. To have something that is physically significant, one must have the whole bracket; that is, a scalar product $\langle u|v\rangle$ which, like all scalar products, is a number (in this case, a complex number).

In Hilbert space we consider linear operators analogous to the symmetrical operators of ordinary space. We call them *Hermitian* operators. We say by definition that for a Hermitian operator, A_{lm} is the complex conjugate of A_{ml}, which means that we can write $A_{lm} = A_{ml}{}^*$. It is possible to show that a hermitian operator has all real eigenvalues.

The eigenvectors of a Hermitian operator are mutually orthogonal and can be normalized. But it is not always certain that they will form a complete base, that is, a base with which we can express any vector (of the Hilbert space) as with equation (4.82), because we have passed to an infinite number of dimensions. We shall assume that the base is complete every time it is necessary.

4.15. The formalism of quantum mechanics

Nonrelativistic quantum mechanics can be based on the following axioms:

    1. Given the state of a physical system, one can (almost always)[65] assign a vector $|\psi\rangle$ of a proper Hilbert space that corresponds to it. This is called a *state vector* and represents all we can know about the system. We assume it is normalized, therefore $\langle\psi|\psi\rangle = 1$.

    2. The state vector evolves according to[66]

$$|\psi(t)\rangle = T(t)|\psi(0)\rangle \tag{4.84}$$

where $|\psi(0)\rangle$ represents the state at time $t = 0$ and $|\psi(t)\rangle$ represents the state at time t. We use $T(t)$ to indicate a linear operator, which we know how to construct starting from the total energy expression (or *Hamiltonian* expression) of the system. We shall not describe the precise rules by which $T(t)$ can be constructed. It is sufficient to know that they exist. Moreover, $T(t)$ preserves the normalization, therefore $\langle\psi(t)|\psi(t)\rangle = \langle\psi(0)|\psi(0)\rangle = 1$.

    3. A Hermitian operator A corresponds to every observable physical quantity of the system (energy, momentum, angular momentum, number of particles, etc.). The eigenvalues a_i, which, as we know, satisfy the equation

$$A|u_i\rangle = a_i|u_i\rangle \qquad (4.85)$$

represent the only possible results of a measurement of A.[67]

4. If by analogy with equation (4.82), we represent the state vector $|\psi\rangle$ in the base of the eigenvectors $|u_i\rangle$ by the expression

$$|\psi\rangle = \sum_{i=1}^{\infty} c_i|u_i\rangle \qquad (4.86)$$

the component c_i, which by analogy with equation (4.83) is calculated by

$$c_i = \langle u_i|\psi\rangle \qquad (4.87)$$

represents, with the square of its modulus

$$|c_1|^2 = P_i \qquad (4.88)$$

the probability P_i that the measurement of A gives the result a_i.

5. If we carry out the measurement and it gives the result a_i, the state vector $|\psi\rangle$ becomes $|u_i\rangle$ immediately after the measurement. Therefore the same measurement, repeated immediately, gives the result a_i with certainty.[68]

To see the connection to Schrödinger's wave mechanics, let us consider the case in which the system is a particle with a state vector $|\psi\rangle$. Let us consider the observable "position of the particle along the x axis" and X, the corresponding operator. The eigenvalues x_i are all the possible positions of the particle. Because a priori, the result of a measurement of the position can give any result, the x_i are all the values that x can assume between $-\infty$ and $+\infty$. If we use $|x\rangle$ to indicate the eigenvector corresponding to the eigenvalue x, we derive from axiom (4) that $|\langle x|\psi\rangle|^2$ represents the probability of finding the particle on the x abscissa. But this is familiar and is recognized when we write

$$\langle x|\psi\rangle = \psi(x) \qquad (4.89)$$

where $\psi(x)$ indicates Schrödinger's wave function. Using the rules for constructing the operator $T(t)$ of equation (4.84), we find that by applying $T(t)$ to $|\psi\rangle$, we get the same result as that obtained by applying Schrödinger's equation to $\psi(x)$. Therefore the two formalisms are equivalent.[69] But naturally, the one outlined here is more general and useful in dealing with observables other than the simple position of the particle.

When we construct the operator A for every dynamic variable, it turns out that the eigenvalues are sometimes all those of a continuous series, as in the case of the position x, although sometimes they have discrete values, as in the case of the total energy of an electron in an atom (energy levels).

The case of angular momentum is important. The corresponding operator has the eigenvalues $l(h/2\pi)$, l being an integer.[70] This is the reason, for example, that the electron in the hydrogen atom can only have an angular momentum that is a multiple of $h/2\pi$, as Bohr had already realized. The component of the angular momentum along the z axis is found to take only values of the type $m(h/2\pi)$, where m is a whole number between $-l$ and $+l$. This is surprising, because the z axis has a direction that can be chosen in a completely arbitrary way.

If we take $h/2\pi$ as the unit of angular momentum, we can say that the orbital angular momentum of the electron is l and its projection on the z axis is m.

Concerning the intrinsic angular momentum or *spin* of a particle, we know that it must be added to nonrelativistic quantum mechanics as an ad hoc hypothesis. We can also assign operators to its components and find eigenvalues that happen to agree with experiment. For example, the electron has $s_z = \pm 1/2$. For a fermion with spin 3/2 we can have $s_z = \pm 3/2$ or $\pm 1/2$. For a boson with spin 1 we can have $s_z = \pm 1$ or $s_z = 0$, and so on, in every case. The possible values of s_z form an arithmetic progression.

Axiom (5) represents one of the most puzzling aspects of quantum mechanics.[71] It is sometimes called the *reduction* or *collapse* of the wave function. It means that the state vector, which generally represents the superimposition of various possibilities with different probabilities, suddenly becomes one well-determined possibility only.

Let us once again consider the case of the two-hole device in Figure 2.50 used with an electron beam. The state vector of an electron is of the type $|\psi\rangle = c_1|\psi_1\rangle + c_2|\psi_2\rangle$ where $|\psi_1\rangle$ and $|\psi_2\rangle$ correspond to the passage of the particle through F_1 or F_2, respectively. This $|\psi\rangle$ gives rise to interference fringes, provided one does not try to determine from which hole the electron came. If instead we carry out a measurement and verify that the electron has come, say, from F_1,[72] the state vector suddenly becomes $|\psi_1\rangle$ and there is no longer interference.

In this way we arrive at one of the most important affirmations made in quantum mechanics, which we have already met in regard to the uncertainty principle. Whatever the objective reality of the physical system may be, the observer cannot determine it without disturbing it in a way that is unforseeable a priori. As a direct consequence of this conception, physical quantities in general do not have well-defined values in a microscopic system, but acquire them only as a result of measurement.

Sometimes physicists have wanted to conclude, probably in a way that is not entirely justified, that the object and the observer form an inseparable and interdependent whole. It seems possible to conclude less drastically that the subject and object can also have an independent reality, but that their interaction is different from what classical physics had assumed. We shall consider this subject again in regard to the theory of measurement.

The disturbance produced by the observer has another important effect. Let us assume that the initial state $|\psi\rangle$ of a system is an eigenstate $|u_i\rangle$ of an observable A. Consequently, if we carry out a measurement of A on the system, we find a well-determined result a_i. But if we measure B, once we have measured A, and obtain the result b_k, we know that the state passes immediately to an eigenstate $|v_k\rangle$ of B. Because generally $|v_k\rangle$ is not an eigenstate of A as well, A no longer has a well-determined value. Therefore it is meaningless to make a precise and simultaneous measurement of A and B. In this case we say that two observables are *incompatible*. They are *compatible* when all their eigenvectors are equal, as is evident. Now one can easily show that when two operators A and B have the same eigenvectors, they commute (and conversely) in the sense that given any vector $|\psi\rangle$, we obtain the same result by applying first operator A then operator B, or by applying first B and then A. We write

$$AB - BA = 0 \qquad (4.90)$$

We conclude that two observables are compatible when the respective operators commute. If instead the operators do not commute, generally, the two observables are not compatible.

If the two observables are the position x and the momentum p_x of the particle, the respective operators do not commute and in fact, as we know, the two quantities cannot be measured at the same time with precision (uncertainty principle).

The *maximum* information on a microsystem is obtained when the values of the maximum possible number of mutually compatible observable are known.

There is an important remark to make here. We shall deal shortly with some conceptual difficulties that quantum mechanics has had, because it is of specific interest for the purpose of this book. However, we must keep in mind that since 1926 until today, quantum mechanics has experienced a notable series of successes. More precisely, we must be aware that at the present time no known experiment exists that definitely contradicts the predictions of quantum mechanics.

We should also note, without going into detail, unfortunately, that quantum mechanics, born to explain the phenomena of microphysics, has recently found great success in interpreting phenomena of clearly

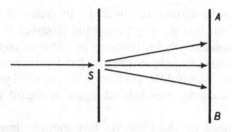

Figure 4.17

macroscopic character, such as *superconductivity, superfluidity,* and *laser* emission.

4.16. Revision of the general scheme of physics

At this point it may be helpful to reread §1.11, which concerns the formulation and meaning of laws in classical physics.

The advent of quantum mechanics is mainly due to the fact that microscopic physics has attained such a high level of precision ϵ_n (revelation of single photons, electrons, etc.) that for some processes we cannot even find a function $\bar{f}$, such that

$$\bar{f}(a_n) = 0 \qquad (4.91)$$

correctly describes the results of experiment, as does equation (1.6).

For example, let us consider a simple device (Fig. 4.17) in which a beam of electrons of well-determined direction is shot at a small hole S which diffracts it, that is, which widens it into a cone of different directions. Afterward we observe the single electrons on a screen by means of a set of counters. The set of all a_n minus 1 represents the numbers that specify the position of the screens and the hole, the energy and direction of the electrons, and so on.

The last a_n represents the position in which the electron is shown on the screen AB. We would like to derive it from the knowledge of the other values a_n, by solving an equation of the type, such as equation (4.91). But we know that this is not possible, because the electron will go once in one place and once in another, in an unpredictable way. Therefore $\bar{f}$ does not exist.

In other words, the set $G(A_n, \epsilon_n)$, which appears in equation (1.7) and to which any $\bar{f}(a_n)$ that agrees with experiment belongs, in general, not only loses elements gradually as ϵ_n decreases, but actually becomes an *empty* set for $\epsilon_n = \bar{\epsilon}_n$. For $\epsilon_n < \bar{\epsilon}_n$ a revision of the general method of physics is necessary.

When a function $\bar{f}(a_n)$ exists for the phenomenon under study, it means that, given all a_n minus 1, the latter is perfectly determined (up to ϵ_n) from $\bar{f}(a_n) = 0$. Obviously, the process is said to happen in a *deterministic* way.

As seen, the orthodox school interprets the nonexistence of $f(a_n)$ below a certain level by saying that the phenomenon is *indeterministic* at that level. We cannot calculate a precise value for one a_n by knowing the values of the others. We can only calculate the distribution of probability.

Let us call a_0 the variable in which we are interested, and indicate symbolically by a_n the other variables $a_1, a_2, \ldots$ which have been measured. What we are looking for is a function $p_{a_n}(a_0)$ which gives us the probability of finding a certain value for a_0, once the a_n values are given.

Quantum mechanics has provided an algorithm for calculating this probability.[73] We shall indicate this algorithm with the symbol $\mathcal{A}_1$.

It has sometimes been said, not without the presumption of making a great revelation, that this algorithm is not the only possible one that agrees with the facts. The great revelation is only a banality.

A set G of algorithms that agree with experimental results certainly exists; $\mathcal{A}_1$ belongs to this set. The set G depends on the class P of phenomena to which we refer and on ϵ_n. We write

$$\mathcal{A}_1 \in G(P, \epsilon_n) \tag{4.92}$$

This equation expresses a physical truth.

By decreasing ϵ_n, for example, by studying the interaction among particles at increasingly smaller distances, or by widening the class P of the phenomena considered, we expect to go to other algorithms, which belong to subsets increasingly smaller than that which appears in equation (4.92). Therefore we should have

$$\mathcal{A}_2 \in G(P', \epsilon_n') \tag{4.93}$$

where

$$G(P', \epsilon_n') \subseteq G(P, \epsilon_n) \tag{4.94}$$

with $\epsilon_n' \leq \epsilon_n$ and $P \subseteq P'$, and so on. But no one can be certain that this will happen in the future precisely.

What is important to note is that the truth of equation (4.93) does not destroy the truth of equation (4.92).

From the functions $f(a_n)$ of classical physics we have gone on to the *algorithms* of contemporary physics. As a matter of fact, it is necessary to point out that analogous algorithms (e.g., solutions of differential equations with given boundary conditions) are widely used even in classical physics. But the meaning attributed to them is different.

In classical physics, algorithms are useful methods for solving problems that would demand the simultaneous or successive application of elementary laws (those of $f(a_n)$) to innumerable contiguous elements. For example, if we study the elastic vibrations of a solid body, we can be sure that the interactions among its contiguous elements are simply governed by the laws of Newton's mechanics and by Hooke's law (by which the deformation in an elastic body is proportional to the force that provokes it). The effect of all these interactions can be synthesized in a differential equation (that of elastic waves), whose solution exempts us from applying the elementary laws an infinite number of times. Thus we can calculate, for example, the frequency of vibration, disregarding the movements of the single particles of the body. But the latter are measurable in principle. We could make analogous considerations for electric and magnetic fields and for Maxwell's equations.

When we proceed in this manner, we often say conventionally that we have *reduced* one theory to a more elementary one. For example, we reduce the theory of elastic waves to Newton's laws and to Hooke's law, or we reduce the theory of the propagation of light to the theory of elementary electromagnetic interactions. The mathematical algorithm is the tool used to carry out this reduction.

The situation in orthodox quantum mechanics is completely different. Schrödinger's ψ is not at all observable and measurable. It obeys a wave equation that does not derive from the interaction, among contiguous elements. This can be seen by observing that when ψ describes a system of many particles, it is a function of the coordinates of all the particles.

Inevitably, the algorithm must replace the functions. It certainly cannot be said that we have reduced the theory of the behavior of atomic particles to something more elementary. Not having evaluated these facts well is behind the venerable question of the duality of waves and particles.[74] Are we dealing with waves or not? According to the considerations already made, we are not dealing with waves, but with entities whose distribution of probability obeys an equation analogous to that of waves. This is not a very singular coincidence in physics; the case of absolutely disparate phenomena that obey analogous laws is very frequent. To say that particles are waves is more or less like saying that a pendulum is composed of an inductance, a capacity, and a resistance put in series.

Then are they corpuscles? Certainly not, if by corpuscle we mean an aggregate of an enormous number of smaller entities, such as those which constitute macroscopic bodies. Besides, the concept would be *contradictory*! Yes, though, if by corpuscle we mean an entity that obeys quantum mechanics. This in turn is *tautological*. But how could it be otherwise?

4.17. Difficulties of quantum mechanics

Since their birth, the orthodox formulation of quantum mechanics and the Copenhagen interpretation have encountered much criticism and opposition. Among the most famous names we have already cited are A. Einstein, M. Planck, L. de Broglie[75] and E. Schrödinger.[76]

The primary motive of this opposition is without doubt constituted by the psychological difficulty of accepting indeterminism: "God does not play dice." But we cannot neglect that such psychological behavior can be justified by several reasons worthy of consideration. Not everything is perfectly clear and satisfactory in quantum mechanics:

First of all, there are the "paradoxes." The one pointed out by A. Einstein, B. Podolsky, and N. Rosen in 1935 is typical. It can take various concrete forms. We shall formulate it in the following way, due to D. Bohm and Y. Aharonov, which makes the central concept comprehensible.

We shall see shortly that many subatomic particles can decay or disintegrate spontaneously, giving origin to other particles. Let us consider a particle with zero spin that, being at rest, decays into two particles, A and B, each with spin 1/2. These particles move apart in opposite directions, for example, along the x axis. Because the initial particle had zero spin, A and B because of the conservation of angular momentum, will have antiparallel spin. This means that if, for example, we measure s_z on A and find it to be $+1/2$, an analogous measurement on B must give $-1/2$. This has been confirmed by many experiments. Now when we measure s_z on A, B can be very far away (even light years!). Therefore Einstein argued, because it is absurd[77] that the measurement carried out on A influences B instantaneously, we must conclude that B *has* the property $s_z = -1/2$. B has it even before the measurement is carried out on A and does not acquire it only in the act of measurement.[78] But if this is true, we must think that by repeating the experiment several times in identical conditions, we shall obtain two distinct sets of cases E^+ and E^-. In the case of the set E^+ the particles A and B have, respectively, $s_z = +1/2$ and $s_z = -1/2$, whereas in the case of the set E^- they have $s_z = -1/2$ and $s_z = +1/2$. But even this is absurd, because z is a direction we chose arbitrarily on the plane perpendicular to x, which has nothing to do with the physical system. If we perform experiments measuring s_y, we obtain analogous

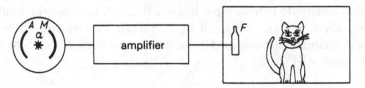

Figure 4.18

results! Therefore it is not possible to say that the particles *have* a direction of angular momentum independent from the measurement.

The "solution" of the paradox given by some physicists is that there is no paradox. The two particles *A* and *B* form a whole which has a state vector and cannot be separated into two independent systems (each with its own state vector). But at this point a serious problem arises: When can we be sure that the system is independent of others and has its own state vector? Any system has certainly interacted in the past with other systems and the measurements carried out on it should be correlated with those carried out on other distant systems, apparently independent. This problem of the *separability* of physical systems is one of the major problems of quantum mechanics and therefore of all physics.[79]

For Bohr, the system *A, B* also is, in a way, nonseparable from the instrument with which it interacts, for example, when we measure s_z. If instead of measuring s_z, we want to measure s_y, we have to change the instrument (or its orientation). The whole system is different and we should not be surprised if the results obtained are apparently incompatible with the previous ones.

For Einstein, however, the difficulty arises from the fact that quantum mechanics is not *complete,* even if it is correct for that part of reality that it describes.

Another interesting paradox was conceived by E. Schrödinger, also in 1935. Explaining it with some nonessential changes, let us suppose we have a spherical cavity (Fig. 4.18) in the center of which is an excited atom *a*. The atom emits a photon, which can strike either the left wall *A*, which is an absorber, or the right wall *A*, which is a photoelectric metal. If it strikes *M*, an electric impulse, suitably amplified, activates a mechanism that breaks a phial *F* full of toxic gas. The gas expands in a box that is close to and isolated from the exterior, in which there is a cat. The cat, when reached by the gas, dies. Now let us consider that the excited atom *a* has an equal probability of emitting the photon toward the right or toward the left. The state of the system, when the atom is no longer excited, is a combination of $|\psi_R\rangle$ = (photon emitted to the right) and $|\psi_L\rangle$ = (photon emitted to the left). But then following the chain, and assuming that the cat can be described by a state vector, we also assume that we have a combination of $|\psi_{LC}\rangle =$ (*live cat*) and $|\psi_{DC}\rangle =$ (*dead cat*)! The "measurement" consists of opening the box and looking to see if the cat is alive or dead. As long as the box is closed, the cat is neither alive nor dead. We are the ones who on opening the box and observing it, kill him or give him life.

This is one of those cases in which, as M. Bunge observes, Hamlet's dilemma, "To be or not to be," would be substituted by "To look or not to look."

The situation is very odd.

Owners of cats will, however, generally agree with Schrödinger that this description is paradoxical, since even before the outside observer comes into action there is at any rate one individual for whom the observable "cat alive" or "cat dead" has a quite definite value, in one half of the cases at least this individual being is, of course, the cat (d'Espagnat, 1976, p. 205).

Thus we can see that it is dangerous, following the chain of interactions, to extend the formalism of quantum mechanics to macroscopic objects. One of the thorniest and most controversial problems of quantum mechanics, known as the *problem of measurement,* arises from precisely this difficulty. We can give a simple idea of the problem in the following way.

We carry out a measurement of an observable A which has the eigenvectors $|u_i\rangle$ and the eigenvalues a_i, on a microscopic system S. If the system S is in a well-determined eigenstate $|u_k\rangle$ before the measurement, the result of the measurement will be a_k and the eigenvector will remain $|u_k\rangle$. The result will be obtained by a macroscopic measuring device M,[80] which will allow us to read the value a_k, say, on a scale.

If, for consistency, we want to deal with M as well, using the rules of quantum mechanics, we must say that it has a set of eigenstates $|v_i\rangle$ and that immediately after the measurement, it is found in the state $|v_k\rangle$, corresponding to the reading a_k on the scale.

Let us suppose that initially M is in the eigenstate $|v_0\rangle$, corresponding to zero on the scale. The whole system $M + S$ is thus initially found in the eigenstate described by $|v_0\rangle$ for M and by $|u_k\rangle$ for S. Symbolically, we indicate this by saying that the system $M + S$ is in the state $|v_0\rangle$ $|u_k\rangle$.[81] The state $|v_0\rangle|u_k\rangle$ evolves according to equation (4.84), in which $T(t)$ is the suitable operator that includes the interaction between M and S. Whatever its precise form might be in the case at hand, we have established that it brings about the passage from the initial state $|v_0\rangle$ $|u_k\rangle$ to the final state $|v_k\rangle|u_k\rangle$. In other words, if τ indicates the (very brief) time of the measurement, we have

$$T(\tau)|v_0\rangle|u_k\rangle = |v_k\rangle|u_k\rangle \qquad (4.95)$$

Clearly, an equation of this form must be valid for any eigenvector $|u_k\rangle$ of S, that is, for any value of the index k.

Let us now assume that S is initially not in an eigenstate $|u_k\rangle$ of A, but in a general state $|\psi\rangle$ represented as in equation (4.86) by

$$|\psi\rangle = \sum c_i |u_i\rangle \tag{4.96}$$

where $|c_i|^2$ indicates the probability that the result of the measurement will be a_i.

If M is initially in the state $|v_0\rangle$, the whole system will be described by $|v_0\rangle|\psi\rangle$. If we apply the operator $T(\tau)$ to this state vector, taking into account equation (4.95) and the fact that $T(\tau)$ is a linear operator (so that to apply it to a sum amounts to the same as applying it to each term and taking the sum), we obtain

$$T(\tau)|v_0\rangle|\psi\rangle = \sum c_i T(\tau)|v_0\rangle|v_i\rangle = \sum c_i |v_i\rangle|u_i\rangle \tag{4.97}$$

So we conclude that the reduction of the wave function has not occurred at all, for we do not find S in a well-determined eigenstate $|u_k\rangle$, and correspondingly, M in a well-determined eigenstate $|v_k\rangle$, which would allow us a reading a_k. The final state vector is still a combination of eigenvectors.

All equation (4.97) tells us is that if we succeed in carrying out a measurement on M with a device M', and if we find that M is in the state $|v_k\rangle$, then S is in the state $|u_k\rangle$. But we shall encounter the same difficulties for M' and must go to a device M'', and so on. Thus we have an apparently endless chain (von Neumann's chain). We can think that this chain passes from the physical devices to the sense organs of the observer, to his nervous system, and finally to his brain. According to J. von Neumann we can formulate the hypothesis that the chain will break at the psychic level, or that the reduction of the wave function will occur when the observer acknowledges the result. Naturally, this interpretation immediately solves the difficulties of the theory of measurement. Some authors consider it inevitable. For example, E. P. Wigner has maintained very effectively that the violation of the linearity of the laws of quantum mechanics in fact happens at the level of consciousness. Nevertheless, not everyone is ready to accept such a bold conclusion.

H. Jauch says rightly:

> The involvement of consciousness at this stage seems hard to reconcile with the fact that the experimental situation can be so arranged that consciousness plays actually no part whatsoever during the act of measurement for instance by storing the permanent record in the memory of a large computer. This information can be recalled at a much later time involving only the observation of macroscopic and classical systems. It is hard to see why this last step of

becoming aware of these macroscopic and classical data in some consciousness should have a decisive influence on the actual measuring process which was completed a long time ago (Jauch, 1973, p. 68).

In this way the idea arises that axioms (2) and (5) of quantum mechanics may not be compatible. Furthermore, the reduction of the state vector seems to be relativistically noninvariant.

The difficulties of quantum mechanics can be confronted in various ways. First of all, we note that in equation (4.92) the class P of the phenomena, to which the algorithm $\mathcal{A}_1$ is applied, is explicitly placed in evidence. Consequently, should some type of experiment, for example, the Einstein–Podosky–Rosen kind, give results that clash with the predictions of $\mathcal{A}_1$, it would be sufficient to exclude it from the class P. One should include it in a larger class P', and look for a new algorithm $\mathcal{A}_2$, like that of equation (4.93), capable of describing it correctly.

Curiously enough, an intense search for $\mathcal{A}_2$ has occurred before any evidence has cast doubt on $\mathcal{A}_1$. An important example is the hypothesis of *hidden parameters* (or *hidden variables*). Initially, scientists had been driven to this hypothesis by the old metaphysical prejudice[82] against indeterminism. But later, research has developed with more useful goals, such as overcoming the present difficulties of quantum mechanics or, at least, of arriving at more general and effective conceptions.

The hypothesis of hidden parameters consists of assuming that microphysical processes can depend not only on the usual observable physical quantities a_n, but also on other parameters ξ_m, which we do not know how to measure. These ξ_m have different values from case to case and, together with the a_n, they determine the phenomenon in an unique way. In this case it is usual to talk of *dispersionless states*. For those of us who cannot know the values of the ξ_m, the process can only have a statistical description.

A famous theorem demonstrated by von Neumann in 1932 seemed to rule out the possibility of dispersionless states and of theories of hidden parameters that gave results totally coincident with those of quantum mechanics. The conditions posed by von Neumann for his theorem, especially the conditions of linearity are very restrictive. One can loosen these conditions; nevertheless, the situation does not become much more promising, at least as long as one looks for a *local* theory. A local theory is one in which the result of a measurement carried out with a device M_1 does not depend on the nature or orientation of another device M_2, possibly located far away from the first. J. S. Bell has shown in 1965 that no local theory of hidden parameters

292   4 Microphysics

concerns an important rule in some aspects of modern physics. To define it, we say that because of the principle of causality, no signal can be received before being sent. More precisely, if an event A sends a signal that is received by an event B, A must precede B.[103] We can see in the following way that this law is closely connected to our sense of freedom of action and to the second principle of thermodynamics, that is, we can influence the events of our future but not those of our past. Assume that I break a leg coming down the stairs. If I could send this information back in time, I could warn myself and therefore avoid going down that stairway! Then an event could happen and not happen,[104] which is absurd. To remove the absurdity we should recall the tragedy of Oedipus and admit that our freedom is only illusory and that in reality we cannot avoid *fate*, even if we know beforehand. But who still believes in these things?

4.21. The inductive inference

Now we have all the facts that we need to analyze the question of induction in physics.[105] This is a central problem, which has been mentioned several times. Moreover, two centuries after D. Hume, this problem not only remains unsolved, but has become increasingly controversial (see, e.g., Salmon, 1966; Hesse, 1974). In fact, natural science has aimed at being more and more *exact* and making objective and undisputable affirmations. Yet this science is based mainly on a process (the inductive) that cannot be justified rigorously in any case![106] In fact, a general opinion is that Hume's criticism is correct. If by justifying we mean deriving with *deductive* process, it is even obvious that inductive inference cannot be justified. On the other hand, to say: "Everything has gone well until now and we have been able to construct science, therefore everything will continue to go well," means clearly to apply that inductive process that we do not know how to justify. This is circular reasoning.

Thus along with K. Popper, we can assume an antiinductionistic attitude (Popper, 1959, 1965, 1978). For Popper, scientific laws and theories cannot be confirmed, but can be falsified, and experimental science proceeds by successive falsifications. At the most, we can speak of corroboration, when a scientific hypothesis resists one or more potential falsifiers.

As mentioned several times, this image of the history of physics as a chain of successive falsifications arises primarily when one chooses to ignore the modern concept by which a theory T has a domain of validity D, whose specification cannot be disjoined from the definition orators since 1962. To treat the macroscopic instrument adequately, we must apply statistical quantum mechanics. Unfortunately, the theory is not at all simple, and it cannot be explained adequately here. According to this theory, the classical macroscopic observables of the instrument can acquire well-defined values (within the precision ϵ) immediately after the measurement, without infringing on the ordinary Schrödinger time evolution.

Not everyone is convinced of the correctness of his approach, which has been subjected to some criticism. Nevertheless, it represents one of the most interesting and promising proposals for solving the difficulties of quantum measurement (see Caldirola, 1974, p. 193). Analogous results have been obtained with different methods by C. George, I. Prigogine, and L. Rosenfeld (see, e.g., Rosenfeld, 1974, p. 431).

Finally, let us mention the fact that according to some scholars, quantum mechanics leads to the introduction of a logic that differs from classical logic. This logic is usually called quantum logic. The idea suggested by G. B. Birkhoff and J. von Neumann in 1934 and by H. Reichenbach in 1944, has been developed by G. W. Mackey, H. M. Jauch, and others.[85] Surprising as this point of view might seem, it is by no means unacceptable. Modern research has revealed that there is not necessarily one logic only, and that different kinds of logic are acceptable (see, e.g., Dalla Chiara, 1974, Chap. VI). According to H. Putnam, logic has an aspect that pertains to natural science and its "necessary truths" can turn out to be false for empirical reasons. Putnam says "Quantum mechanics itself explains the *approximate* validity of *classical* logic in the large, just as non-Euclidean geometry explains the *approximate* validity of *Euclidean* geometry in the small." (Putnam, 1969). Not everyone agrees with this point of view (see, e.g, Heelan, 1974).

We can give an intuitive idea of the starting point of quantum logic in the following way. Let us consider a physical system whose state vector $|\psi\rangle = \Sigma c_i|u_i\rangle$ is *not* one of the eigenstates $|u_i\rangle$ of the Hermitean operator A. It is not true that the physical quantity has the value a_i (eigenvalue corresponding to $|u_i\rangle$), nor is it true that A does not have the value a_i. Therefore the law of the excluded middle is not valid in the strong form, which affirms: for every proposition p, either p is true or its negation is true. Nevertheless, one can reasonably introduce a new type of disjunction (p or q) which validates a weak principle of the excluded middle; in this way "p or not p" remains always a true proposition. This must be interpreted by saying that $|\psi\rangle$ can always be expanded into a vector parallel to $|u_i\rangle$ and one orthogonal to $|u_i\rangle$. J. Bub has shown that a logic different from the classical one can make quantum mechanics complete in Einstein's sense.

4.18. Microphysics and reality

The reasons that have shown the general form of quantum mechanics to be unsatisfactory have prompted some scholars to try to complete or render the present theory more precise, or even to conceive a new and more general one.

But unfortunately, these reasons often entail various psychological, ideological, and conceptual demands that are much less justified and which only cause confusion. Some people want to discuss modern science, so deeply structured and diversified in its problems, by using simplistic and worn-out key words, which sometimes are used as banners around which *to fight and die*. One often ends up fighting about too general terms, used improperly, and creating oppositions that are artificial, or, at least, not altogether clear.

As a typical example, we observe that the opposition of *realism* to *positivism* has become popular among some physicists who analyze the foundations of quantum mechanics. But what kind of realism, and what kind of positivism? Is it really true that these terms are opposites? If we consult any philosophical dictionary we see that this is not the case. It is more or less the same kind of error that one makes by identifying *agnostic* with *atheist*.

More correctly, many oppose realism to *idealism*.[86] But one should remember that Kant, in the first edition of the *Critique of Pure Reason* (fourth paralogism), said: "The transcendental idealist . . . can be an empirical realist, therefore a *dualist,* and admit the existence of matter, without going beyond the pure consciousness of himself. . . ." I say all this, only to show how dangerous it is to comment on science, with its precise language, by committing ourselves to general philosophical terms that are abstracted from various contexts in which they were used historically and by taking for granted that they always have the same meaning.

Whole volumes can be, and are, written on questions such as those of the metaphysical implications of science, but this task is not usually very profitable. Accordingly, let us just mention a few points that need further clarification.

One of the most serious concerns of today is that of the *reality* of the external world, which some think is "endangered" by the present theory of quantum mechanics.[87] Sometimes, their reasoning is subtle and refined, or too critical and hasty, and is involved with purely emotional demands, which are ill-suited to scientific investigation.[88] To explain this better, I will present an extreme and somewhat idealized case of unacceptable reasoning. Someone says: (1) For the world to be *real,* it must be made as those who have not studied imagine it[89] to be. (2) Quantum mechanics shows us a world that is different from that mentioned in (1). (3) Therefore quantum mechanics postulates that the world is not real.[90] Formally, this reasoning (substantially an application of the *modus tollendo tollens*, as logicians say) would be flawless, but premise (1) is unsustainable.[91]

More often than not, the lay person may think of atoms as small balls having *all* the properties of macroscopic balls, except for their size. But we know this is not the case. For example, atoms do not have a temperature, even though they are quite real. Why then, in order to be real, must they have a well-defined position and momentum?

Dependence on the observer disturbs many people. They say that things must exist independently of the observer. After all, the world existed before the origin of humankind! This is quite a reasonable thesis, maintained by all realists (in particular, the dialectic materialists, Engels and Lenin). Is quantum mechanics really contrary to this thesis?

First, let us mention that it is doubtful today that we must really think of the observer as a conscious subject. It is probably sufficient to speak simply of interaction with a macroscopic system of measurement. But let us analyze this more closely.

In logic the properties or qualities of individuals are expressed by *predicates* which can be *monadic* (or one-place predicates) or *polyadic* (or many-place predicates). A monadic predicate, for example, "blonde," can only be applied to one individual; for example, "John is blonde." A two-place predicate, for example, "husband," can only be applied to an ordered pair of individuals; for example, "John is Mary's husband." Now the property of being blonde is possessed by John, independently of the existence of other individuals; but John cannot possess the property of being a husband if there is no woman who has married him. Now, would it make sense to say that because John cannot have the property of being a husband independently of the existence of a wife, John is not real or does not even exist? Evidently, it would be foolish to say so. Yet this conclusion is analogous to that derived by saying that a particle, because it does not have a well-defined s_z, independently of the observer or of the instrument of observation, is not real or does not even exist!

These considerations help us to understand that some of the psychological, or ideological, difficulties created by quantum mechanics are readily overcome when we accept this simple statement: "The modern study of microphysics has taught us that not all the properties of objects, which we believed to be expressed by monadic predicates or by relations among the objects themselves, are such; some are relations with the observer or with the macroscopic device of observation.

One can choose whether or not to accept this statement, but one cannot reasonably say that it has anything to do with the reality of objects![92]

Probably, the representatives of the Copenhagen school have sometimes taken imprudent positions in regard to the *objectivity* or the *reality* of physical objects. However, I suggest that these are not essential to the interpretation of the theory. For example, many (too many) times the "irrealism" of the Copenhagen school has been attacked by quoting Heisenberg's opinion, according to which quantum mechanics does not represent particles but, rather, our knowledge of the particles.[93] In my opinion, we can summarize the situation in less dramatic terms by examining two assertions:

1. Quantum mechanics describes our knowledge of particles.

2. Quantum mechanics *not only* describes our knowledge of particles, but *also* does something else.

Assertion (1) is a safe affirmation, whereas (2) is, to say the least, a matter of opinion. Scientific honesty should then suggest that we limit ourselves to *affirming* (1), without excluding the *possibility* that (2) is true as well.

If one writes to me about his activity, I can safely affirm that I know *what* he writes to me about his activity. In contrast, I am much less justified to affirm that I know his activity with certainty. But this does not at all imply that I think he is unreal!

In conclusion, let us recall a passage by Francis Bacon, which said: "The universe should not be narrowed down to the limits of the understanding, as has been man's practice up till now, but rather the understanding must be stretched and enlarged, to take in the image of the universe as it is discovered" (Francis Bacon).[94]

Therefore the reasonable, but very old position, shared by scientists, is that we should not attempt to prescribe to reality how it must be made and suggests instead that we *extend our understanding* to include those forms of reality that we learn through study, and which perhaps we might have considered absurd a priori.[95]

4.19. Determinism and indeterminism

Determinism–indeterminism is a much debated dilemma which is almost too tiresome to discuss again.

Humankind has always known that the world was not created in a random way. We obtain information about the universe that surrounds us through our senses, sometimes aided by instruments. This information in turn, allows us, to derive other information about non-observed facts. Therefore it is superfluous to try to gain knowledge of the latter by direct observation. We can say then that the capacity of the channel of sense data is redundant in respect to the messages we receive from the outside world.

Let us clarify this concept with a simple example. Suppose we want to know the movement of a stone thrown in the air (for simplicity, omitting the resistance of the medium). A priori it would seem necessary to determine experimentally the space coordinates of the stone as a function of time. This would represent considerable information. But physics teaches us that much less is needed: To be precise, it is sufficient to know the initial position and momentum. In this way, through the knowledge of the equations of motion, information is coded and redundancy is eliminated.

Galileo probably thought that this was true for any precision of measurement ϵ, no matter how small. If this is true, then the more precisely initial data are fixed, the more precisely the trajectory is known. When we make the initial information tend to infinity (by having ϵ tend to zero), the redundancy that can be eliminated through our code (the classical law of free fall) also tends to infinity. Nothing could have guaranteed the correctness of this extrapolation. It could have been right; but instead it was wrong!

Quantum mechanics teaches us that both the initial information and the redundancy that we eliminate can only be *finite*, because of the uncertainty principle and the probabilistic interpretation of $|\psi|^2$, respectively. That is all there is to the famous indeterminism!

What, then, is conceptually difficult? Has anyone ever stopped to reflect that it is conceptually difficult, instead, to accept the classical vision with its infinities?[96] One must always be suspicious when meeting infinity (even though potential) in physics.

The physicist investigates the behavior of nature and keeps to the results that this investigation suggests. It is simply absurd for him to want to prescribe the modality of this behavior a priori, particularly so that he should *want* nature to be deterministic or indeterministic.

What a great mind like that of Einstein thinks about time, space, gravitation, and so on is of exceptional interest to me as a physicist. But what Einstein thinks about the possibility of God playing dice or not, is of interest only as an anecdote, or at most as history. Is this presumptuous on my part? No, because the facts that Einstein has for knowing God's habits are exactly the same as mine – none! I do not think it presumptuous to say that if Einstein and I would bet on the drawing of lots tomorrow, we would have exactly the same probability of winning.

All we know today is that we possess an indeterministic theory T_n, which works very well (perfectly for an enormous class of phenomena).

To say that it is not the final theory is trivial, as it is to say that someone can work out a more general and more coherent theory T_{n+1}, which can predict new phenomena. This T_{n+1} could also be deterministic. We shall see. What is certain is that if T_{n+1} has the characteristics of greater generality, coherence, and fertility, physicists will accept it.[97]

Let us conclude with Dirac's recent affirmation: "It may be that in some future development we shall be able to return to determinism, but only at the expense of giving up something else, some other prejudice which we hold to very strongly at the present time." (Dirac, 1973, p. 7).

4.20. Causality

Finally, let us consider the *principle of causality* and the affirmation that it too has been jeopardized by quantum mechanics. In making this affirmation, we are in effect referring to Leibniz's concept – nothing happens without a cause. This concept is tightly bound to the principle of sufficient reason, which states there must be a reason for everything that exists to be as it is and not otherwise. Clearly, this principle is nothing but a form of strict determinism. Therefore, to note that it is negated by indeterminism (even merely probabilistic) is simply trivial.

However, one wants to see more than determinism in the concept of cause. If we accept this position, in my opinion it is wrong to "blame" quantum mechanics, simply because a certain naïve concept of physical cause had already been challenged a long time before.

Naturally, we cannot ignore the fact that the literature concerning the concept of cause is extensive and ranges from Aristotle to Leibniz, to Hume, to Kant and to a whole series of modern scholars[98] (who more and more had to consider the results of physics and science in general). Because I do not think the subject, at least in its traditional form, is very significant in modern physics, I shall only discuss a few considerations. Nevertheless, I believe B. Russell a bit exaggerated in affirming that: "The law of causality, I believe, like much that passes muster among philosophers, is a relic of a bygone age, surviving, like the monarchy, only because it is erroneously supposed to do no harm" (Russell, 1963, p. 132).

Starting from Laplace's statement: "We must consider the present state of the universe as the effect of its preceding state and the cause of that which follows," we can immediately see that the concept of cause is of little use. Naturally, Laplace meant to say that there are precise (deterministic) laws that tie the prior state of the universe to

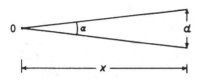

Figure 4.19

the posterior one. But to say this, it is not necessary to disturb the venerable concept of cause!

This becomes extremely clear with general relativity, which represents essentially an attempt to interpret the physical becoming of the universe in terms of laws concerning its geometric structure (space–time structure). If the program had completely succeeded and quanta had not appeared, each particle would have had its own precise world line and the set of world lines would have formed a geometric structure, subject to strict laws. To speak of cause in such a universe would have been quite silly. Let us give a very simple example in ordinary space. Consider two rays issued from O (Fig. 4.19), at a very small angle α. Who would think of saying that the configuration existing at point O is the *cause* of the fact that the distance d between the two rays at the abscissa x has become αx? Moreover, if in a configuration of this kind in space–time x were to represent time, we would not understand why there should be asymmetry between cause and effect and why we should insist that cause necessarily *precedes* effect.

As already discussed if for *causality* we mean strict *determinism*, then it is tautological and not very enlightening to affirm that modern microphysics and the current interpretation of quantum mechanics have threatened the causality principle. Some physicists wanted to point out, almost to excuse themselves (for what?) that ψ evolves in a "causal" way (i.e., it obeys Schrödinger's equation), whereas *events* can be predicted only in a probabilistic way. Why this consideration should "console" those who adhere to the old principle of cause, is a mystery.

Actually, I believe, these questions are really impossible to understand, if one wants to ignore at all costs the fact that the concept of cause has essentially an anthropomorphic origin and only applies figuratively to the physical universe.[99]

We know that our actions can change at will the structure of the world lines that surround us, within certain limits. That our freedom of action is real or illusory is not important here. The fact is that each of us feels psychologically sure of having this freedom.

When following an act of will, we change something in the surrounding world, we say that we have *caused* that change. Without us and our act of will, the change would not have occurred. Generally, things happen in this way. Our will (through processes that we shall not analyze here and which, moreover, are almost totally unknown to us) brings about some physical processes in our bodies, which are tied by physical laws (known even on an intuitive level) with other processes of the outer world, which we want to realize. For example, if we want to break the glass of a window, we move our arm and hand in a position to throw a stone, thus using the laws of the motion of falling bodies and the fragility of the glass to reach our goal. Humans have believed in animism for thousands of years[100] and easily assumed that there is exactly the same relationship between the stone and the breaking of the glass as there is between our will and the movement of our hand. The stone that moves becomes the *cause* of the breaking of the glass. Indeed, we even imagine a *causal chain* between the different physical processes that follow one another and give this concept a meaning independent of the existence or nonexistence of an act of will that started the sequence of events. But if the situation is like this, is Russell not right in affirming that the concept of cause is a relic of the past?

Here I would like to distinguish between two uses of the word cause; one absolutely useless, the other containing a valid and useful concept. The first is the use of Laplace, for which the prior state of the universe is cause for that which follows. It is sufficient to say that physical laws exist. The second is exemplified by saying that the arrival of the stone is cause for the breaking of the glass. This use appears to be perfectly reasonable and cannot be given up. How can we dispose of it if we want to decide who has to pay for the glass? To illustrate this second use we must make a more in-depth analysis.

As mentioned, we have the impression of being able to change freely (within certain limits) the world lines that surrounds us. What does this mean? Does it perhaps mean that we can make the world around us no longer obey physical laws? No, with a little reflection we are convinced that this is not so. And then?

The key statement to understanding how to deal with this problem is *cause must always precede effect*. This condition of antecedence leads us immediately to surmise that the anisotropy of time must enter into the question, as well as the second principle of thermodynamics and the growth of entropy. In fact, this is so (see Grünbaum, 1973, Chap. VII).

Humans represent a notable source of negentropy for the surrounding world. Naturally, we derive negentropy ultimately from solar light and through food, but this is not important. When the world lines of the bodies that surround us enter into interaction with the world lines of the parts of our body, the former can undergo a conspicuous *injection* of negentropy. Interactions are in perfect agreement with the laws of physics and involve no violation. An interaction takes place during the time in which an external body (or system of bodies) can be considered as part of one and the same system together with the human body. Afterward, the external body branches off from the overall system and gradually increases its entropy. This is the case of the stone that is thrown.

Our body is provided with parts that are admirably suited to interact with the outer world and to transfer to it great amounts of negentropy. It is mainly the hand that interacts with great efficiency.[101] Therefore it is not surprising that generally a requisite of *low entropy,* or *improbability,* is associated with the concept of cause. Cause is a more improbable event than all the others that are happening around us.[102] Consider two successive phenomena A and B, linked by physical laws. The more improbable A is (in the part of the universe being examined), the more likely we are to recognize that A is cause for B. Furthermore, the more improbable B is, the more we tend to ask, what is its cause – that is, what is system A, even more improbable than B, with which B has interacted. We do this because assuming that the low entropy of B is a spontaneous fluctuation is much less reasonable (much less probable) than assuming the existence of a low entropy system A, with which B has interacted. (Recall Schlick's example of the footprints in the sand, §3.16).

On the basis of these considerations the concept of cause cannot have a precise definition, and acquires a value that is mainly practical. For this reason the individuation of cause is often a matter of opinion or convention. In fact, in the case of the stone and the glass, let us assume that stones often approach the glass (some boys in the neighborhood usually throw them for fun), which is protected by a grate. If one day the grate is removed and the glass is broken, we can say that the removal of the grate was the cause of the breaking of the glass. If the removal of the grate is more improbable than the approaching stones, we can presume to say just that.

Moreover, some examples derived from medicine and its diagnoses are very clear. For instance, let us assume that I have a toothache and I affirm that the cause of my pain is a cavity in my tooth. Why do I not say instead that the cause is some particular nerve that carries a certain signal from the tooth to my brain? – simply because it is very probable (virtually certain) that I have that nerve and it functions in that way, whereas having a cavity is a much more unusual event.

Finally, let us consider a more technical aspect of causality which concerns an important rule in some aspects of modern physics. To define it, we say that because of the principle of causality, no signal can be received before being sent. More precisely, if an event A sends a signal that is received by an event B, A must precede B.[103] We can see in the following way that this law is closely connected to our sense of freedom of action and to the second principle of thermodynamics, that is, we can influence the events of our future but not those of our past. Assume that I break a leg coming down the stairs. If I could send this information back in time, I could warn myself and therefore avoid going down that stairway! Then an event could happen and not happen,[104] which is absurd. To remove the absurdity we should recall the tragedy of Oedipus and admit that our freedom is only illusory and that in reality we cannot avoid *fate*, even if we know beforehand. But who still believes in these things?

4.21. The inductive inference

Now we have all the facts that we need to analyze the question of induction in physics.[105] This is a central problem, which has been mentioned several times. Moreover, two centuries after D. Hume, this problem not only remains unsolved, but has become increasingly controversial (see, e.g., Salmon, 1966; Hesse, 1974). In fact, natural science has aimed at being more and more *exact* and making objective and undisputable affirmations. Yet this science is based mainly on a process (the inductive) that cannot be justified rigorously in any case![106] In fact, a general opinion is that Hume's criticism is correct. If by justifying we mean deriving with *deductive* process, it is even obvious that inductive inference cannot be justified. On the other hand, to say: "Everything has gone well until now and we have been able to construct science, therefore everything will continue to go well," means clearly to apply that inductive process that we do not know how to justify. This is circular reasoning.

Thus along with K. Popper, we can assume an antiinductionistic attitude (Popper, 1959, 1965, 1978). For Popper, scientific laws and theories cannot be confirmed, but can be falsified, and experimental science proceeds by successive falsifications. At the most, we can speak of corroboration, when a scientific hypothesis resists one or more potential falsifiers.

As mentioned several times, this image of the history of physics as a chain of successive falsifications arises primarily when one chooses to ignore the modern concept by which a theory T has a domain of validity D, whose specification cannot be disjoined from the definition of T. The physicist is *sure* that T is valid within D, but does not commit himself to what can happen outside of D. An experiment outside of D, which does not agree with T, does not falsify the assertion that T is valid within D. Nevertheless, this certainly does not solve the problem of inductive inference. How can the physicist be sure that T is always valid within D, having performed only a finite number of experiments?

At this point I think that for our purposes, it is worthwhile to set aside the question "*quid iuris?*", which concerns the legitimacy of the inductive process, to consider the question "*quid facti?*", which concerns instead the analysis of how physicists use induction today. We realize that physicists have long given up some classical forms of induction, which they know are not valid. They have learned at their own expense, by repeatedly encountering unpleasant counter examples, the limited reliability of those processes.

A typical process that physicists have abandoned is that of inference *from many to all*. Let us examine *many* objects of the same class. We find that they have a certain property; we then conclude that *all* the objects of that class have that particular property. We can describe this process using more precise symbolic language.

The assertion that a physical object belongs to a certain class (e.g., that of stars, of solid bodies, or of electrons, etc.) means that it has a certain property or a set of properties A. Let us assume that we have ascertained many times that objects of class A also have the property B and that we have not found any counterexamples. Then we are tempted to formulate a general law. To do this we shall use A and B as *monadic predicates*, meaning that Aa_1 and Ba_2, respectively, indicate that the object a_1 had the property A and that the object a_2 has the property B. We shall also use the *universal quantifier* $\forall$ (which is read *for all*) and the implication sign $\rightarrow$ (which is read *if . . . , then*). The general law is written

$$\forall x(Ax \rightarrow Bx) \tag{4.98}$$

In words we read: for all x, if x has property A, then x has property B. Let us assume to have found an object a_1 which has both property A and property B. Using the symbol of *conjunction* $\wedge$ (which is read *and*), we write

$$Aa_1 \wedge Ba_1 \tag{4.99}$$

We say that a_1 represents a *positive instance* of expression (4.98). The simple rule says that the more positive instances we find,[107] the more the general law, expression (4.98), is confirmed.

Now the modern physicist knows that no matter how numerous the positive instances are, it is imprudent to assert the general law, equation (4.98). This is true for classical physics as well, but becomes particularly evident for microphysics. Let us give an example.

Let us assume that A indicates the predicate "is a uranium atom"[108] and B the predicate "does not decay within an hour." Further, assume that we are able to observe one uranium atom at a time and to check that it has the property B, that is, that it does not undergo radioactive decay during one hour of observation. If we observe one hundred, one thousand, one million of such atoms, we are virtually sure of finding all positive instances of the kind of expression (4.99). Yet expression (4.98) is false!

In fact, the mean lifetime of uranium is extremely long. It takes 4.5 billion years for any quantity of uranium to halve itself through radioactive decay. This means that less than one in fifty thousand billion atoms decays within an hour. If we observe successively one million atoms, we are practically sure of finding all positive instances of type, expression (4.99). We would have to repeat the experiment millions of times (for a million atoms every time) to have a nonnegligible probability of finding a negative instance!

One might think that the radioactive decay of uranium is a very particular and exceptional case in physics. On the contrary, it is quite within the rule. Much as it might seem paradoxical, modern physics has accustomed us to think that very rare cases are quite common. Moreover, a paradox discovered by C. G. Hempel[109] suggests that something is wrong with the *many-to-all* rule. Let us illustrate this with an example derived from physics.

Assume that A indicates the predicate "is an electron" and B the predicate "is negatively charged." Then expression (4.98) states the general law that all electrons are negatively charged (which everyone believed to be the case before the discovery of Dirac and Anderson). Every time one observes an electron and ascertains that it is negatively charged, one has a positive instance, expression (4.99) and a confirmation of the general law. But expression (4.98) is logically equivalent to the assertion that all that is not negatively charged is not an electron. By introducing the symbol of *negation* $\neg$ (which is read *not*), we can write

$$\forall x(\neg Bx \to \neg Ax) \tag{4.100}$$

According to expression (4.99), this law is confirmed every time we find an object a_2, such that

$$\neg Ba_2 \wedge \neg Aa_2 \tag{4.101}$$

that is, an object that is not negatively charged and is not an electron.

This object, for example, could be a photon. In this case the observation of a photon would confirm expression (4.100) and necessarily expression (4.98) as well, which is its logical equivalent. So the observation of a photon would confirm that all electrons are negatively charged! Although several writers have tried to overcome the paradox, no physicist is prepared to accept such an absurdity. But physicists do not need to *overcome* the paradox at all, because they know very well that the *many-to-all* rule is incorrect.[110]

But then we wonder how does the physicist proceed, or how is science in general possible? The physicist recognizes only one rule of induction as valid, which is represented by the space–time invariance of laws, already discussed in §1.4. It tells us that nature acts *here and now* as it does *there and then* (both in the past and in the future). It is a law *from one to all*. This law, like all laws of induction, is not logically justified. Yet at least in a limited region of space–time, of human dimensions, it is, as we have already noticed, a necessary condition for living and thinking. Therefore it becomes a postulate, or better, a way of reasoning, even at an intuitive level. Scientific investigation extends the validity of the postulate to much wider ranges than the human one.[111]

In macroscopic physics it is sufficient to observe a phenomenon once in order to conclude that provided that the conditions are identical, it will always occur in the same way.

Naturally, things are different when dealing with: (1) a macroscopic process,[112] under conditions chosen at random within a certain domain, and (2) a microscopic process, under conditions fixed or chosen at random within a certain domain.

In both these cases we must repeat the experiment a large number of times and calculate probabilities. We have only to consider probability as a physical quantity and apply the space–time invariance to it as well. The distribution of probability for a certain physical quantity, measured in a series of experiments of type (1) or (2), is the same wherever and whenever we perform the experiments.

Let us see how these rules are applied in some concrete examples. Suppose that an astronomer, examining n stars chosen at random, finds that n_1 of them have the property B, while $n_2 = n - n_1$ do not. He will say that the probability that a star will have property B is (n_1/n) $\pm$ ϵ, ϵ being on the order of $1/\sqrt{n}$. If he finds $n_1 = n$, he will not conclude that *all* the stars have property B, even if n is very large; he will say instead that the probability that a star does not have property B is on the order of $1/n$ or less.[113]

Taking an example from microphysics, let us assume that we observe a certain type of unstable particle n times under identical conditions.

If the particle decays n_1 times in a way M_1, n_2 times in a way M_2, and n_3 times in a way M_3, we say that M_1, M_2, M_3 have the respective probabilities $(n_1/n) \pm \epsilon$, $(n_2/n) \pm \epsilon$, $(n_3/n) \pm \epsilon$. But we do not exclude that there are other modes of decay whose total probability is on the order of $1/n$ or less. In particular, observing a million atoms of uranium for one hour, and seeing that none of them decay, we do not conclude that uranium is *stable*. We say that the probability that an atom of uranium decays within one hour is less than one millionth.

Now, from the confirmation of simple propositions of the type already examined, let us go to the confirmation of more general hypotheses or theories. To *confirm* a theory means to find that it has a domain of validity that is "fairly wide." We do not and should not confirm any theory absolutely, independently of the domain of validity. For the same reason, to give an absolute *falsification* or *confutation* of a theory means to establish that its domain of validity corresponds to an *empty* class of phenomena. This is a very difficult task which it is doubtful has a well-defined operational meaning.

It is interesting to observe that even when dealing with a well-determined phenomenon O, it is sometimes difficult (or even impossible) to falsify a scientific hypothesis H. The logical scheme would be the following: Let us assume that we have proven that from the hypothesis H there follows a fact O that is experimentally observable.[114] We write

$$H \to O \qquad (4.102)$$

Let us also assume that we have observed $\neg$O, that is, a fact that excludes O; from it one derives (*modus tollens*) $\neg H$, the falsification of H.

But as early as 1906, P. Duhem had observed that the physicist can never subject an isolated hypothesis to verification. We always deal with a group of hypotheses and experiment cannot tell us *which* of them is refuted. In other words, together with H, we consciously or unconsciously assume one or more auxiliary hypotheses A, so that what we subject to verification is the assertion

$$H \wedge A \to O \qquad (4.103)$$

To have found $\neg$O can contradict H, but it can also contradict A without concerning H. For example, H can be Thomson's hypothesis: "electrons are particles" and A, the auxiliary hypothesis, "particles move according to classical mechanics." Electron diffraction ($\neg$O) does not falsify H, as we might think at first glance,[115] but it falsifies A!

W. V. Quine has even affirmed that: "Any statement can be held true come what may, if we make drastic enough adjustments elsewhere in the system" (Quine, 1953, p. 43). This affirmation is probably too strong, as A. Grünbaum has maintained (Grünbaum, 1971, p. 69). Nevertheless, the numerous examples (especially in contemporary physics) in which this situation arises should be sufficient to make us very careful about *falsifying* scientific hypotheses or theories. It is part and parcel of the methodology of the physicist to keep his eyes open and to make sure every time that he has refuted H and not A.

Incidentally, note that in the case in which expression (4.103) is valid, the position of a Popperian is desperate. In fact, to say that $\neg O$ falsifies H, one must be sure that A is true. But that is impossible, given his assertion that hypotheses can only be falsified or not falsified![116] For example, let us assume that we are about to verify the results of general relativity by observing a star near the sun during an eclipse. Let us also assume that we do not find the deflection predicted by Einstein. Have we falsified relativity H? Yes, if we assume the auxiliary hypothesis A that our telescope works in the same way in which all telescopes have worked up until today. No, if we do not assume the validity and certainty of the inductive process, and so are not sure of A.

The important case of quantitative relations of the kind $f(a_1, a_2) = 0$, represented in Figure 1.5 deserves particular attention. Let us assume that we are dealing with a completely new process whose course is unknown. One samples at random[117] n values of a_1 and measures the corresponding values of a_2, drawing the crosses as in Figure 1.5. One then chooses a two-variable function f, such that the curve $f(a_1, a_2) = 0$ cuts through all the crosses. The property B common to all the crosses is that of meeting the curve $f(a_1, a_2) = 0$. We say that the probability that a new measurement (with a_1 chosen at random in its interval) does not have property B, that is, a new measurement does not agree with the law $f(a_1, a_2) = 0$ is less than $1/n$.

As a matter of fact, the methodology previously explained is a bit abbreviated and simplified. To be really rigorous, we must start from a hypothetical curve *preexistent* to the n experimental trials. It seems like a vicious circle, but we can proceed as follows.

First, we carry out a certain number of measurements and propose a hypothetical curve C that agrees with them. Next, we proceed to the n measurements (n being very large), with abscissas chosen at random in the interval. Two cases are possible:

1. All the n measurements agree with the curve C. We then assume this curve to be the graphic image of the law we are looking for, and say that the probability that a new measurement does not agree with C is of order $1/n$ or less.

2. Some of the n measurements do not agree with C. We then look for another hypothetical curve C' that agrees with all the measurements made. Next, we proceed to make n' new measurements.

Two cases are now possible:

1'. All n' measurements agree with C'. We then assume that this curve is the graphic image of the law we are looking for and say that the probability that a new measurement does not agree with C' is of the order $1/n'$ or less.

2'. Some of the n' measurements do not agree with C'. We then look for another hypothetical curve C'' that agrees with all the measurements made; and so on.

This procedure may appear very complicated, but it represents the scheme of what we should do to be really sure; in practice some abbreviations are tolerated. Those who are accustomed to the *safe* laws of classical physics may be surprised. But let us see what this procedure really entails.

For classical laws, n is very large, not only because an enormous number of direct measurements have been made, but also because the number of indirect confirmations is enormous. A phenomenon that does not proceed in the predicted way would have detectable repercussions on many other phenomena. When n is very large, the physicist is perfectly accustomed to neglect $1/n$. This is what we do, for example, when we say that the entropy in an isolated system *always* increases.

But modern physics necessitates being careful when n is not very large. If, for example, one measures the luminosity of the sun as a function of wavelength, point by point, with a spectroscope, one very probably will begin by finding the continuous spectrum, that is, a Planck's curve (Fig. 4.5). But it is a mistake to conclude that this is the correct curve for *all* the points of the spectrum. One fully realizes this when hitting on a Fraunhofer line! Analogous situations present themselves with the *resonances* of today's high energy physics (see §4.26).

In conclusion, it is very dangerous to *regularize* an experimental curve and to say with certainty that between the observed points there are no narrow peaks.[118]

Before ending these brief observations on induction, we should note another attempt to link the procedure of inductive inference to probability. This is *inductive logic*, founded by R. Carnap and developed gradually in various forms (Carnap, 1950, 1952; Hintikka, 1966; Carnap and Jeffrey, ed., 1971; Niiniluoto and Tuomela, 1973).

According to the definition of R. Carnap and W. Stegmüller, inductive logic is characterized by these fundamental concepts:

1) Every inductive inference, in the general sense of a non-deductive and non-demonstrative conclusion, is an inference based on probability. 2) Therefore inductive logic, as a theory of the principles of inductive inference, is the same thing as the logic of probability. 3) The concept of probability, which must serve as the foundation of inductive logic, is a logical relation between two propositions or assertions, and precisely, is the degree of confirmation of a hypothesis on the basis of certain premises. 4) The so-called frequentistic concept of probability, as it is used in statistical research, is in itself an important scientific concept, but cannot be used as a basis for inductive logic. 5) All the principles and theorems of inductive logic are analytical. 6) Therefore the validity of inductive inference does not depend on a synthetic presupposition, as for example, on the much debated principle of the uniformity of the world (Carnap and Stegmüller, 1959, p. 1).

Generally speaking, inductive logic attempts to establish a *confirmation function* $\mathscr{C}(H/E)$, which represents the probability that the hypothesis H is true when we have the evidence E. The probability is calculated by complicated rules of computation of linguistic combinations, which cannot be described here. It is clear that it cannot be a matter of *frequence* but rather of a *logical* probability, or a *rational belief*.

To explore these subjects in depth would be digressing, instead let us point out two facts. First, as we have seen, the physicist tends to arrive at the *assertion of a probability*, whereas the inductive logician tends to give the *probability of an assertion*. That the latter approach can be successful in the case of universal assertions is made extremely doubtful by the fact, as mentioned, that in physics very rare cases are very common. Second, the application of inductive logic, even using an extremely simple and elementary language, requires the computation of an astronomical number of combinations, which make its utilization practically impossible (see, e.g., Suppes, 1966, p. 21).[119]

4.22. Quantum electrodynamics

In the second half of the 1920s quantum mechanics had obtained a decisive foothold as the theory most likely to interpret all atomic phenomena correctly. But the role of electromagnetic fields was not clear, for their interaction with electrons led to the emission or absorption of quanta $h\nu$ according to some mysterious mechanism. This was the question: Is the electromagnetic field a classical entity or can it be subjected to the rules of quantum mechanics as well?

It was discovered quite soon, by M. Born, W. Heisenberg, P. Jordan, and in particular, by P. A. M. Dirac (1927) that the *quantization of electromagnetic fields* was possible. To do so we must divide the field into its degrees of freedom (e.g., into plane waves as already done many times), and consider for each degree of freedom the physical quantities E, H or energy, momentum of the field as operators in a convenient Hilbert space.

Taking into account a degree of freedom of the field, we find that the mathematical form of the equations which it obeys is perfectly analogous to what is valid for a *linear oscillator* (such as the pendulum or the tuning fork). We know perfectly how to quantize the linear oscillator and obtain the remarkable result that its energy levels are *equispaced*. If v is the frequency, hv is the spacing.

Thus photons derive naturally from the general rules of quantum mechanics, without ad hoc hypotheses. A degree of freedom of the field can pass from one energy level to another in the same way as an atom can; but given the level spacing, the jump can only be a multiple of hv.

The theory of the quantization of electromagnetic fields was completed by N. Bohr and L. Rosenfeld in 1933, with a fundamental work on the measurement of the quantities of the field, which cast light on limitations analogous to those of Heisenberg's principle.

The *quantum theory of electromagnetic radiation* can be described as follows. Let us first consider a system of atoms and an electromagnetic field, apart from their interaction. Each atom will be in a well-determined energy level and the same will be true for each degree of freedom of the field. This situation can go on indefinitely and constitutes a *stationary state* of the system. But in reality, the electromagnetic field and the electrons interact, and this interaction can be dealt with as a disturbance of the stationary state. After precise calculations, we find that the disturbance has the effect of occasionally making an atom pass from one energy level E_m to another E_n, at the same time, making a degree of freedom of the field pass from one energy level to an adjoining one, conserving the total energy. Therefore we have $E_m - E_n = hv$, where the photon is emitted if $E_m > E_n$ and is absorbed if $E_m < E_n$. We can calculate the probability with which each of these processes occur (transition probability), agreeing perfectly with experience.

We recall that wave mechanics was born from a need for symmetry between the electromagnetic field, of which a particle aspect had been discovered, and ordinary particles, in which, conversely, one expected to find a wave aspect. Now the quantization of the electromagnetic field seems to break this symmetry once again. If photons are quanta of the electromagnetic field, should not electrons in turn be quanta of their wave field ψ?

This idea might seem strange if we think that although the electromagnetic fields E and H are observables (i.e., are measurable), ψ is a mathematical quantity that enters into the algorithm of quantum mechanics but is not observable. Yet P. Jordan and E. Wigner succeeded in demonstrating that ψ, which enters into Dirac's equations, can be quantized in a way analogous to that of the electromagnetic field. In this case we speak of *second quantization*.[120] In this way electrons and positrons somehow become the quanta of the field ψ and, as happens to photons, they can be emitted or absorbed (but only in pairs) in the processes of interaction. We shall soon see how a very evident and intuitive interpretation of these processes can be given.

The science whose object is the theory of the electromagnetic and electronic fields quantized and interacting is called *quantum electrodynamics*. Through several improvements brought about by E. Fermi, W. Pauli, S. Tomonaga, J. Schwinger, R. Feynman, F. Dyson and many others, it has become one of the most precise and highly perfected instruments of modern physics.

In order to see what phenomena quantum electrodynamics deals with and how it describes them, let us make two important preliminary comments. The first concerns antiparticles. In the equations of quantum theory that govern the motion of particles, energy and time always appear associated. To be precise, it is their product that appears. This explains equation (4.58). If we change the signs of energy and time, simultaneously, the product remains unchanged. In regard to Dirac's equations we have already encountered some negative energy solutions and have observed that they are associated with the antiparticles of the ordinary positive energy particles. By following these indications and developing the theory, one can show that an ordinary particle, which moves forward in time, is described by the same mathematical formulas that describe an antiparticle which moves *backward* in time.[121] Without entering into the difficult question of the physical reality of this remarkable conception, we shall content ourselves with the *mathematical* equivalence of the two phenomena. An electron that follows a given path is equivalent to a positron that follows the same path in the opposite direction in time.

The second observation we must make concerns the conservation of energy. This conservation is respected in quantum mechanics as shown, for example, by equation (4.31). But if we want to verify precisely that energy is really conserved, we must have a long time at our disposal. In fact, equation (4.58) tells us that if we have a very little time Δt, the inaccuracy ΔE with which we can determine the energy is very large. In light of the current interpretation of Heisenberg's principle, we say that the energy of a particle or a system *is* indeterminate and that the conservation of energy is valid only over long periods of time. For very brief times we can also find notable fluctuations. Recalling the relativistic equivalence between mass and energy $E = mc^2$, we see that even the mass of a particle can fluctuate and that the smaller the interval of time we consider, the more conspicuous is the fluctuation. As a result, particles can be emitted or generated in a system, even when the system does not possess the necessary energy, provided that such particles exist for very brief times and are immediately reabsorbed by the system. We then speak of *virtual particles*.

A static charge can with this mechanism generate and absorb photons and thus produce its own electromagnetic field. But these photons are not destined to appear in experimental reality because, as stated, they must be reabsorbed by the charge itself, or by another charge, immediately after they are generated. These are called virtual photons. Those that are most capable of moving away from the charge are the least energetic ones. Those with greater energy instead have a shorter life and will be confined to the immediate surroundings of the charge. Each charged particle thus appears surrounded by a cloud of photons whose extreme borders are constituted by low energy photons which can go very far before being reabsorbed, whereas the center part is formed by high energy photons which are limited to extremely short times and distances. Developing the calculations, we find that this virtual photon cloud is equivalent to the well-known Coulomb field which exists around the particle and decays as $1/r^2$.

Going from the electromagnetic to the electron field, we can also speak of the creation and annihilation of electron–positron pairs in a virtual sense. As we know, a photon γ with sufficient energy can give rise (near a nucleus) to a pair. If the $h\nu$ energy of the photon is less than $2m_0c^2$ (energy at rest of the pair),[122] the *real* creation cannot come about, but in analogy with what we just stated for an electron, the virtual creation can occur for a sufficiently brief time.

A graphical illustration, suggested by R. Feynman in 1949, makes the various types of electrodynamic interaction particularly evident.

A *Feynman diagram* has three types of elements: external lines, internal lines, and vertices, which are the points in which the lines meet. The external lines, each of which has only one vertex in common with the rest of the diagram, represent the initial (or *incident*) particles and the final (or *generated* or *scattered* particles).[123] The internal lines each join two vertices of the diagram and represent the virtual particles. The vertices represent elementary processes of interaction.

A single elementary electrodynamic process is always represented by a diagram such as that of Figure 4.20. There are three external lines that converge in a vertex. The wider lines correspond to electrons; the wavy line, to a photon. The electron lines are provided with an arrow:

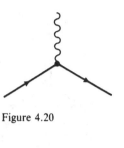

Figure 4.20

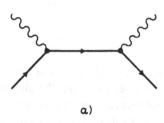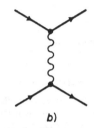

a)   b)

Figure 4.21

the one directed toward the vertex indicates the incident electron; the one pointed away from it indicates the scattered electron. But, according to what we said about the mathematical description of particles and antiparticles, each wide line can represent, instead of an electron, a positron that moves in the opposite direction (in this case, from right to left). The wavy line does not have an arrow because, as we shall see, photons are their own antiparticles. Thus the line can represent either an incident photon (from above to below), or an emitted photon (from below to above). This elementary diagram could by itself represent indifferently, according to how it is read:

1. An electron that changes state by emitting or absorbing a photon.
2. A positron that changes state by emitting or absorbing a photon.
3. An incident photon that creates an electron–positron pair.
4. A pair that is annihilated generating a photon.

Actually, these isolated processes cannot occur (because they would not conserve momentum), and we have to associate more than one process, or elementary cell, of the type described.

Thus, for example, Fig. 4.21(a) represents the collision of an electron and a photon and the consequent scattering or change of direction of both (Compton's effect). As we see, the process is interpreted as follows: the incident electron absorbs the incident photon (left vertex) and is converted into a virtual electron (internal line). This emits another photon (right vertex), becomes a real electron again and moves away. The experimenter does not realize that the final products are a

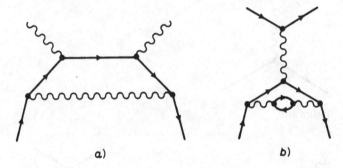

Figure 4.22

different electron and a different photon from the incident ones (which besides would be meaningless). He observes only that the electron and photon are in states that differ from the incident ones. It is obvious then that the same diagram (read from right to left) can represent the collision of a positron with a photon. But it can also represent the annihilation of a pair, with the creation of photons. An electron comes from the left and a positron from the right. The electron at the left vertex emits a photon and becomes a virtual electron, which in turn is annihilated at the right vertex with the positron, emitting a second photon. Figure 4.21(b) represents the collision of two electrons (or of two positrons). The first electron emits a virtual photon and changes state. The virtual photon is absorbed by the second electron which changes its state, too.

The processes we have illustrated are very simple. But the same processes can occur in more complicated and less probable ways. Let us give some examples. The phenomenon described by Fig. 4.21(a) can occur as in Fig. 4.22(a). Here the incident electron, before emitting a real photon, emits a virtual photon, which is then absorbed by the diffused electron. The phenomenon of Fig. 4.21(b) can occur as in Fig. 4.22(b). One of the incident electrons emits a virtual photon preliminarily. This gives rise to a virtual pair which is annihilated creating a new virtual photon, which is finally absorbed by one of the final electrons.

From these examples we see that many increasingly complicated Feynman diagrams can correspond to the same physical phenomenon. Fortunately, the electrodynamic interaction is sufficiently weak. Because of this, the higher the number of vertices, the smaller the correction that the process represented by the diagram brings to the fundamental one, Figure 4.21(a) or (b). We say that the more the *order of the perturbation* increases, the more negligible it becomes. More precisely, the characteristic parameter of electrodynamic interactions is the *fine structure constant*

$$\alpha = 2\pi \frac{e^2}{hc} \simeq \frac{1}{137} \tag{4.104}$$

which tells us roughly how much smaller each order of perturbation is than the preceding one. As can be seen, the decrease is sufficiently rapid. In many cases one can limit oneself to considering the smallest order, that is the Feynman diagram that describes the phenomenon with the minimum number of vertices.

The phenomena for which Figure 4.22(a) and (b) differ from 4.21(a) and (b) are phenomena of interaction of an electron with the field it has generated. We say that these phenomena bring *radiative corrections* to the simple behavior that electrons would have in their absence.

The continuous creation and annihilation of virtual pairs that occur wherever there is an electromagnetic field is particularly interesting. Some effects of this process are analogous to those brought about by the polarization of a dielectric medium which we saw in §2.13. For this reason we speak of *vacuum polarization*. Because of its effect, a charged particle in motion carries not only an electromagnetic field but also a whole cloud of virtual particles. Thus the charge of the particle is partially screened and from a distance appears different from what it really is. But the *naked* particle is not observable. It is worthwhile therefore to take the measured value for the charge and to disregard the screening. We then naturally have to build a theory in which the constants of the naked particle do not appear but, rather, those of the particle accompanied by its field.[124]

The radiative corrections also affect the energy levels of atoms. The displacement of some levels of hydrogen detected by W. E. Lamb and R. C. Retherford in 1947 is a typical example. Modern quantum electrodynamics accounts perfectly for this, as well as for a slight difference that occurs between the real magnetic moment of an electron and the theoretical value that derives from Dirac's equation (*Bohr's magneton*).

Before leaving the subject let us make some general considerations. The amazing success of quantum electrodynamics has been recognized for a long time. Yet those who comment on or criticize the principles of quantum mechanics rarely refer to it. Instead they are likely to remain bound to the perplexities and battles of the twenties or thirties. How is it possible to continue to point out as *indigestible* the fact that a particle does not simultaneously have a well-defined position and momentum, when we have to accept facts much further from simple intuition, for instance, as the fact that every region of space swarms with (infinitely many) ghost particles, which appear and disappear without being able to locate them in any way?[125] On the other hand, if we refused to accept their existence, we would place ourselves in the position of those phenomenists who, in spite of Boltzmann's theoretical successes, refused to accept the existence of atoms!

As H. Mehlberg rightly points out (Mehlberg, 1967, p. 46), we can extend the argument set forth by Poincaré for molecules to the particles of electrodynamics. Particles are real[126] because they can be counted (with Feynman diagrams). Once more, the objects of physics are associated with natural numbers.

We can now state with greater precision what we saw in §1.9, concerning measurements. We said that the result of a measurement can never be a real number (infinitely precise). We then ask: what can be said in the case of *counting* particles? Are not integral numbers real numbers (infinitely precise) as well? To answer this question, we must consider that in order to *count* particles we must have them interact with a given system and that the interaction is only probabilistic! In this way we shall never be sure of the exact number or, in other words, we shall obtain a number $n \pm \epsilon$, in analogy to what we obtain in other types of measurements.

Finally, regarding the *intension* and *extension* concepts of §1.6, we can ask, for example: What is the extension of the concept, electron? By now it is difficult to think that this extension is constituted by a well-determined set (finite or infinite) of individuals. First of all, the lack of permanence of these individuals would make it necessary to change the set from instant to instant. But should we admit only real electrons to the set, ignoring virtual electrons? This seems very arbitrary if we think that there are virtual electrons that, with an exchange of very little energy with the environment, can become real and vice versa. Furthermore, real and virtual electrons seem to have practically the same functions in Feynman diagrams. On the other hand, if we admit all virtual electrons to the set, we admit an infinite number of possibilities, some of which are extremely far from coming true, and are quite different from our concept of real physical objects.

Because this same reasoning can be applied to all the particles that constitute the world, we ask if some logical distinctions made in the past did not originate in the erroneous presupposition that physics is necessarily classical physics. In particular, we can even ask if, besides intensions, extensions exist in physics. Perhaps it is excessive to give a drastic negative answer to this question. Nevertheless, the interpretation of the logical concept of extension may definitely need a profound revision in modern physics.

4.23. The atomic nucleus

Once the structure of the atom, composed of nucleus and electrons, was clarified, it was natural that scientists began to ask themselves how the nucleus is made. It was sufficiently evident from radioactive phenomena that the nucleus was not *elementary* but, in turn, composite. It was known that the nucleus of some elements could emit α particles (i.g., nuclei of He), β particles (i.e., electrons) and photons γ. It would have been difficult for an elementary particle to have these properties.

Because a nucleus of atomic number Z has a positive charge equal in absolute value to Z times that of the electron, the following idea could arise spontaneously: The nucleus of H, called *proton*, is the heavy *elementary* particle, and has a positive charge equal in absolute value to that of the electron; the nuclei of the successive elements are formed by a certain number A of protons plus $A - Z$ electrons. We thus obtain a total charge of $Ae - (A - Z)e = Ze$ and an approximate weight equal to A times the weight of a proton (given the smallness of the mass of the electron).

But there are some theoretical reasons that stand in opposition to this conception. A very important one comes from the uncertainty principle $\Delta p_x \Delta x > h/4\pi$. The nucleus turns out to have dimensions on the order of 10^{-13} cm. If an electron is confined within the nucleus, the uncertainty of its position is not greater than 10^{-13} cm. Then Δp_x cannot be less than $h/4\pi 10^{-13}$, that is, it must at least be on the order of 10^{-14} in CGS. Now it is fairly evident that a quantity will never be smaller that its uncertainty. Therefore the momentum p of the electron will not be less than 10^{-14}. On the other hand, the energy of the electron will be given according to equation (2.72) by $E = c \cdot \sqrt{p^2 + m^2 c^2}$. Recalling that the mass of the electron is about 10^{-27} g, one can easily calculate that the energy cannot be less than about 10^{-4} erg or 100 MeV. This energy is too high; making it very difficult to understand what keeps the electron within the nucleus and, what is more, not agreeing with the fact that in the β emission the electrons come out with much lower energies, on the order of a MeV.

The matter was clarified in 1932, when J. Chadwick discovered the existence of the *neutron*, a *neutral* particle that has almost the same mass as the proton and spin 1/2. From then on there emerged an image of the nucleus which was valid.

A nucleus is made up of Z protons and N neutrons or, as we say, of $A = Z + N$ *nucleons*. Therefore the nucleus has a positive charge Ze and a mass number A, that is, a weight equal to about A times the weight of the hydrogen atom. The Z number of protons is equal to the number of peripheral electrons of the neutral atoms, and so characterizes the chemical species. But, Z being equal, because we can have different values for the number N of neutrons, we are led to the conclusion that *isotopes*, or atoms of the same chemical species with different weight, can exist. To indicate these characteristics of the nucleus we can use the symbolic notation $^{A}_{Z}X_{N}$, where X is the chemical symbol of the atom. Let us give some examples.

The nucleus of ordinary hydrogen is made up of only one proton and we indicate it with $^{1}_{1}H_{0}$; but heavy hydrogen or *deuterium* (discovered by H. C. Urey in 1932) also exists. This consists of a proton and a neutron, called a *deuteron*, and is indicated by $^{2}_{1}H_{1}$. Ordinary helium (or α particle) is indicated by $^{4}_{2}He_{2}$; but $^{3}_{2}He_{1}$ also exists (*triton*) and has one neutron less. Oxygen has several isotopes, for example, $^{15}_{8}O_{7}$, $^{16}_{8}O_{8}$, $^{17}_{8}O_{9}$, $^{18}_{8}O_{10}$; and so on, until we arrive at uranium, whose most abundant isotope is $^{238}_{92}U_{146}$. Usually, we are satisfied with a more synthetic notation, like $^{16}_{8}O$, or even ^{16}O, since the chemical symbol establishes Z unequivocally, and from it we derive N.

The constitution of the nucleus so established agrees with the laws of radioactive displacement seen in §4.4. In α emission the nucleus frees two protons and two neutrons bound together; in β emission a neutron of the nucleus emits one e^{-} and becomes a proton. We shall speak again of these phenomena.

We can also become aware of the disintegrations and artificial transmutations of nuclei, which are brought about by bombarding them with suitable projectiles of sufficient energy. The first of these transmutations was discovered by Rutherford in 1919 by bombarding nitrogen with α particles emitted by radium. The corresponding nuclear reaction can be written as:

$$^{14}_{7}N + ^{4}_{2}He \rightarrow ^{17}_{8}O + ^{1}_{1}H$$

The α particle (^{4}He) is captured by the nucleus, the isotope ^{17}O is formed and a proton (^{1}H) is emitted. Today the number of nuclear transmutations obtainable in this way (by bombarding with α particles, with protons, neutrons, deuterons, γ rays, etc.) is not even countable. It has been said very often that the alchemists' dream has been realized.

The nucleus that is formed after the inclusion of the particle that strikes it may not emit another particle, but decays after a certain time, thus giving rise to *artificial radioactivity*. The latter was discovered by Irène Curie and F. Joliot in 1934. E. Fermi and his collaborators succeeded in making the majority of known nuclei radioactive by bombarding them with neutrons. The neutrons had the advantage of reaching the nucleus easily, because they are not charged, they are not repelled by the electric field of the nucleus. Slow neutrons, or neutrons slowed down by making them pass through water or paraffin until they reach a kinetic energy on the order of the thermal energy kT, were especially useful for inducing transmutations and for studying the atomic nucleus in general. In artificial radioactivity there is also a β emission phenomenon which was not known in natural radioactivity. To be exact, a proton can emit one e^+ and can become a neutron. Naturally, the element then falls back one step in the periodic system.

All these studies led, among other things, to the establishment of a very important result. The mass of a nucleus is always *less* than the sum of the masses of its component nucleons taken separately (*mass defect*). To what is this due?

To understand the phenomenon, one must think only that each nucleon is kept inside the nucleus by a binding energy E_b. This is the energy necessary to remove the nucleon from the nucleus. On the average it is equal to 8 MeV of energy for each nucleon.[127] But to furnish the nucleon with the energy E_b is equivalent, according to Einstein's relation equation (2.69), to furnishing it with a mass $m = E_b/c^2$. Therefore the nucleon will weigh more[128] when separated from the nucleus than when included in it. More precisely, because the mass of a proton is equivalent to 938 MeV, we see that the mass increase when it is released is about one one-hundredth of its mass, that is, a perfectly measurable quantity. If we carry out the opposite process and reassemble the nucleus, beginning with the component nucleons, its mass will decrease, and we must expect an energy equal to the binding energy to be freed. This is the principle of the production of nuclear energy.

In reality, the binding energy of the nucleon is not the same for all nuclei.[129] If we plot the mass number A along the abscissa and the binding energy per nucleon E_b along the ordinate, we obtain a graph such as that of Figure 4.23 (it is an average curve, in which some small oscillations are omitted). As we see, the curve climbs very quickly for light elements; it reaches the maximum of about 8 MeV per nucleon around iron ^{56}Fe, and thereafter decreases slowly up to uranium.

If we split a nucleus of uranium into two nuclei of approximately the same atomic number, the binding energy per nucleon in the two resulting nuclei is greater. Therefore in the process of division (fission of uranium) some of the binding energy will be liberated. This energy can appear in several forms: γ rays, neutrons, and kinetic energy of the two final nuclei. The process of fission can be spontaneous or induced by particle bombardment. The yield is especially high when we bombard the isotope ^{235}U with neutrons. The neutron liberated by the fission can induce other fission processes. In fact, when a neutron collides with a ^{235}U nucleus, it can be absorbed, giving rise to a more unstable nucleus which, before long, breaks, emitting other neutrons

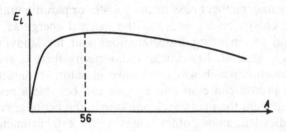

Figure 4.23

and we obtain a *chain process*. In every single act of this succession, a remarkable quantity of energy is liberated.

The fission of U was observed by O. Hahn and F. Strassmann in 1939 in the course of a search for *transuranic elements* (elements with $Z > 92$ obtained through neutron bombardments). Actually, many transuranic elements have been discovered, among them plutonium $^{239}_{94}$Pu, which is also very useful for the obtainment of nuclear power: ^{239}Pu can be produced in a *nuclear reactor* or *atomic pile*, in which the chain process of uranium fission is exploited. The first atomic pile was built by Fermi in 1942. These controlled fission plants can be used not only to produce atomic bombs, as unfortunately happened then and still happens today, but also to produce energy and a whole series of radioactive isotopes which are valuable in medicine and in industry. After World War II the attention of many physicists, who were searching for ever greater energy sources (and of the military, searching for ever more potent weapons), was directed toward the first part of the binding energy curve. Passing from a light nucleus like H to the He nucleus, the binding energy increases much faster than in the fission process of U. The inverse process of fission, nuclear *fusion*, appeared much more promising as a high-yield energy source. There is a difficulty, though; roughly, it is that of succeeding in bringing together H nuclei (to be precise, 4H nuclei, two of which become neutrons by emitting two e^{+}), in order to bring the nuclear forces into play. To overcome electric repulsion, we need temperatures on the order of 10^7 K. So whereas in the case of fission the problem concerned the dimensions of the fissionable material (it was necessary to reach the *critical mass* in order to obtain the starter for a self-sustained reaction), in fusion a strong initial energy was necessary to push the protons against one another. Once the fusion was started, the enormous energy freed would have increased the temperature (i.e., the energy) of the system further, and the reaction would have continued.

Unfortunately, the use of these ideas (the H bomb) was reached quite soon in the war, although we have not yet been able to obtain controllable energy from fusion. Oddly enough, this happened notwithstanding the fact that nuclear fusion is really the physical phenomenon that furnishes us with the energy we need to live. In fact, the reaction that occurs within stars, particularly in the sun, is what supplies its radiative power. We shall discuss this point further in Chapter 5.

Now we should ask what holds the nucleons of the nucleus together. Evidently, it is not electromagnetic forces, for they are too weak and cannot act on the neutrons. More considerable forces, called *nuclear forces*, are needed. We know only a few characteristics of these factors and not all that we want to know or would need to describe them adequately. These are *short-range* forces because they cannot act beyond distances on the order of 10^{-13} cm, that is, on the order of the size of the nucleus, Within the range, they are about one hundred times more intense than electromagnetic forces. One very important property is their *charge independence*. Derived from this property is the fact that among the pairs $p-p$, $n-n$, and $p-n$ the *nuclear* interaction has the same qualitative and quantitative structure.

Nuclear forces will be discussed later. Here we observe that in spite of our imperfect knowledge of them, we have succeeded in building some models of the nucleus, which agree in many ways with experimental data.

The oldest model is that of the liquid *drop* proposed by C. F. von Weizsäcker in 1935, which starts from the observation that the nucleus is apparently composed of homogeneous and incompressible *nuclear matter* like that of a liquid. This model was used by N. Bohr and J. A. Wheeler to explain some peculiarities of fission.

The most important model of the nucleus is the *shell* model proposed by M. Goeppert-Meyer and H. Jensen in 1949. This concept is analogous to that of the electron shells of the atom. Whereas electrons are held in the atom by the electrostatic attraction of the nucleus, nucleons are held in the nucleus by their mutual interactions. We assume that the total effect of these interactions is equivalent to a uniform *potential well*. This means that, in practice, each nucleon is included in a spherical container of the same dimensions as the nucleus. By applying the laws of quantum mechanics to this model, we find that each nucleon moves inside the container in orbitals of quantified energy and momentum. We thus find orbitals s, p, d, and so on as for electrons in the atom. Taking into account Pauli's exclusion principle, we can calculate the number of protons or neutrons needed to fill the successive shells. We find the *magic numbers*: 2, 8, 20, 28, 50, 82, 126. A nucleus having a magic number of protons or neutrons is particularly stable. When the nucleus is excited by bombardment, a nucleon can pass from its orbital to a higher unoccupied one. The nucleus can then return to the ground state, emitting a γ photon.

This simple model accounts already quite well for many properties of the nucleus, but the agreement with experiment can be improved by taking into account that rotation deforms and causes it to deviate from the spherical shape.

The α particles have a magic number, too, both for protons and neutrons, and are particularly stable so that they retain some individuality even inside a heavy nucleus. When they strike the walls of the box, that is, the potential barrier that separates them from the exterior, they have a nonzero probability of getting through by *tunnel effect* (Fig. 4.10). In this way α radioactivity was explained by G. Gamow, as early as 1928.

4.24. The second crisis of classical physics

It is usually said that classical physics underwent a crisis at the end of the nineteenth century, and that it was overcome through quantum mechanics. In reality, I would distinguish two very different phases of this crisis; the first was practically surmounted and the second is still in full and dramatic development.

To the modern observer it appears that quantum mechanics was only a partial break with classical physics. The different rules of *correspondence*, whose discovery began with Bohr and ended with Dirac, placed classical physics in a central position. To give a known example, let us observe that the potential energy E_p which appears in equation (4.49), and therefore in Schrödinger's equation, is that of classical forces: for example, that of a Coulomb force. By the same token, the electromagnetic field quantized in quantum electrodynamics is the classical electromagnetic field of Maxwell. Again, the momentum that is quantized is initially the classical one. It is true that afterward we find that the spin must be added, but also that the spin must obey laws perfectly analogous to ordinary quantized momentum.

Thus we have the impression that at this stage the physicist still speaks the language of classical physics and deals with classical quantities. The only changes are (1) the way of dealing mathematically with these quantities (or if we want, the set of laws that the quantities themselves obey), and (2) the possibility of measuring them simultaneously with precision (the uncertainty principle).

The second crisis, more serious than the first, began with nuclear and subnuclear physics. Until today we have the impression that quantum mechanics is valid even at this level. But now classical expressions like E_p of equation (4.49), that is, those expressions that we have to use in quantum mechanics so that it would not be an empty skeleton, seem to escape our attempts at precise definition! Whereas by using classical mechanics and electromagnetism in quantum mechanics, we can sometimes attain an amazing degree of accuracy, we do not know what to use for nuclear forces in a way that permits us to escape from a semiqualitative stage.

This situation depends on many factors, one of which is represented by the short range of nuclear forces. Coulomb's electrostatic force has practically an infinite range. It is therefore possible to study it even at enormous distances from the charged particles that generate it. In other words, it is possible to study the force's classical limit, which has a well-defined meaning. To study nuclear forces, in constrast, we must be very close to the nucleons, where classical concepts are no longer even approximately valid.

Second, nuclear forces are too strong to be studied with perturbation methods. Nevertheless, we can construct Feynman diagrams, as for electrodynamics. But the constant that now corresponds to the fine structure constant of expression (4.104), is on the order of unity instead of 1/100. Consequently, the successive perturbative orders do not diminish to the point of being negligible. On the other hand, we do not know of an alternative mathematical way of dealing with the problem. The situation is really paradoxical. We do not know what to use as initial data in a problem that we would not know how to solve mathematically anyway!

Let us recall that in electrodynamics the successive orders of perturbation represent, as we might say, the successive levels of reality at which the particles involved in the interaction arrive. The more we proceed toward the higher orders, the more their reality becomes remote. But what can we say if all the orders are present at the same time and with the same right? Can we still speak of particles?

In atomic physics we often use a classical language and construct *models* that somehow claim to visualize the situation. We know that such models are fictitious, but we obtain the correct description of the phenomena by applying quantum mechanics to them. Even this possibility of intermediate visual models ceases in subnuclear physics. The possibility of a detailed description of interactions in particular, ceases. For this reason, particle physicists, when speaking of models, almost always refer to *mathematical models*, that is, to semitheories or heuristic hypotheses formulated in very abstract terms (see also §1.15). In a world in which the transitory and elusive nature of particles is even more pronounced than what we saw in quantum electrodynamics, it seems that the permanence of the substance loses all meaning. It appears that Heraclitus definitely wins the battle against Parmenides.

But really, it is not so. Parmenides reappears stronger than ever. How could we speak of reality if there was nothing permanent about it? There must be some invariant, as noted in §2.6. As a matter of fact, in contemporary physics the loss of meaning of the classical concept of substance and detailed interaction, accompanies the placement of the principles, or laws, of *conservation* in a central, almost exclusive, position.

The things conserved in particle physics are already numerous; some are old acquaintances, but others are absolutely unknown and unimaginable in classical physics. We could perhaps say that they are the substance with which physics deals. To understand the importance of this, it is sufficient to say that many physicists are even convinced that any process that does not violate a conservation law *can* occur in reality.[130]

Sometimes, we speak of *symmetries* instead of laws of conservation, because some symmetries or invariances of physical laws appear necessary and we can demonstrate that a quantity that is conserved corresponds to each of them. Thus, for example, the conservation of momentum corresponds to invariance in respect to translation in space. The conservation of energy corresponds to invariance in respect to translation in time. The conservation of angular momentum corresponds to invariance in respect to orientation.[131] But it has not been demonstrated that this relationship is as close and universal as many assume. Besides there are quite a few conserved quantities of whose symmetry correspondences we are uncertain.

The laws of physics concern phenomena. Symmetries concern laws, and so they are *superlaws*. E. P. Wigner says:

> If we had a complete knowledge of all events in the world, everywhere and at all times, there would be no use for the laws of physics, or, in fact, of any other science. . . . They might give us a certain pleasure and perhaps amazement to contemplate, even though they would not furnish new information. [Similarly:] if we knew all the laws of nature, or the ultimate law of nature, the invariance properties of these laws would not furnish us new information. They might give us a certain pleasure, and perhaps amazement to contemplate, even though they would not furnish new information (Wigner, 1967, p. 16–17; see also Salam, 1972, p. 69).

We spoke before of the limits of validity of physical laws. Now we must develop the same argument for the superlaws or symmetries. Symmetries have limits of validity as well, in that they are generally respected for certain classes of phenomena and violated for others. But we shall come to this in due time.

When we discussed Galileo's method, we observed that the questions that classical physics asks of nature are in some way tendentious. The experimenter guesses at least part of the answer; this is why the experiment can be sensible. But today things have changed dramatically. The objects studied are anything but directly guessable. We reach them (if we reach them at all) through a long and complicated mathematical and experimental chain. Nothing appears immediate anymore; everything is mediated, in large measure. The questions to ask nature become ever more uncertain and difficult.

It is useful to observe, finally, that the refined complication of the questions posed by physicists, corresponds to an ever higher *cost* of the experimental apparatus. On the other hand, the a priori uncertainty regarding their adequacy increases. This leads many to formulate a not unreasonable doubt: Is it worthwhile?

4.25. Particles multiply

After the discovery of the neutron in 1932, one might have thought that the list of *elementary particles* was exhausted. There were as many as seemed necessary and sufficient to construct the known physical world. Protons and neutrons formed the nuclei; by adding electrons, atoms were made; and the picture was completed by adding photons and possibly *gravitons*, the quanta of gravitational fields.

But β radiation remained rather mysterious. Here a neutron (proton) of the nucleus emits an electron (positron) and transforms itself into a proton (neutron). A close examination of experimental results revealed that the principle of the conservation of energy (as well as those of the conservation of momentum and spin) was apparently violated in the reaction of nuclear decay. Thus in 1931 Pauli formulated the ingenious hypothesis that an additional particle was emitted in β decay. Fermi baptized this particle the *neutrino* and was able in 1934 to construct a very satisfying theory of β emission with it. The neutrino has zero mass and charge (like the photon) and spin 1/2.

The neutrino interacts in a very *weak* way with the other particles. As a result, its reactions, when they can occur, are extremely rare. Thus a *free* neutron takes about a quarter of an hour (mean lifetime) to decay into a proton and an electron, emitting a neutrino. For reasons that we shall soon see, the process today is written as:

$$n \rightarrow p + e^- + \bar{\nu}_e$$

and the emitted particle $\bar{\nu}_e$ is called *antineutrino* rather than neutrino.

For the energy balance, we note that the neutron has a mass equal to 939.54 MeV, whereas the proton has a mass equal to 938.25 MeV, which when added to the mass of 0.51 MeV of the electron gives 938.76 MeV. Thus 0.78 MeV are left over, which become the kinetic energy of the electron and proton and the energy of the antineutrino. But these relations are valid only for the free neutron, which decays into a free proton (plus electron and antineutrino). But when the nucleons are in the nucleus, we know that they lose their individuality; moreover, we must also take into account the nuclear binding energy and the energy of electrostatic repulsion between protons. In the majority of cases the energy necessary to bring about the reaction is not available; the nucleus then is stable. But where there is still sufficient energy, there is β radioactivity. For the same reasons we see that the reaction

$$p \rightarrow n + e^+ + \nu_e$$

which cannot occur with free particles (because the proton is lighter than the neutron) can sometimes occur in a nucleus. In that case we have (artificial) β^+ radioactivity.

The existence of neutrinos was accepted by physicists for a long time, simply because the principles of conservation demanded it and because the success of Fermi's theory confirmed it. But direct detection of the particle is extremely difficult, due to the smallness of the weak interaction. A neutrino can cross millions of kilometers of dense material without having an appreciable probability of being absorbed (or better yet, of interacting)! Only in 1956 were C. Cowan and F. Reines finally able to reveal (using a powerful beam of antineutrinos produced by a reactor) the inverse reaction of β decay

$$\bar{\nu}_e + p \rightarrow n + e^+$$

in which a neutron and a positron are born from the collision between an antineutrino and a proton.

Having settled the question of β decay with the neutrino, one could now think he had exhausted all the elementary particles. But again, this was not the case. In reality, the particle *zoo* (as physicists call it) was constantly increasing.

This was due both to the availability of increasingly more powerful accelerators and to a number of bold and successful theoretical ideas.

We have already seen how it is possible to conceive the electrostatic field in terms of photon theory. The electric interaction between two charged particles was interpreted on the basis of an exchange of virtual photons. Once this point of view was accepted and its fertility was verified, it was natural to ask if other force fields, too, could be treated in an analogous way. Particular attention was focused on nuclear forces and the possible existence of a *quantum* of the nuclear field, responsible, through virtual exchange, for nucleonic interaction. This theory of the nuclear field was proposed in 1935 by H. Yukawa.

The quantum of the nuclear field was called *meson*, because its predicted mass was halfway between that of the electron and that of the proton. The order of the mass of the meson can be deduced from the uncertainty principle. In fact, we know that the range of the nuclear field is on the order of 10^{-13} cm. Consequently, a virtual meson of the cloud that surrounds the nucleus will not go farther than 10^{-13} cm away. If its speed is on the order of c, it will subsist for a period of time on the order of 10^{-23} s. By applying the uncertainty principle in the form equation (4.58) with $\Delta t = 10^{-23}$ s, we find the order of 10^{-4} erg for the uncertainty of the energy ΔE.

The energy of the particle cannot be less than ΔE. On the other hand, we imagine that, roughly speaking, the particle comes and goes; at the maximum distance from the nucleon it will have zero kinetic energy and all the energy will be attributed to mass. Then, dividing by c^2, we find a mass on the order of 10^{-25} g, or a hundred times the mass of the electron.

We know that photons are not always necessarily virtual; when there is sufficient energy they can be real. The same must occur for mesons.

Actually, a year after Yukawa's proposal, physicists discovered certain particles produced by cosmic rays, which have a charge of $+e$ or $-e$ and a mass approximately 207 times that of the electron. It was natural to think of Yukawa's meson and these particles were initially called μ *mesons*. Later, when this interpretation was found untenable, they were called *muons*.

Serious difficulties arose quite soon. As the particle responsible for nuclear interactions, Yukawa's meson had to interact strongly with the nucleons. But an experiment by M. Conversi, E. Pancini, and O. Piccioni demonstrated that this was not the case. Besides, as we recall from our discussion of relativity, the muons of cosmic rays arrive on earth in great abundance. This fact, besides confirming the relativistic dilation of time, demonstrates that muons do not interact very much with nucleons in the atmosphere.

The muon has an average life of 2.2×10^{-6} s and decays to form an electron (positive or negative) and two neutrinos, according to the scheme

$$\mu^- \rightarrow e^- + \bar{\nu}_e + \nu_\mu$$

$$\mu^+ \rightarrow e^+ + \nu_e + \bar{\nu}_\mu$$

For reasons suggested by experiment we have arrived at the conclusion that the electron and the muon each has its own neutrino. We speak here of different particles, which we indicate, respectively, by ν_e and ν_μ. The neutrino of the muon was actually observed as different from ν_e in 1962.

For a long time e, ν_e, μ, ν_μ have formed the *lepton* family (from Greek λεπτός meaning *light*). The electron and the muon were identical in all respects, except for the mass and lifetime. This fact was a puzzle for the physicists who asked: Why must μ exist? But this in turn gives rise to a philosophical question: Why do physicists ask this question? Why is it easier to accept the existence of two totally different particles than of two particles equal in every way except in mass? Naturally, this depends largely on the fact that they did not know how to include the two particles in a reasonable framework.

But in recent years the problem has shifted to a different level, when experimental evidence has been accumulated pointing to the existence of a new lepton called τ. Because the mass of τ is about 1800 MeV it is also termed *heavy lepton* (a curious contradiction in terms). Its charge is e, its lifetime less than 5×10^{-12} s and it is believed that its spin is $\frac{1}{2}$. There is general agreement that there should be a ν_τ neutrino. As a result, today the lepton family has six members. Leptons are subject to weak interactions but are insensitive to the strong interactions of nuclear forces.

Returning to history, a particle having the characteristics of Yukawa's meson was finally discovered in 1947 by W. M. Powell and G. S. Occhialini. This particle was given the same π *meson* or *pion*. Charged pions have a mass 273 times that of the electron and a charge of $+e$ or $-e$, and are indicated respectively by π^+ and π^-. Around 1950 a neutral pion π^0, with slightly lesser mass, was discovered as well. Pions interact strongly with nuclei, as Yukawa required. Today all strongly interacting particles are called *hadrons*.

Charged pions have a mean lifetime of 2.6×10^{-8} s and decay to form muons and neutrinos according to the scheme:

$$\pi^+ \rightarrow \mu^+ + \nu_\mu$$

$$\pi^- \rightarrow \mu^- + \bar{\nu}_\mu$$

The π^0 has instead a shorter life (0.9×10^{-16} s) and decays into two photons according to the process:

$$\pi^0 \rightarrow \gamma + \gamma$$

Shortly after pions, other mesons, called K-mesons or kaons, with mass 966 m_e, lifetime 1.2×10^{-8} s and charge of $\pm e$ were discovered in cosmic rays. Later the neutral meson K^0 was discovered as well. This particle possessed the extremely curious property of being a combination of two different states or particles, with two different lifetimes.

K-mesons provide an excellent example of particles that decay in different ways, with different probabilities. For example K^+ decays in 64 percent of the cases into $\mu + \nu_\mu$, in 21 percent of the cases in $\pi^+ + \pi_0$, in 5.6 percent of the cases into $\pi^+ + \pi^+ + \pi^-$, in 4.9 percent of the cases into $e^+ + \nu_e + \pi^0$, and so on until we arrive at very rare modes of decay.

When the introduction of increasingly more powerful particle accelerators made it possible to make highly energetic protons and pions interact, it was seen that another neutral unstable particle of great mass (1115 MeV) was produced at the same time as the K-meson. This particle was given the symbol Λ^0. A new class of particles, called *hyperons*, with masses greater than that of the nucleon, was thus introduced. Two typical reactions producing Λ^0 (at the threshold energy) are:

$$p + \pi^- \rightarrow \Lambda^0 + K^0$$

$$p + p \rightarrow \Lambda^0 + K^+ + p$$

In the collision of two energetic protons other hyperparticles, for example, Σ and Ξ, can be produced. Typical reactions are:

$$p + p \rightarrow p + \Sigma^+ + K^0$$

$$p + p \rightarrow p + K^+ + K^0 + \Xi^0$$

and so on.

Nucleons and hyperons together form the class of *baryons* (or heavy particles, from Greek βαρύς, meaning *heavy*).

As we have seen, the attempt to make Schrödinger's equation relativistically correct led Dirac to the introduction of the concept of *antiparticles* (in his particular case, of the positron). The general application of the quantum theory of fields requires the existence of an antiparticle for every particle, the properties of which are obtained from those of the particle by applying a particular transformation, called *charge conjugation* (which we shall discuss later), to the mathematical entities associated with every physical quantity.

The term, antiparticle, originates from the fact that every time a particle and its antiparticle come into contact, they annihilate one another, delivering energy in the form of other particles. A very important observation concerning charged particles is that a charged particle does not always have the antiparticle we would expect at first sight. For example, the hyperon Σ^+ differs in mass from Σ^-, which therefore cannot be its antiparticle.

Only two neutral particles, the photon and the meson π^0 (besides the possible graviton) coincide with their antiparticles, therefore we say that the γ and π^0 are antiparticles of themselves.

According to the laws of conservation, which we shall soon explain better, antiprotons can only be born at the same time as protons, thus it is necessary that, in the collision $p-p$, the incident proton have an energy higher than 5.6 GeV, so that (in the *center-of-mass* frame) an energy higher than $2m_p c^2$ is available. Then the following process occurs

$$p + p \rightarrow p + p + p + \bar{p}$$

The existence of $\bar{p}$ was experimentally demonstrated by O. Chamberlain, E. Segrè, and others in 1955, when protons of such energy became available. Similarly, an antineutron can be generated according to the following process:

$$p + p \rightarrow p + p + n + \bar{n}$$

Today there is no doubt that each particle has its antiparticle.

4.26. Interactions and conservations

The particle family is much more numerous than the brief list so far discussed. But that list is enough to show that the goal of having a few elementary bricks out of which to build all matter was far from being attained. The situation was analogous to that existing for the chemical elements about the middle of the last century. In a sense the particle Mendeleev was still to come.

Until a decade ago the situation was somewhat confused. Although some partial classifications were known, a complete and satisfactory theory appeared to be far away. But in recent years there have been some very interesting openings and physicists can now hope that a unitary view, based on a comparatively small number of elementary components, may be attained in the near future. Before discussing these recent achievements we want to return for a while to the historical development and describe briefly the picture that was available until a few years ago.

A first subdivision can be made on the basis of the type of interaction that a particle exhibits in the vicinity of another particle. Fortunately, the types of interaction that have been verified until today number only four: gravitational, electromagnetic, strong, and weak.[132]

These interactions manifest themselves with widely varying characteristic intensity. By attributing the value 1 to the intensity of electromagnetic interactions, we obtain the following table:

interactions:	strong	10^2
	electromagnetic	1
	weak	10^{-11}
	gravitational	10^{-36}

Note the extremely low intensity of gravitational forces in respect to electromagnetic ones. In this context it is surprising that humans should daily experience gravitational more than electromagnetic effects. This is due to the perfect compensation between negative and positive charges that occur in any material object of macroscopic dimensions. If two men, 1 percent of whose electrons had been removed, were to meet, they would repel one another with an enormous force, capable of lifting a weight equal to that of the entire earth! Because analogous compensation does not exist in the case of masses (which are all positive), it is clear that when these masses reach the size of heavenly bodies, they can make their effects felt.

Both *electromagnetic* and *gravitational* interactions have an infinite range and are felt at any distance from the source, although attenuated by the factor $1/r^2$. By contrast, *strong* and *weak* interactions have extremely short ranges of about 10^{-13} and 10^{-15} cm, respectively. Weak interactions are responsible for some processes with comparatively long reaction times, such as β radioactivity and μ decay, whereas strong interactions have typical reaction times on the order of 10^{-23} s. The latter represents in a sense the natural time scale to be used in judging whether a subnuclear process is fast or slow.

Electromagnetic forces are the best understood of the four interactions. For this reason all the efforts made (since Yukawa's time) to explain the other forces have been based on analogies with the electromagnetic forces.

Experimental evidence has shown that leptons are insensitive to strong forces, although being subject to weak forces, to gravitation and, if charged, to elecromagnetic forces. Only mesons and baryons have strong interactions and are accordingly called *hadrons* (from Greek αδρός, that is, *strong*).

After this first subdivision based on the type of interaction shown, let us now discuss another criterion of classification based on the laws of conservation. Here particles have surprised us in a very interesting way. In fact, we have already noted that in any reaction the four classical laws of conservation (energy-mass, momentum, angular momentum, and electric charge) are always satisfied. To repeat, so far no reaction has been found for which the set of these four principles is not satisfied. In studying an increasing number of particle reactions, keeping in mind the existence of antiparticles, one has realized that there are many processes in nature that are acceptable by virtue of the four principles previously mentioned. But others, *equally possible*, did not occur. This observation has required the introduction of other conservative entities that do not appear in the everyday world. These entities have sometimes been given fancy names.

First of all, from the observation of the nonoccurrence of certain reactions among particles, physicists have been induced to attribute a *baryonic number* to each of the baryons. More precisely, the number +1 was attributed to these particles and the number −1, in a symmetrical way, to their antiparticles. The following regularity has never been disproven by facts: In all the processes observed, the algebraic sum of the baryonic numbers on the left always equals that of those on the right. Therefore a new principle, *the conservation of the baryonic number*, exists. Many processes that would be possible, because they respect the first four principles, do not occur in nature for they would violate the constancy of the baryonic number.

This new conservation principle is at the moment considered a postulate, because we are not capable of fully understanding the symmetry of the universe, whose constancy is empirically translated as the constancy of the baryonic number. For example, the reaction

$$p + p \rightarrow p + \Lambda + \Sigma^+$$

cannot occur, even when the energy is available, because the baryonic number on the left equals 2, whereas that on the right equals 3.

Contrary to what occurs for baryons, an analogous conservation law does not exist for mesons. The number of these particles is not conserved in the universe. The same is true for photons.

Two analogous laws were known for leptons: the *conservation of the muonic number* (muon and its neutrino) and the *conservation of the electronic number* (electron and its neutrino). The electronic number of e^- and ν_e is +1; that of e^+ and $\bar{\nu}_e$ is −1. For example, β decay must be written in the form $n \rightarrow p + e^- + \bar{\nu}_e$, for because there are no leptons on the left, the number is 0, and must be the same on the right as well (naturally, the baryonic number is +1 on both the right and left). Similar properties apply to the muonic number (+1 for μ^- and ν_μ and −1 for μ^+ and $\bar{\nu}_\mu$). This explains why, for example μ^- decay should be written $\mu^- \rightarrow e^- + \bar{\nu}_e + \nu_\mu$ with an electronic neutrino and a muonic neutrino on the right. For similar reasons π^- decay is written as $\pi^- \rightarrow \mu^- + \bar{\nu}_\mu$. The muonic number of the left is zero, and so it must be zero on the right.

Of course, since the heavy lepton τ is known physicists think that an analogous conservation law should exist for it and its neutrino.

On the basis of these conservation principles, we understand why the transformation

$$p \rightarrow e^+ + \gamma$$

which would be possible from the point of view of energy and charge, does not take place. It cannot occur because the baryonic and leptonic numbers are not conserved.

Similarly, the transformation of p into other particles of lesser mass does not occur because they all have the baryonic number zero. Thus the stability of the proton is dependent on this law of conservation, and analogous reasoning leads to the same conclusion for the electron. In this case it is the conservation of charge that gives rise to stability; below the electron there are only particles with zero charge.

Now we naturally ask ourselves which law forbids a photon to decay to form a certain number of photons with lower energy. This law does not exist. But let us observe that if the photon decays, however brief its mean life in its system may be, it lasts an eternity for us, since the photon travels at the speed of light! Besides, because of the conservation of momentum and energy, the several photons produced would travel one on top of the other and would not be distinguishable. This may be the explanation of stability. An analogous consideration can be made for neutrinos.

As we have seen, the combination of the theory of relativity with that of quantum mechanics indicates the existence of an intrinsic angular momentum or spin of well-determined value for every particle. If a particle has spin 1/2, the projection of the spin in a predetermined direction (e.g., the third component s_z) can only be $\pm$ 1/2. If the spin is 1, the results of a measurement can only be $\pm$ 1; 0. If instead the spin is equal to 3/2, the possible projections are: $\pm$ 3/2; $\pm$ 1/2; and so on. In the case of 1/2 there are two possibilities ($\pm$ 1/2), and we therefore speak of a *doublet*. In the case of 1 we speak of a *triplet*, and so on.

Starting from the charge independence of nuclear forces, Heisenberg found an interesting analogy between the possible states of the spin of a particle with spin 1/2 and the possible states of a nucleon in the nucleus. Because of the equality of strong interactions among nucleons and because of the near coincidence of proton and neutron masses, it is possible to consider the proton and the neutron as a doublet of particles that are like two aspects of the same entity: the nucleon. This is evidently true if we disregard the difference of magnetic moment, the lifetime and the essential difference that one is charged and the other is neutral.[133]

In this vision, which has shown itself to be full of consequences, another doublet is constituted by the particles K^+ and K^0, which must also be regarded as two aspects of an entity, different from the nucleon and equally unknown to us. Another doublet is formed by the Ξ^- and Ξ^0; Λ^0 can instead be considered as a *singlet*; the three pions and the

Σ present themselves as *triplets*. Naturally, we do not take into account the differences of mass, which in reality are small, and the differences of lifetime, just to mention the most important parameters.

The essential element which is placed in the foreground of this purely empirical classification is electric charge. This is zero for singlets; is given by $+1,0$ or $-1,0$ for doublets; has the values $+1$, 0, -1 for triplets, and so on. Thus we see that it is possible to introduce a new quantity connected with charge and assign it the value 0 for singlet states, 1/2 for doublet states, 1 for triplet states, and so on. Analogous to the case of spin, we are led to admit that it is a vector whose projections in a privileged direction, necessarily quantized, are 0 for singlets; $\pm$ 1/2 for doublets; 0, ± 1 for triplets, and so on. The space in which the vector can be orientated is evidently not physical space, but an abstract space whose significance we do not substantially know.

Because of these analogies, this vector has been called *isotopic spin* or *isospin*, even though it has nothing to do with spin. The abstract space in which the vector is represented is similarly called *isotopic space*.

Isotopic spin I is therefore a vector with a characteristic value for each type of particle. When we consider in particular a set of protons and neutrons, that is, a given isotope of a nucleus, we have a total value of the isotopic spin. A variation in the value of the total isotopic spin of a set of particles or, more precisely, a variation in the value of the chosen component (let us say, the third component I_3) represents a variation in the number of positive charges, and thus the passage from one nucleus to another.

In regard to symmetries we can also say that the charge independence of strong interactions represents a symmetry and that a conservation law must correspond to these interactions. What is conserved? Isotopic spin. In all strong interactions, isotopic spin remains unvaried; thus we are presented with a new mysterious property of particles.

We can also state the following rule: When considering reactions in which strong interactions come into play, we can vary the orientation of the vector representing isotopic spin in its abstract space without changing the laws that describe the process.

As we have seen, stability is reduced to the inviolability of certain conservation principles. For protons and electrons, the relevant conservation principles are those of baryonic number and electric charge.

On the other hand, the particles generated in strong interactions (which therefore imply very brief formation times, 10^{-23} s) can have lifetimes that seem enormous when compared to the time necessary for their birth. This behavior becomes difficult to explain for hyperons.

For instance, the particle Λ^0 has a lifetime on the order of 10^{-10} s, or 10^{13} natural time units. It is slow to die ($\Lambda^0 \rightarrow p + \pi^-$); once generated ($\pi^- + p \rightarrow \Lambda^0 + \cdots$), its behavior is strange, and we can try to explain it by appealing to the existence of another entity whose conservation prevents the decay of Λ^0 in periods of less than 10^{-10} s. This entity is not conserved in weak decay processes, and so the strange particles can decay in this way, with long lifetimes. But in strong interactions, responsible for the rapid formation of the particles themselves, it is conserved.

The unknown quantity, which is introduced to explain the behavior of the Λ^0 and other particles produced in strong interactions, was quantized by M. Gell-Mann and K. Nishijima, by affinity with other quantities, and was called *strangeness*, precisely to underline our present incapacity to understand its meaning. Strangeness can have the values, 0, ±1, ±2, and so on.

As an example, let us consider the process

$$p + p \rightarrow p + \Lambda^0 + K^+$$

This process is possible because the stangeness is zero both on the left and on the right. In fact, the stangeness of protons is 0, that of Λ^0 is -1 and that of K^+ is $+1$. In contrast, any process in which a single strange particle is produced by initial particles that are not strange, is prohibited. Sometimes instead of strangeness, we use *hypercharge*, the sum of the strangeness and the baryonic number.

We might be a bit skeptical about the necessity of proceeding in this way, constantly introducing new entities, in a research field that seems already so obscure. But as already observed, conservation laws are the strongest supports we have today. We must humbly learn from nature what is conserved and under what conditions.

We have laid our hands on some profound entities of the physical universe, but we have been able to comprehend only certain of their characteristics, which had led us to call them particles. These entities actually have many determinations subject to conservation laws. Some of these are found even in our macroscopic world, whereas others are peculiar to the subatomic world only. We can only acknowledge this.

We naturally ask: What happens if a particle formed in about 10^{-23} s in a strong interaction is not restrained by any conservation law from decaying the same way? Evidently it will decay in a time that is also on the order of 10^{-23} s. But in which sense, then, will we be able to speak of particles?

To understand the situation, let us give a classical (and purely indicative) example. Consider an acoustic wave of determined frequency ν which encounters the chord of a musical instrument tuned to the

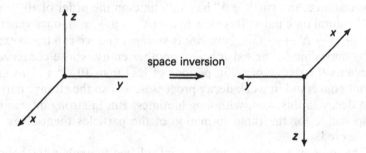

Figure 4.24

frequency v_0. Generally, when v is not very near v_0, the chord absorbs some energy from the acoustic wave and immediately returns it to the surrounding air. But if v is very near v_0, the chord enters into *resonance*; it absorbs a remarkable quantity of energy from the wave and, when the latter has passed, it continues to vibrate at the frequency v_0 for a certain time τ, which depends on the various characteristic parameters of the system. We can use Fourier's analysis to show that the longer τ is, the *narrower* the resonance is, that is, the smaller the difference $\Delta v = |v - v_0|$ must be, so that the wave can excite the chord.

Very similar things take place with particles when we speak of energies instead of frequencies (which amounts to the same, because $E = hv$). The kind of process that occurs is a function of the energy E of the incident particle (or, more precisely, of the total energy available). There can be a particular energy E_0 at which resonance occurs. In this case we have a product of the reaction, whose lifetime τ is longer than we would normally expect. The narrower the resonance, that is, the smaller the admissible $\Delta E = |E - E_0|$ (uncertainty principle), the longer is τ.

Actually, a large number of extremely ephemeral particles of this kind have been found, and they are in fact called resonances. Their (mean) lifetime is usually on the order of 10^{-23} s.

We have already discussed the principal space–time symmetries and the corresponding conservation laws. Another very important kind of symmetry is that of *reflection*. It occurs when two phenomena are the mirror image of each other. If the mirror occupies the xy plane, the description of the second phenomenon is obtained from that of the first by changing z into $-z$.

More generally, we can consider *space inversion* (Fig. 4.24), which occurs when the direction of all three coordinate axes is inverted.

When a physical law is expressed through a certain coordinate function, $f(x, y, z)$, we say it is invariant for space inversion if the function

f is an even function (i.e., if it contains only arguments of even degrees, x^2, y^2, z^2, xy, xz, etc.); in this case, in fact, if we change x to $-x$, y to $-y$, z to $-z$, the value of the function remains unchanged.

From the first elements of quantum mechanics, we derive that the wave functions of the various particles have a well-determined mode of behavior in regard to these inversions of axes. More precisely, ψ turns out to be either an even or odd function, that is, to have a well-determined *intrinsic parity*. Since the quantity that we usually measure is $|\psi|^2$, we are led to think, by extrapolation, that in all physical phenomena the laws do not change when the space is inverted.

The importance of this conclusion resides in the consequent physical possibility for any known phenomenon to occur in inverted space, or in mirror-image space. We say in this case that physical phenomena *conserve parity*. The consequence of the law of the conservation of parity is that if a physical phenomenon occurs in real space, the phenomenon seen in the mirror is perfectly possible. If the observer were not aware of the mirror, he would judge the phenomenon as being physically correct.

Let us imagine a fictitious world in which *lefthanded* people do not exist. Seeing a person writing with his left hand, one would immediately think of a phenomenon seen in a mirror. Until 1955 it was thought that, in physics, situations of this kind did not exist, or that nature did not distinguish right from left. The fact had already been verified for electromagnetic and strong interactions. Many workers even thought that this was necessary, on the basis of the principle of insufficient reason.

But in 1956 there was an extraordinary turn in the line of development of these concepts: The law of parity was violated in weak interactions. In that year, in fact, T. D. Lee and C. N. Yang, following certain anomalies observed in K-meson experiments, suggested the violation. The following year C. S. Wu and co-workers demonstrated the violation of parity without doubt. The experiment consisted in a study of the decay of ^{60}Co into ^{60}Ni:

$$^{60}\text{Co} \rightarrow {}^{60}\text{Ni} + e^- + \bar{\nu}_e$$

Introducing the radioactive Co preparation in a low temperature environment (absolute zero) and subjecting it to a very intense magnetic field (Fig. 4.25), we obtain this condition: practically all magnetic moments and therefore the spins of the Co nuclei are oriented in the direction of the field, because the very low temperature eliminates thermal fluctuations. It was found experimentally that the electrons coming from the decay are preferentially emitted to the side indicated in Figure 4.25(a), that is, in the direction opposite to that of the spins of the nuclei.

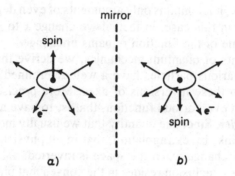

Figure 4.25

In the mirror image Figure 4.25(b) of the phenomenon, the direction of the spin changes (think of a top reflected in the mirror), but the direction of emission of the electrons does not change. Therefore the emission of electrons in the direction of the spin of the nucleus should be equally probable. But the evidence shows that it is not so, and we conclude that in the processes in which weak interactions occur, parity is not conserved.

Moreover, it results that neutrinos are only lefthanded; that is, their spin is opposite to that of a corkscrew that advances in the same direction. Note that this fact could not have universal validity if neutrinos did not travel at the speed of light and therefore did not have zero mass. In the opposite case, in fact, an observer could travel in the same direction with greater speed and would see them rotating in dextrorotatory sense. In a letter to V. Weisskopf at the time of these discoveries, Pauli wrote: "I am struck not so much by the fact that God is left handed, but by the fact that he is symmetrical when he expresses himself in a *strong* way."

Incidentally, we observe once more how the danger of the principle of insufficient reason becomes clear. The fact that we do not have a sufficient reason for preferring one thing to another does not imply that nature does not have it! With the discovery of the violation of parity in weak interactions, then, followed the possibility of distinguishing right from left in nature. For instance, if we wanted to communicate with hypothetical inhabitants of other worlds and to indicate to them which side we call right, it would be sufficient to indicate the correct phenomenon and the relevant instructions.

Right–left symmetry is restored by performing the mirror image and at the same time changing particles into antiparticles. The latter operation is called *charge conjugation*. In this way physical phenomena become symmetrical,[134] and the possibility of distinguishing left from right disappears again. Our distant interlocutors, in fact, have no way of communicating whether they are matter or antimatter, because the photon is its own antiparticle, and therefore light and electromagnetic signals could not carry this information to us.

Thus we have found a more complex symmetry than that of mirror images; it is indicated by CP (C for charge conjugation, and P for parity). Neither P symmetry nor C symmetry exist always in nature, but CP symmetry does exist. Or so we believed for some time, until certain experiments on the decay of K-mesons disproved even this hypothesis. The problem is presently filed in the profound mysteries folder.

Although the discovery that weak interactions possess less symmetry than electromagnetic or strong interactions was a great surprise for its time, we currently accept this fact today. For reasons still unknown, the various fundamental interactions have different symmetries. Precisely, the stronger the interaction, the more *constrained* it is by symmetries.

There is yet another kind of transformation to which we can imagine subjecting any physical process: time inversion (T). This can be defined as follows: if we change the sign of time in all equations of a phenomenon, we obtain the *inverted-time phenomenon*. If, for instance, we project the film of any process backward, the sequence of images is an inverted-time process. If this process actually occurs in nature, we can say that it is *invariant for time inversion*.

In the field of classical mechanics, as we have already noticed, phenomena obey the principle of time reversibility. The irreversibility of thermodynamics is merely a statistical phenomenon, due to special initial conditions, in which our world is found. In the particle domain we must assume a different position. It is not possible to affirm or negate invariance a priori, but it is necessary to study in particular all particle phenomenologies. We have to see if we can assume the invariance valid or not valid for the various particular manifestations. The present situation is the following. We have a single phenomenon on the basis of which we can doubt the invariance of all the manifestations of the microscopic world in respect to time inversion. Nevertheless, we can by no means say that the problem is completely closed.[135]

From the first principles of relativistic quantum mechanics, it is possible to derive a theorem that is valid for all physical systems. It says that if we apply the three transformations: matter → antimatter (i.e., charge conjugation C), space-inversion (parity P), and time inversion (T) to a certain physical process in any order, we obtain a process that can certainly occur in nature. This means that all physical processes must be invariant for *CPT* transformation.

From this theorem one can derive, for example, that both the mass and the lifetime of an antiparticle are always identical to those of the corresponding particle. Experimental evidence confirms this point.

As stated, with very high probability, we can hold time inversion to be valid in all processes. If the invariance in respect to *T* is valid, and in addition there must be invariance in respect to *CPT*, it follows that there must always be *CP* invariance. But *CP* appears violated, and we must assume that *T* is violated to the same degree. And then? – Again, these questions are still very mysterious.

The large amount of data and unanswered questions that have accumulated on the new particles has kept theoretical physicists very busy in the last decades. The main point is to understand why some configurations of properties (spin, mass, charge, etc.) can persist for long or short periods of time, giving rise to those entities that we call particles, whereas other combinations are excluded; to understand why they interact and decay in the way we know; and to understand why the various interactions have different symmetries, and so on. Whether or not a connection exists among these questions, both the possibility of any traditional form of answer, and the adequacy of the structure of quantum mechanics to provide a response, must be included in the actual problematics.

As our discussion has clearly shown, the concept of *elementary particle* had gradually become quite fuzzy and questionable in the last decades.

Experimental evidence has shown that provided that sufficient kinetic energy E_k is put into play, it is possible in a collision to have reversible particle transformations. For instance, we can have

$$\text{particle } A + E_k + \text{target} \rightarrow \text{particle } B + \text{other particles}$$

and conversely,

$$\text{particle } B + E_k + \text{target} \rightarrow \text{particle } A + \text{other particles}$$

Particle *A* therefore does not have greater rights to elementarity than particle B. Hence one can speak of particle *democracy*.

In this vision, the world appears therefore to be composed *not* of particles but of a complex of entities which we see to granulize into what we call particles. Each particle in turn, because of what we saw in field theory, has the possibility of virtually emitting all the particles with which it can interact. A proton, for instance, will emit and absorb virtually all the series of strongly interacting particles (pions, kaons, Λ_0 particles, etc.). A physical or *clothed* proton, as it really exists and

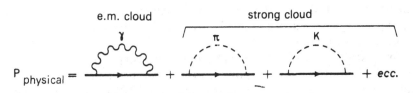

Figure 4.26

is observed in the laboratory, must be thought of as the superimposition
of the *naked* proton with a *strong cloud* and an *electromagnetic cloud*.
This is indicated schematically in Figure 4.26, where the solid line
indicates the naked proton.

But every particle of the various clouds can in turn create its own
virtual electromagnetic and strong (and weak) clouds, and the process
is repeated infinitely. In this vision every particle assumes an extremely
complicated configuration whose description requires the simultaneous
intervention of all the entities with which we know it can interact. This
situation has been given the name *bootstrap* to signify an entity that
sustains itself (in fact, it is said that the Baron of Münchausen could
lift himself from the ground by pulling on his bootstraps).

4.27. Toward the grand unification

The analogy existing between the various states in which particles are
found and their modes of decay, on one hand, and the atomic states
and their modes of decay, on the other hand, is quite obvious. The
possibility that the various particles are states of a *fundamental system*
that decay to form other states (just like the various levels of the hydrogen atom are excited states of the proton electron system) has been
studied in depth. From this point of view, the emission of light in an
atomic transition that can, for instance, be written as (see Figure 4.11):

$$2p \rightarrow 1s + \gamma$$

does not differ in principle from the transition:

$$\Lambda_0 \rightarrow p + \pi^-$$

We therefore ask if it is not possible to consider the entire particle zoo
as composed of various levels of one or more fundamental systems.

A mathematical instrument called *group theory,* much more potent
than the application of the elementary symmetries seen until now, is
very useful in confronting this problem.

Starting from a study of energy level distributions for mesons and
baryons, and basing their investigations on the application of group

theory, M. Gell-Mann and Y. Ne'eman indicated in 1961 the possibility of explaining these distributions by a type of symmetry named SU(3) in the technical jargon.

Interest in this type of conception became very strong, especially after the confirmation of its power, which took place in 1964 with the discovery of the Ω^- particle, postulated by Gell-Mann and Ne'eman on the basis of their theory, as the last of a series of ten levels (of which nine were known).

As was independently proposed by M. Gell-Mann and G. Zweig in 1963, the SU(3) symmetry can be thought of as derived from the combination of three subnucleonic and fundamental objects. Gell-Mann has called these hypothetical objects *quarks,* with a term used by James Joyce in *Finnegan's Wake.* Different combinations of the three quarks and of their antiparticles, the *antiquarks* were believed to explain all the properties of the hadrons.

The three original quarks, today called u (*up*), d (*down*) and s (*strange*), had some surprising properties. They had spin 1/2 as all elementary fermions, but their electric charges $+2/3$, $-1/3$, $-1/3$ (in e units) and their common baryon number 1/3 had fractional values. Moreover, the fact that they had never been isolated seemed to require that their masses should be very large.

The ordinary hadrons are combinations of the u, d quarks and of their $\bar{u}$, $\bar{d}$ antiquarks. Mesons are a quark–antiquark pair. For instance, the pair u$\bar{d}$ has the charge $2/3 + 1/3 = 1$ and the baryon number $1/3 - 1/3 = 0$ and can represent π^+; as a consequence, $\bar{u}$d will represent π^-. The pair u$\bar{u}$ represents π^0. Nucleons are instead composed of three quarks. Thus uud with charge $2/3 + 2/3 - 1/3 = 1$ and baryon number $1/3 + 1/3 + 1/3 = 1$ represents the proton, whereas udd with charge $2/3 - 1/3 - 1/3 = 0$ and baryon number $1/3 + 1/3 + 1/3 = 1$ represents the neutron.

The quark s, of course, takes part in the formation of *strange* particles.

The quark hypothesis had some very attractive features. It could also account for the experimental results found by bombarding protons with high energy electrons (or protons or neutrinos). The outcome was more similar to the situation depicted in Figure 4.3 than to that of Figure 4.2. As a result, protons could not be elementary particles or have a homogeneous structure. Within a proton there must exist some smaller particles that give rise to large scattering angles. These particles, initially called *partons,* are believed to coincide with the quarks.

But the three original quarks turned out shortly to be insufficient and their number had to increase. A fourth quark c (*charm*) was proposed by L. Glashow, J. Iliopoulos and L. Maiani in 1970 in order to explain the absence of some decays otherwise permitted. An experimental confirmation of its existence came in 1974 when a very narrow resonance (10^{-20} s) was discovered, corresponding to a large mass (3.1 GeV) particle called ψ that could be interpreted as the pair $c\bar{c}$. This combination was termed *charmonium* by analogy with *positronium*, an exotic form of hydrogen atom where a positron forms the nucleus and an electron turns about it. Several excited states of charmonium have also been found. The lifetime of the ψ particle is comparatively long because the particle cannot very easily decay into the lighter hadrons composed of the early three quarks and not containing c. The detection of *charmed* particles (both mesons and baryons), or of particles containing the c quark in their structure, brought final confirmation to the existence of the charm.

But the quark family was to grow further. In 1977 a new heavy particle (9.46 GeV) of comparatively long life was discovered. This particle, called Y, must consist of a fifth quark b (*bottom*) and its antiquark $\bar{b}$, forming *bottomonium*. Several excited states have also been revealed, giving rise to a new hadron family, the heaviest so far known. It is currently believed that at least a sixth quark t (*top*) must exist. *Toponium* has not yet been detected, probably because its mass exceeds 30 GeV.

Where has all this led us? Why do we believe the present situation to be a little more satisfactory than that of a few years ago? The answer is that we have apparently succeeded in passing from a maze of hundreds of different and certainly not fundamental particles to a set of twelve fundamental fermions, divided into two groups with a certain symmetry. On the one hand, we have the six leptons e, ν_e, μ, ν_μ, τ, ν_τ, behaving as pointlike particles (in that no experiment has revealed any internal structure at least down to 10^{-16} cm). On the other hand, we have the six quarks u, d, s, c, b, t, also believed to be fundamental and without internal structure. The twelve fundamental fermions can be grouped into three generations. The first generation, represented by the pairs e, ν_e and u, d is all that is needed to build *ordinary* matter. The second generation s, c and μ, ν_μ as well as the third generation b, t, and τ, ν_τ appear only at high energies. In a sense, their role in nature is a puzzle. This may suggest the question: Is our list of leptons and quarks complete? Today no one can tell.

Let us return to the properties of quarks. What keeps them together? It is thought that besides having six possible *flavors* (u, d, s, c, b, t) quarks come in three different *colors* (*red, blue, green* and their anticolors *antired, antiblue, antigreen*). Needless to say, these fancy names have nothing to do with the flavors and colors of ordinary matter. Color is the analog of the electric charge and keeps the quarks together in the hadrons. A proton is made up of three quarks of different colors, hence it is *white*; a meson contains a quark–antiquark pair and is also white.

Quantum electrodynamics, as we know, describes very well the electromagnetic interaction. The attraction or repulsion between two charges arises through the exchange of photons, which represent the quanta of the electromagnetic field. *Quantum chromodynamics* describes in a similar way the interaction of colored quarks. The particles having the role analogous to photons are massless bosons of spin 1, called *gluons,* because they glue the quarks together. Quarks are kept together by exchanging gluons. But whereas there is only one kind of photon, there are eight different kinds of gluons. Moreover, whereas the photon is neutral, so that a particle emitting or absorbing it does not change its charge, gluons are colored – in that each gluon carries one color and one anticolor – and make the colors of quarks change all the time. Quantum chromodynamics still encounters some difficulties, but it has nonetheless obtained considerable success and looks very promising today.

We can ask, Are these quarks and gluons really existing? Have they ever been seen? The answer to the last question is, Yes and no. It was remarked in §1.7 that when we *see* an object, our perception represents the end link of a long chain of physical processes. In the case of the particles that were known before the quarks, the concept of seeing something had already undergone a considerable evolution, with respect to the concept of seeing a macroscopic body. But in the case of quarks and gluons we must content ourselves with still a more abstract notion.

If quarks were easily separable from the hadrons they form, it should be possible to reveal them by an experiment similar to that made by Millikan for electrons, for the fractional charge of a quark would be a clear sign of its presence. As a matter of fact, some experiments of this kind have been attempted, but so far their results have been negative. It is widely believed today that quarks may be *inseparable*. This would depend on the nature of the forces that keep them together. Consider first the case of a proton and an electron bound together in a hydrogen atom. In order to separate the electron, we have only to give it a finite energy (the ionization energy); this is so because the electric force that binds the two particles together decreases sufficiently rapidly with distance. But in the case of quarks we are led to surmise that the binding force may not decrease or may even increase with distance. As a result, when we attempt to separate two quarks, the energy we must supply soon becomes sufficient to create new quark–antiquark pairs; these particles in turn combine with the original quarks and as a net result we obtain a number of hadrons (π, K, ψ and so on) and no isolated quark.

It is impossible today to say a final word on this subject. But many physicists think it probable that quarks may be inescapably *confined* in the hadrons. If this is true, quarks would never exist as single particles. Yet there may be several indirect methods to *see* them.

A spectacular phenomenon observed in some high energy collisions is represented by the appearance of two or more *jets*. A jet is a bunch of high energy particles flying apart in very close directions and filling a narrow cone. When two moderately high energy particles, say, an electron and a positron, collide head-on, they give rise to various particles that fly away from the impact point in all directions. But at very high energy something new may happen; two particle jets are generated that fly in exactly opposite directions. This is explained by the assumption that a quark–antiquark pair is produced at the point of collision. Both particles start in opposite directions, but for the reasons already mentioned, end up by each generating a bunch of hadrons forming a jet. Moreover, at still higher energies three coplanar jets may be noticed. The process can be interpreted as follows. The quark and the antiquark newly produced in the collision travel at high velocities and one of them can emit a gluon (in the same way as an accelerated electric charge can emit a photon). The gluon in turn generates a jet.

These findings, together with a number of other experiments and theoretical elaborations, have rendered the quark and gluon hypothesis something more than a mere hypothesis. Physicists today cannot help but smile when they read Gell-Mann's statement of 1963: "It is fun to speculate about the way quarks would behave if they were physical particles . . . instead of purely mathematical entities." Yet no one has ever seen an isolated quark and perhaps no one will ever see it.

It is expedient now to return to the leptons and to the weak interactions. For a long time physicists had tried to go beyond Fermi's phenomenological theory, attributing weak interaction to the exchange of an *intermediate boson* W; this particle played the same role as the photon in electromagnetism or Yukawa's meson in nuclear interaction. Due to the very short range of weak forces, Yukawa's type of argument (§4.25) seemed to lead to the conclusion that the intermediate boson was extremely massive. No wonder then that it had not yet been observed. But the intermediate boson must also be positively or negatively charged ($W^{\pm}$). To show this let us represent the weak decay of the neutron $n \rightarrow p + e^- + \bar{\nu}_e$ with the Feynman diagram of Figure 4.21(b). Now n enters from bottom left, p leaves at bottom right, ν_e enters (or equivalently $\bar{\nu}_e$ leaves) at top left and e^- leaves at top right; the wavy line stands either for a W^- that transfers the charge -1

upward or for a W^+ that transfers the charge $+1$ downward. We say that a *charged current* has occurred. However, it was shortly realized that a *neutral current* was also necessary. This current carried by a *neutral* boson Z^0 is required to account for scattering reactions such as $v + e \rightarrow v + e$ or $v + p \rightarrow v + p$, where no charge exchange occurs. Such events were experimentally observed in 1974. The three intermediate bosons have all spin 1 – the same as the photon.

Thus we have been led to conceive the physical world as formed by twelve fermions (six leptons and six quarks) cemented by a number of bosons. The situation is definitely better than that of two decades ago when particles started to multiply wildly. But obviously we would like to reduce further the number of independent elements. Moreover, we ask, why should there be four different interactions?

It is natural to wonder whether the analogy between electromagnetic and weak forces is accidental or due to a deep reason. Cannot both interactions be accounted for as different aspects of one and the same phenomenon? The unification of electromagnetic and weak forces has a long history, starting from an interesting theory proposed by C. Yang and R. Mills back in 1954 and a later contribution by S. Glashow in 1960. A special type of symmetry, called *gauge symmetry*, has represented a fundamental tool for these investigations. It is a form of *local* mathematical invariance – well known in the case of Maxwell's electromagnetic fields – which can be generalized to weak interactions. Unfortunately, it cannot very easily be explained in nontechnical terms. This generalization seemed at first to require that intermediate bosons should have zero mass, which was contrary to experience. The difficulty was overcome by means of a new concept, that of *spontaneous symmetry breaking*. The concept was introduced by W. Heisenberg in the case of ferromagnetic materials, where the theory is perfectly symmetrical, without any preferred direction, whereas each sample presents a well-defined direction of magnetization. A second classical example of broken symmetry is crystallization of a saturated liquid about a few initial germs that arise in an unpredictable way in the liquid phase; a third example is turbulence that spontaneously starts in the otherwise regular flow of a fluid. In any case, an asymmetry of the physical system arises without an apparent cause, starting from perfectly symmetric conditions. When a gauge symmetry is spontaneously broken, the zero mass bosons associated with it can become massive (*Higg's mechanism*). Leaning on arguments of this kind – which are described here only in a sketchy and inadequate way – S. Weinberg and A. Salam in 1967 proposed a model of unification of electromagnetic and weak interactions that has since met with ever-increasing success.

The unification of two physical forces that are apparently very different is not a new occurrence and had taken place in the last century when electric and magnetic forces were found to be aspects of a unique physical entity. As long as the electric charges are at rest, the electric and magnetic forces may not seem to have much in common. However, a moving charge is equivalent to a current and generates a magnetic field; a second charge moving in that field experiences a force. It follows from first principles that this force mediated by the magnetic field becomes on the order of the electric force when the speed of the particles is on the order of that of light, that is, when the particles have considerable energy. An analogous fact is believed to occur in the case of weak and electromagnetic forces.

But as soon as weak forces and electromagnetism are essentially unified we naturally also want to unify strong forces. Many physicists are working on this problem, mainly using gauge theories and broken symmetries. We can think of the existence of three initial and identical forces binding together massless leptons and quarks. Then spontaneous symmetry breaking causes the forces to diversify and lends nonzero masses to some leptons and to the quarks. The three forces should converge toward one and the same intensity at very high energy (about 10^{15} GeV).

In the form that is mainly investigated today this *grand unification* seems to require the possibility of transitions from quarks to leptons. The most spectacular consequence of this decay would be that the proton, one of the fundamental constituents of our universe, would not be stable. In actuality, its mean lifetime (about 10^{30} years) would be so much longer than the age of the universe (about 10^{10} years) that there is no danger that the world should vanish before our eyes. However, the decay of a few protons should be observable by means of properly designed experiments. Some laboratories are working in that direction.

Finally, once on the road of the grand unification, why should we not think of the fourth interaction, that is, the gravitational force? In this case, of course, the problems become more difficult and involved, but many theorists are working on them. The hypotheses become bolder and the technical jargon acquires new terms. Today physicists talk of *supergravity, supersymmetries, gravitons, gravitinos.* It is too early to predict a complete success, but some promising results have already been obtained.

4.28. Materialism and mechanism in contemporary physics

One may ask why, until now, we have not mentioned *materialism* (except in passing), in spite of the fact that materialism represents a philosophical conception and a problematic apparently closely linked to subjects dealt with by the physicist. The answer to this question is in fact Lapalician. To speak of materialism (at least according to some traditional conceptions), we must have a clear concept of matter. And today any concept of matter that does not take into account the results of contemporary physics, is absolutely inadequate.

Naturally, we do not claim to confront (and still less to resolve) the *metaphysical* problem. We merely want to ask what kind of materialism is maintainable today by the physicist, and what kind must be abandoned, in so far as it presupposes a concept of matter already obsolete.

What is matter? If we omit the ancient Greek conceptions of matter as an undifferentiated principle capable of taking on form, or as power, we find at the beginning of the scientific revolution the Cartesian definition of matter as *extended substance*. According to Descartes, the principal and characterizing property of matter is that of occupying space. Is this conception still valid?

In reality, today we know that the matter of microscopic bodies is made up mainly of empty space and that its *extension* represents much more the distance between particles than the volume of the particles themselves. And what is meant by the volume of a particle? One of the most reasonable meanings we can give to this expression is that of a portion of space inside which the interaction of the particle with other particles can be felt. But beside the fact that this is not a univocal definition, because it depends on the type of interaction, we are very far from the classical concept of volume of a body, not to be occupied by other bodies.

The particle property closest to what was classically called the *impenetrability* of bodies (or *antitipia* according to Leibniz) is represented by Pauli's exclusion principle. But unfortunately, not all particles obey Pauli's principle. Only particles with semiintegral spins obey it. Should we say that they are matter and the others are not? It seems absurd.

It is instead interesting to remember at this point that even the concept of empty space has changed. Empty space, as we have seen, is not *nothing* and does not only have geometric properties, but has physical ones as well in that it swarms with virtual particles. Paradoxically, we could say that in modern physics, although matter has lost its extension, extension has acquired materiality.

A concept affirmed largely following Newton's work is that of matter as *mass*. After the discovery of the conservation of mass, even in chemical reactions (Lavoisier), this concept seemed more and more reasonable. But after Einstein and Dirac, once the fact that particles can appear and disappear (one particle transforming itself into another) was accepted as normal and not at all strange, and after we recognized the existence of particles of zero mass, like the photon and neutrino, mass was less and less regarded as a good candidate to the name of matter.

One can object that if mass is not conserved, energy (or mass + energy) is conserved. Can matter be identified with energy? This identification, under various forms, has been maintained in the past by several scientists such as H. Helmholtz, W. Rankine, and W. Ostwald, and is known as *energetism*. If it is only a question of definition, it can certainly be accepted. But energy is only *one* of the physical quantities that are conserved. Why can we not call momentum or angular momentum, matter? And what should we say regarding those conservative quantities discovered more recently, such as isospin, leptonic numbers, baryonic number, and so on? If we can speak of matter, is it not a bit naïve to consider it mainly on the basis of its macroscopic properties, that is, of those properties whose only merit is that of not being canceled when many particles join together?

It seems that the traditional concept of matter includes *divisibility* into parts; divisibility that can either (1) continue to infinity or (2) lead to the discovery of elementary blocks (the real *atoms*). But the quark and gluon *confinement* gives rise to a peculiar situation. The ultimate constituents of matter might be nonseparable from one another. Is this an acceptable property? Does it not clash with a reasonable definition of matter? And anyway should not the parts of a material body be smaller than the whole? Quarks and intermediate bosons confront us with striking counterexamples to this rule; heavier particles can assemble to form lighter particles. It is not important that quarks have not yet been shown with certainty. The mere possibility of conceiving such a theory is sufficient to demonstrate how far we are today from traditional ideas of matter.

In my opinion, the most reasonable thing we can do is to recognize (without regret) that matter in its traditional meaning of permanent, divisible, corpulent, hard and extended substance, has dissolved in our hands and no longer exists. But it would be a mistake to think that, after this acknowledgement, *materialism* is dead and buried.

All that has been written about materialism in all its meanings and connotations could fill several libraries, therefore I will not be so ingenuous as to try to state an original thought on materialism in a few words. Rather, I shall limit my task to discussing those aspects that strictly concern physics. I ask the question: In what sense is modern physics, or can modern physics be, materialistic?

If we consider the traditional concept of matter (which, as we said, has already shown itself to be obsolete) as an indispensable presupposition of any materialism, then it is clear that a modern physicist cannot be a materialist. But everyone knows that materialism is not necessarily so crude.

Let us take, for instance, one of the fundamental theses of materialism (dialectic or nondialectic), by which things outside of us exist independently from our consciousness and sensations. With few exceptions, a physicist, as he pursues his profession, accepts such a postulate, whether consciously or unconsciously. I do not exclude that some physicists can have an absolutely idealistic vision of the world. But it seems very unlikely to me that this vision can have a direct relation to the actual activity of research and to the epistemology that informs it. Instead, we must observe that this thesis deserves the name, *realism,* much more than that of *materialism.* It is not a matter of knowing if things exist outside of us, but of knowing if they are material, and of what matter they are made. Unless, naturally, things are material by definition.

It is more difficult that a physicist should declare himself in favor of the other thesis, which denies any difference of principle between phenomena and things and which affirms that the content of consciousness faithfully reflects external reality. The physicist will probably make us realize that today our knowledge of reality is mediated by a long chain of connections and abstract mathematical relations, to which it is very difficult to attribute the character of a reflection.[136]

We could go on, but as I said, I do not intend (nor am I competent) to write a treatise on materialism. The idea that I want to put forward briefly is the following. At first sight it seems obvious that physics has a materialistic foundation, but if we analyze the meaning of this statement and try to give it a precise content, we remain very perplexed. We have the impression of having said with words much more than we can justify with concepts.

It is also interesting to examine the position of modern physics in regard to *mechanism.* As we know, a restricted meaning of mechanism concerns physics only. This is the theory that all physical phenomena depend only on the *motion* of bodies or, at the most, on the laws of mechanics. This theory seemed to collapse with the fall of the elastic hypothesis of light and with Maxwell's work. But, paradoxically, we cannot deny that there is a tendency in modern physics toward a return to a certain type of mechanism. To be exact, it has seemed possible to reduce everything to the movements and interactions of particles. It is a matter of definition, whether the forces that characterize these interactions are mechanical or nonmechanical. Therefore it is perfectly maintainable that physics today is much more mechanistic than it was at the turn of the century!

It is another matter when, abandoning the restriction to physics, one considers materialism and mechanism in a larger sense. Precisely, we can refer to the *monistic* or *reductionistic* hypothesis, according to which all natural phenomena (including *biological* phenomena) depend only on the laws of physics. But we do not always realize the whole meaning of this hypothesis. More exact, it can mean two quite different things.

First of all, we may want to hypothesize that the biological world depends only on the laws of physics known today. But I really do not think that any physicist wants to affirm something of this kind with certainty. As an example, we could discover one day that various complex structures of thousands of atoms give rise to new phenomena,[137] which today we may not even imagine. But if these phenomena were unequivocally describable, measurable and predictable, even if with probabilistic laws, it would be absurd to say that they do not obey physical laws.

Then the only nonnaïve meaning that the hypothesis in question could have would be the assertion that biological phenomena exist which *will never be* tractable and predictable with the method of physics. Here we enter the realm of what cannot be falsified, at least in a reasonably imaginable future. Everyone can think what he wants.[138]

Nevertheless, an important fact must be emphasized. In the last decades biology has progressed immensely, applying the methods of physics and showing that some of the fundamental processes of living organisms depend on molecular structures subjected to physical laws. This method has enjoyed great success, although no alternative method has been found. It would be foolish to stop proceeding along this road, only because no one can ensure us that it will not be barred one day.

I am a physicist, not a prophet. But I have serious doubts about the prophetic virtue of others.

5   The universe

5.1. General laws and historical facts

According to a traditional conception, physics should deal with general laws that govern the universe, rather than with single objects and historical facts. We have already mentioned the problem of being perfectly faithful to this distinction. Indeed, it is difficult to decide in what degree general laws would appear different from those we know, if the physical world were factually determined in a different way from what it actually is.[1] Arguments of this kind can be drawn, for instance, from general relativity and thermodynamics.

But that is not all. We can in fact ask: Is it reasonable to study general laws without considering the concrete objects to which they apply and the structure of the universe that contains these objects? In modern research there are many factors that suggest the opposite. The physicist's interest in the historically determined universe that surrounds us increases constantly. Therefore the scope of this book would be considered too limited if it did not contain at least some mention of the environment in which we live and the modern investigation of it. Naturally, we shall deal with *points of reference* necessary to shed light on the epistemological implications, rather than with information, which could not be given systematically and completely in such a limited space.

In microphysics we have encountered *nomological objects* (atoms, molecules, nuclei, particles), that is, objects whose constitution is strictly governed by general laws. After all, we could think (and some people have affirmed) that they *are* nothing but laws or complexes of laws. For example, the electron could represent the law by which a mass of 0.511 MeV and a charge e are always accompanied by spin 1/2, a Bohr magneton, and so on. In this case the interest in the *class* of objects is much greater than that in the individual. Indeed, we can say that the physicist is totally disinterested in the individual electron.

But if we leave microphysics, we realize that in the environment that surrounds us, there are also *quasi-nomological* objects. It is sufficient to look up at the night sky. All the stars we see have in common some peculiar characteristics that we must assume to be due to laws. Never-

342

theless, they are not all equal! What we say for stars, can be repeated for galaxies, planets, comets, and so on. But it can also be repeated for terrestrial objects, for biological objects, for example. A given human chromosome can have an enormous number of different molecular structures. Yet from a certain point of view, it is always the same element, with the same characteristics in the cellular structure.

We could think that in the cases mentioned, these are classes of objects with *some* (but not all) common characteristics, due to laws. But as yet they are not particular cases. What should we say, then, about the planet Saturn, surrounded by rings? Today we know (see §5.8) that other planets, too, are accompanied by similar rings. Therefore we are certainly confronted with a nomological situation.[2] But until a few years ago, when it was believed that Saturn was the only planet with that feature, one could only conclude that it is a class with only one element! Already we have passed almost without interruption from strictly nomological objects to unique ones, which we must accept as a historical fact. The more we move away from strictly nomological objects, the more our interest in the class decreases and that in single objects increases. For example, the planet Mars is of great interest, even as an individual object. What we said for classes of *objects*, can be applied equally well to classes of *phenomena* (eclipses, explosions of stars, biological mutations, and so on).

If those who deal with general laws cannot completely leave out of consideration the historical fact of the structure of the environment, those who deal with it (geologists, naturalists, astronomers, and so on) certainly cannot ignore general laws. Besides, this is so obvious that humans have spoken of the *cosmos* since antiquity, using a word that includes by itself the concept of ordered structure governed by laws, to indicate the universe.

For a useful terminological reference (absolutely conventional), we shall here call *physicists* those who deal with general laws and *astronomers* those who deal with the structure of the universe. The astronomer is in a position of inferiority in respect to the physicist, in that he can only observe the universe rather than *experimenting* with it, that is, asking the universe the questions he wants. We certainly cannot construct a galaxy with desired initial conditions! Therefore the astronomer must resort to the general laws with which the physicist provides him by experimenting in the laboratory. But the physicist in turn is in a position of inferiority with respect to the astronomer because of the same factual circumstances. The astronomer has the galaxies at his disposal, whereas the physicist in the laboratory does not! Certain questions we would want to ask the universe would require a much louder voice than our own; in other words, with human means we cannot perform the experiments that we would like to do. Our only resource is to that immense natural laboratory, the universe. In this case we cannot *ask questions*, but must only *listen*. By listening well, however, and by analyzing what we hear, we can deduce many important facts.

What does *deduce* mean in this context? The term deserves a more formal examination. But we immediately encounter great difficulty. We have spoken of *laws* and of *particular facts*; but a strict and formal distinction between these two concepts is complex, and indeed, even today has not been found!

Let us take expression (4.98: $\forall x(Ax \rightarrow Bx)$), for example, with its universal quantifier. It seems a typical general law. Let us then consider with Quine this particular fact: "Suppose I have lost my key. Then immediately we have this trivial generalization: It is true of everyone x without exception that if x is I, x lost his key" (Quine, 1966, p. 50).

Here a very particular fact such as "I have lost the key," has acquired the form of a general law like expression (4.98). Should we then say that laws do not exist? Or should we say that single facts do not exist? But both physical laws and individual physical facts exist. Everyone uses these concepts, and there must be a way to discuss them! Therefore, giving up that strict formal rigor that seems unreachable, we shall try to give a semiintuitive description, as precise as possible, of how we can deal with laws and particular facts. Thus we start by assuming that we know intuitively (and this is the weak point) what a historically determined physical situation or a particular physical fact is.

A situation or fact of this kind can be expressed by a proposition p. There exists, by hypothesis, the set P of all the propositions that express possible situations or particular facts, and we can recognize when $p \in P$. Let us consider such a p. Even without knowing any physical law, we can define the set Q_0 of all the logical consequence of p; for any proposition $q \in Q_0$ we find that $p \rightarrow q$ (*if p, then q;* or *p implies q*) is a logical law. In particular, p itself will belong to Q_0.

A physical law L widens, by way of a new set Q_L, the set of *consequences* of p and turns it into $Q = Q_0 \cup Q_L$ (i.e., the *union* of Q_0 and Q_L, or the set that contains all the elements of both Q_0 and Q_L). Evidently, it is always possible to choose the new set Q_L so that it does not have elements in common with Q_0. We make this choice and write symbolically, $Q_L \cap Q_0 = \emptyset$ (i.e., the *intersection* of Q_L and Q_0 is empty).

Once we know the physical law L, the set Q_L is a function of p and we write

$$Q_L = f_L(p) \qquad p \in P \tag{5.1}$$

The fact that every proposition of the set Q_L is a consequence of p can be conventionally written

$$q \in Q_L \quad : \quad p \xrightarrow{L} q \tag{5.2}$$

But this is not a logical implication; q does not follow logically (or analytically,[3] if we like) from p. The meaning of expression (5.2) is instead what we discussed in regard to induction: given p, the probability that q does not occur is absolutely negligible.

Let us illustrate with an example. Let p be the proposition: "At such and such a time, in such and such a place, I release a stone (in a vacuum)." By the law of the fall of masses, we obtain as consequences of p, the propositions q_1: "After 1 s, the stone has crossed 490 cm"; q_2: "After 2 s, the stone has crossed 1960 cm"; and many other propositions that form the set Q_L. But the law of free fall makes the empty set correspond to proposition p: "At such and such a time, in such and such a place, I strike a tuning fork." In fact, this law cannot draw any consequence from p (besides logical ones).[4]

Note that according to equation (5.1), a law L can be associated with any particular fact expressed by $p_1 \in P$. It is sufficient to define the function $f_L(p)$ corresponding to this case. Precisely, when p_1 is a logical consequence of p, $f_L(p)$ represents the empty set; in the other cases,[5] it represents the set that has as elements only p_1 and all its logical consequences (which are not yet included among the logical consequences of p). Do not be deceived into thinking, however, that we have strictly defined what a particular fact is. It would indeed be a circular procedure, because we began by assuming that we can recognize when $p \in P$.

Generally, we think that a law deserves such a name and that it is worthwhile to state when, for *many* propositions p, not logical consequences of one another, the set $Q_L = f_L(p)$ has *many* elements, not logical consequences of one another. And obviously, we cannot give a strict definition of what we mean by *many* in this case.

The physicist in fact is only interested in the general law, equation (5.1) and not in knowing if p and which p actually occurs in reality. The physicist is not interested in knowing if at such and such a time, in such and such a place, I have actually released that stone.

The case of the astronomer is different. He is certainly interested in general laws, for example, Kepler's laws or the universal law of gravity; and in this he is similar to the physicist. But the astronomer is also interested in the particular fact that the planet Mars exists, that it is made in a certain way, and that at such and such a time it is found in such and such a position. To deduce these facts, he uses the laws of physics. For example, he observes the stars with a telescope and applies the laws of optics. Then, to derive the structure and composition of the heavenly bodies, he applies the laws of mechanics, thermodynamics, microphysics, and so on.

Let us suppose that the astronomer is interested in ascertaining the fact q_1. The simplest but, unfortunately, least common case is when it is possible to make an observation p from which q_1 follows, through a law (or a complex of laws) such as equation (5.1). Then we must consider ourselves fortunate, and accept q_1 as acquired.

But almost always, the possible observation p *underdetermines* the physical system studied with respect to q_1. To be exact, a q follows from p which says that, q_1 occurs or q_2 occurs or q_3 occurs, and so on. In symbols we write ($\lor$ is the symbol of *disjunction* and is read *or*)

$$p \xrightarrow{L} q; \qquad q = q_1 \lor q_2 \lor q_3 \cdots \qquad (5.3)$$

where L indicates the overall set of known physical laws. The situation is ambiguous, and all the astronomer can conclude is that q_1 is *possible*.

The physicist in the laboratory is generally (but not always) able to avoid this unpleasant situation by manipulating the conditions of the experiment, for example, by realizing the situation $\neg q_2 \lor \neg q_3 \cdots$ artificially. In this case, because

$$p \lor \neg q_2 \lor \neg q_3 \cdots \xrightarrow{L} q_1 \qquad (5.4)$$

follows from expressions (5.3), we can ascertain q_1 by way of the observation p.

The astronomer does not have access to the universal laboratory and must follow a much more complicated (and sometimes rather ingenious) process. Precisely, he tries to exclude the alternatives to q_1 through other suitable observations (generally of a type totally different from p). To give an example of this process, let us suppose that expressions (5.3) offers only three alternatives, and therefore is written as:

$$p \xrightarrow{L} q_1 \lor q_2 \lor q_3 \qquad (5.5)$$

Let us suppose then that the astronomer is able to make two other observations p' and p'' such that

$$p' \xrightarrow{L} q_1 \lor q_2 \lor \neg q_3$$
$$p'' \xrightarrow{L} q_1 \lor \neg q_2 \lor \neg q_3 \qquad (5.6)$$

Let us suppose that q_1 is false. Then if we hypothesize that q_3 is true, the second observation p' requires that q_2 be true as well, whereas the third observation p'' requires that q_2 be false. Thus the hypothesis is absurd. We have thus eliminated the alternative that q_1 is false and q_3 is true. In a similar way, with two other well-chosen observations, we can eliminate the alternative that q_1 is false and q_2 is true. Then we can only conclude that q_1 is true.

This scheme is very idealized compared to what often occurs in reality. The several alternatives often live together for years, decades, or even centuries! Sometimes we tend to ignore some alternatives because we think them not very *probable* for some reason. The advent of astronautics has represented a great step forward, for in many cases it has allowed us to eliminate a great number of alternatives. Concerning the solar system we can use the most effective process of all: to send instruments *in loco* or even to go to *see* in person!

The ascertainment of the factual structure of the environment in which we live constitutes *cosmology*[6] in a broad sense. *Cosmogony* has always been associated with this science, especially today. Etymologically it refers to the *birth* of the universe. This supreme problem is certainly always present in the minds of scientists, and in due time we shall see in what sense and with what limitations it is handled.

But today very great interest is also focused on the problem of the birth of a number of much more particular systems or structures. The examples are many and varied: galaxies, stars, the solar system, mountains, animal species; hence by analogy, such cultural facts as writing, language, and so on.

Why do we pose these kinds of problems? Why do we presume that such questions can have sensible answers? In my opinion, this is not merely an unjustified psychological attitude or an infantile curiosity. As we shall see in many cases, the facts themselves present connotations which lead us out of necessity to formulate those questions.

It is suggested that the reader carefully reread what we said concerning the time arrow (§3.16) as well as causality (§4.20). We find many objects or systems of objects in the universe that bear clear traces (in a broad sense). These constitute formations or situations so unprobable that the formula, equation (3.93), leads us to dismiss the hypothesis that they are due to a spontaneous fluctuation of the system. On the basis of equation (3.94), however, we are much more likely to assume that there has been an interaction with another system of high negentropy. We naturally ask ourselves what this system or interaction was in the same way as, when we see footprints in the sand, it is impossible not to ask ourselves who has passed there.

From this point of view we are convinced that the prescientific attitude of the biblical Genesis (or of analogous geneses of other religions or mythologies)[7] is anything but naïve. If there are footprints, someone must have passed by. The most obvious hypothesis is that it is someone very similar to us, but much more powerful, who has created things as they are.

Only a more mature philosophical reflection distinguishes first the concept of *creator* from that of *builder*, the latter being the forger and organizer of a *pre-existent* matter. Physics then leads us to recognize that pouring negentropy in a material system does not necessarily require the intervention of a thinking subject, but merely interaction with another system far from equilibrium.

5.2. Form and movements of the earth

The study of the environment begins in earliest infancy when the child through his sensations begins to know the surrounding world. Already at that phase a mixed process takes place: a study of the environment and a successive study of the laws that govern it. Here *successive* is intended in a logical sense more than in a strictly temporal one. In fact, we immediately have a strict interdependence of the two aspects: To study the world around us we must know the general laws, and the general laws derive from the world around us.

Let us begin our examination of the environment by dealing with the earth; first, because the earth is the planet on which we live and second, because this study effectively illustrates general problems such as why we are often led to imagine a *teleology* of the universe.

There is no doubt that when we begin to reflect for the first time on the surrounding environment, all things seem to have been ordered in this way with a precise goal; and we can easily convince ourselves that *we* are that goal! In many religions, particularly in those of the Judeo–Christian tradition, this fact is recognized and even sanctioned. Examining the environment in which we live with a critical eye can also help us to investigate if this supposed teleology can have an origin and explanation different from that of divine will. As might easily be expected, this is very possible; but paradoxically, we shall see that the scientific study of the environment can also bring new and valid arguments to corroborate the teleological hypothesis – arguments that the author or authors of the Bible certainly did not possess. Now let us begin by considering two concepts that are sometimes erroneously confused: that (1) the earth is round and not flat and (2) the earth rotates around the sun and spins around itself.

The idea that the earth is spherical was already present in classical antiquity[8] and known by many, even if not universally accepted. Only the *direct* proof of this is relatively recent. This proof dates back to Magellan and the first circumnavigation of the world.

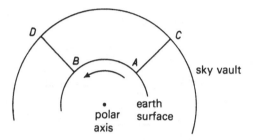

Figure 5.1

What were the suppositions on the roundness of the earth due to? First, the finiteness of the surface was suggested by the observation of the heavenly vault that revolves around us (it does not *flow*, but *rotates*, because the same constellations appear to us periodically). This rotation is meaningful only if we suppose the earth to be *finite*, and not constituted by an unlimited flat surface.

An observer sees the heavenly vault revolving about an axis, the *polar axis*, which passes very near the polar star and its antipode, that is, the diametrically opposed point relative to the center of the earth. The great circles that pass through these two poles are the *celestial meridians*. The observer sees all the meridians passing successively above him.

If the celestial vault did not rotate, then from two points situated at different longitudes, that is, on different terrestrial meridians, we would always see the same two different aspects. But because the celestial meridians rotate, to ascertain this difference, *simultaneous* observations in different places would be necessary. This would require the use of high precision clocks, which would have to be transported hundreds of kilometers apart, maintaining the exact time. This was certainly impossible in antiquity.

To schematize these ideas, let us consider Figure 5.1, in which the direction of the polar axis is perpendicular to the plane of the drawing, and we assume that the rotation of the earth is counterclockwise. If an observer at A sees a certain constellation C at his zenith, an observer at B sees another constellation D at his zenith at the same *instant*. Because we cannot define the simultaneity of the observations, it follows that the observer, initially at A, sees C at a certain time, and after a certain time, sees D. Therefore it is impossible to affirm that the sky is different when seen from two different longitudes. But from the fact that the various constellations observed from any terrestrial point rotate with uniform angular velocity, we can deduce that the celestial vault is at a very great, nearly infinite, distance. Another type of ob-

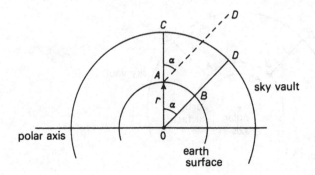

Figure 5.2

servation that could be made, even in antiquity, is that represented in Figure 5.2 concerning two observation points on the same meridian, but at different latitudes. An observer at A sees constellation C pass through the zenith, which is different from constellation D, which passes through the zenith of B, and can also measure the angle $\alpha = CAD = AOB$. If the length of the arc of terrestrial surface AB is known, at least approximately, it is possible to derive the approximate size of the earth, that is, its radius $r = AB/\alpha$ (Eratosthenes of Cyrene, about 250 B.C.).

We can present these facts in a simplified way, as follows. First, let us consider the statements:

q_1 = the earth is spherical

q_2 = the earth is flat and finite

q_3 = the celestial vault is at a finite distance

q_4 = the celestial vault is at an infinite distance.

The longitudinal observation p tells us

$$p \xrightarrow{L} (q_1 \vee q_2) \wedge q_4 \tag{5.7}$$

whereas the latitudinal observation p' tells us

$$p' \xrightarrow{L} q_1 \vee (q_2 \wedge q_3) \tag{5.8}$$

Now from expression (5.7) we have q_4, one of whose logical consequences is $\neg q_3$. But then, from expression (5.8) we obviously derive q_1.[9] Clearly, in this way we have neglected a number of alternatives that we consider not very probable.

It is true that the movement of the earth around the sun and around itself has been discussed since antiquity (particularly by Aristarchus of Samos in the third century B.C.); but the theory was not very successful. The Ptolemaic geocentric system (Claudius Ptolemy of Alexandria, second century A.D.) predominated throughout the middle ages; and the heliocentric system was not scientifically acknowledged until the sixteenth century, with Copernicus. We shall discuss this later.

In examining our planet directly, we note that the earth is not exactly spherical; it is what we call a *geoid*. Roughly speaking, it is a ball 12,600 km in diameter, slightly flattened at the poles. The polar diameter is smaller than the equatorial one, but the difference is very small (about 43 km). The flattening is due to the spinning of the earth, which gives rise to a centrifugal force and thus a tendency to widen at the periphery (i.e., at the equator). One consequence is that we are nearer the earth's center at the pole, and thus the terrestrial attraction is stronger here than at the equator. Then if we add the effect of the centrifugal acceleration present at the equator the result is that the acceleration of gravity measured at the pole is greater than that measured at other points on the terrestrial surface, particularly the equator.

In giving this brief picture of the morphological and astronomical characteristics of the earth, we inevitably repeat information that is well known. Nevertheless, on this matter one can make a rather strange remark: There are often elementary questions concerning our immediate environment, which are not adequately known by everyone. This lack of knowledge often occurs even among so-called knowledgeable people. It is evidently a cultural deformation, by which one is interested in knowledge in the broad meaning of the term; but the more abstract knowledge is – the further it is from us, our surrounding environment, and our everyday needs – the more it seems noble and desirable. This is perhaps the reason that an educated person can sometimes tell you what *pulsars* or *quasars* are (heavenly bodies that we shall discuss later) but cannot explain why the weather in the temperate zone is worse in the winter (in the sense that it rains more, besides the fact that it is colder).

We return to our original argument that the earth moves in two major ways in the solar system – rotation around its own axis and revolution around the sun. The earth's orbit, or the course described in the motion of revolution, is elliptical, as are all the planetary orbits. But it is almost a circle. The orbital plane is also called the plane of the *ecliptic*.

In addition, the earth is spinning around its axis, which is not perpendicular to the ecliptic plane: Its inclination with respect to the normal to that plane is about 23°. A very important consequence of this inclination is the phenomenon of seasons. If the axis of rotation were perpendicular to the ecliptic plane, the great circle that marks the boundary between the lighted and unlighted hemispheres would always coincide with a terrestrial meridian. The sun's rays would always fall perpendicularly on the equator at midday and the length of the day would always be equal to that of the night. Instead, because of the inclination, this situation occurs only two days a year, during the *equinoxes*. At about three months' distance from an equinox, there occurs instead the maximum difference between the lengths of day and night. The sun's rays are always inclined in respect to the equator, and we have a *solstice*. The culmination of each season does not correspond to one of these points, but comes a bit later. This is due to the fact that the earth's surface and the atmosphere possess a certain thermal inertia.

The major causes of the variations of the mean temperature from one season to another are the different length of day with respect to night and the different inclination of the sun's rays. In summer the insolation at not too distant points to the north and south of us is not very different. In winter, on the contrary, the same distance involves a remarkable difference in insolation. The masses of air found at not too distant latitudes are heated in quite different ways. This gives rise to violent air circulation. A warm air mass, which rises and moves rapidly toward a colder zone, carries along a certain quantity of water vapor, which condenses when it reaches the colder zone and causes rain.

Meteorology is a science of enormous practical interest, but even today it is far from being in a sure and definitive shape. We do not think it involves unknown physical laws. Rather, the impression is that very small variations in the relevant parameters are sufficient to cause conspicuous variations in the effect. There is then a double order of problems: On one hand, we cannot measure all the minimal variations with the necessary precision; on the other hand, the parameters we must consider are so numerous that we cannot control them easily.

Meteorology is an emblematic science that leads us to an important consideration. A world can exist that is subject to iron laws, all quite simple, which does not however lend itself to be described and predicted, except in a very broad way. Fortunately, the macroscopic world rarely presents these conditions. Usually, we can investigate it with remarkable precision on the basis of a certain number of more important parameters.

5.3. The earth's structure

When we begin to probe our knowledge of the earth, we realize that our information is quite limited. To think that we are so near to the internal regions of our planet but yet know much more about regions of the universe that are very far away! Actually, we know very little about the inside of the earth through *direct* experience, because we are only able to scratch the surface. However, whatever information we are able to ascertain reveals the power of the application of the general laws of physics.

First of all, by analyzing the way in which the earth gravitates around the sun, the moon gravitates around the earth, and bodies fall on the terrestrial surface, we have been able to deduce the total mass, and thus determine the average density of the planet, that is, the average content in grams for every cubic centimeter. The average density is about 5.5 g/cm^3 or 5.5 times the density of water. However, if we consider the rocks found on the earth's surface (granites, basalts, and so on), we see that their average density is about 2.2 g/cm^3. This means that a heavier part must exist inside the earth, with greater average density.

For other information we can use other methods, still indirect, which take advantage of the propagation of seismic waves. For instance, every time an earthquake occurs, elastic waves called *seismic waves*, propagate from the focus or *hypocenter*, that is, from the point from which the phenomenon originates (from a few km to 700 km deep). Some of these waves propagate toward the surface and others toward the interior. By placing a network of *seismographs* on the surface of the earth, it is possible to detect the waves and to determine the time taken to cover the distance from the focus to the various detection stations. Elastic waves in a solid are both longitudinal – they vibrate in the same direction in which they propagate, like acoustic waves in the air – and transversal – they vibrate in a direction perpendicular to that of propagation, like light waves. Longitudinal waves can propagate in both solids and liquids, whereas transversal waves cannot propagate in liquids. In the latter case, in fact, each layer of the medium that vibrates in a direction transversal to that of propagation would have to transmit its movement to the next layer. This is possible only if the medium is capable of transmitting what we call a *shearing* stress, that is, if the medium is solid.

Let us assume, then, that we have a network of seismographs distributed over the earth's surface (Fig. 5.3). Let us also assume, to begin with, that the vibrations are transmitted in a straight line (which they are not, as we shall see). As long as we do not go very far from the focus O, we see that the stations receive both the longitudinal waves (*l*) and the transversal ones (*t*), even if the latter arrive with a certain delay because their speed is lower. But at a certain distance the seismographs begin to show only longitudinal waves. Further on, when

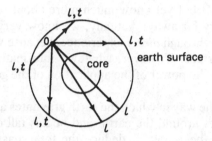

Figure 5.3

the circuit is completed and we return to examining stations nearer the focus, we again find transversal waves as well. This series of observations leads to an extremely important conclusion: that a central, liquid core must exist in which the propagation of transversal waves is impossible.

In reality, there is more than one important transition, and moreover, the nonuniform density prevents the propagation from following a straight line (variable refractive index). Taking all this into account, we are able to deduce the density of the various layers by experiment and calculation. As a first approximation, the image we can deduce of our planet is the following. Very close to the surface is a *crust*, about 30 km deep, made of very light material. Below this is a *mantle*, which as a first approximation, can be considered solid, and which is composed of heavier rocks (silicon and magnesium). Finally, there is a *core* of heavy metals (iron and nickel) in a liquid state. This is because the temperature increases from the surface toward the interior, and is high enough in the core to melt metal.

In regard to the mantle, we must say that it is correct to consider it solid, where fairly rapid movements (like the oscillations of seismic waves) are concerned. But it is also considered a very viscous fluid, a plastic medium that can only be deformed very slowly in a layer from 100 to 250 km deep (this is called the *asthenosphere*). An example of this kind of behavior is found in glass, which is not really a solid. To verify this, affix one end of a long glass rod to the wall and observe it after a few months or years, depending on its length and thickness. We see that the rod bends. This occurs because it does not have the capacity of resisting strain inherent to solids; but its viscosity is so high that a very long time is necessary before the deformation becomes noticeable. The crust and the first layer of the mantle up to 100 km of depth, constitute the *lithosphere*, which behaves substantially like rigid matter.

Now what considerations can we make on these hypotheses of the earth's structure from the cognitive point of view? There are two possible logical positions. A very cautious position is the one of recognizing that we cannot be sure that the internal configuration is as we have described it. But we can affirm that *if* the laws that are valid in the rest of the known universe are valid inside the earth as well, such a configuration would produce those effects (the earth's movements, gravitational acceleration, seismic waves) on the outside which can actually be detected today. The second position, the one that we actually accept, is that when we are capable of observing the interior of the earth directly, we will find that its structure coincides with that surmised today. This situation is quite different from that in the other fields of physics, in which we make suppositions that are not verifiable a priori. Here instead we are confronted by an impossibility which there is no reason to think will last forever.

The gaseous envelope that surrounds the earth and extends for several hundred kilometers in altitude, with a density that decreases progressively, constitutes the *atmosphere*. The existence of the atmosphere was determined with certainty in ancient times, for the movement of winds unequivocally suggests the presence of something material around us. The fact that this natural element in which we live is no different from other elements, at least in the sense that it gravitates with a weight of its own on the earth's surface, has been clarified only in relatively recent times. The uncertainty of this matter resulted from classical theories, in which the earth was regarded as consisting of four elements: earth, water, air, and fire, of which the latter two supposedly rise. These theories are credited to Anaxagoras, but were taken up again by others (even by Aristotle) and reached Galileo's time almost unvaried.

The first to prove that the atmosphere has a weight like all other bodies, even if it is difficult to imagine because we live within it, was E. Torricelli, one of Galileo's pupils. He understood that because there is pressure exerted on all parts of our bodies, in all directions which is counterbalanced by the interior pressure of our bodies, there must consequently be an atmospheric pressure that we do not detect. Using the mercury barometer, a very simple instrument, he revealed its existence and measured it as well (in 1643). Today we usually note the effect of atmospheric pressure on the eardrum, for example, at high altitude. But we must consider that in antiquity, and even in Galileo's time, this effect was negligible, for fast vehicles did not exist and one gradually got accustomed to the change in passing from plain to mountain. The decrease in atmospheric pressure with the increase in altitude above sea level was in fact discovered even later, when B. Pascal brought a barometer to the top of a mountain and repeated the measurement there.

Today we know that in going from sea level to an altitude of 8,000 meters, the atmospheric pressure diminishes by 50 percent. When we arrive at levels on the order of 100 km, the pressure becomes extremely low. Therefore we normally say that the atmosphere is only a few hundred kilometers high. Actually, it is difficult to define a precise limit; the atmosphere does not end abruptly but, rather, dissolves in the interplanetary void.

The atmosphere is conventionally divided into layers, the lowest of which is the *troposphere*, which extends from ground level to an altitude of approximately 10 km. This is the stratum in which clouds are found and in which the most evident atmospheric disturbances (winds, precipitations) take place. When we speak of the composition of the air, we generally refer to data measured in this zone only. Remember that the main components of air are nitrogen (about 78 percent) and oxygen (about 21 percent). Then, in much lower percentages, there are also noble gases, water vapor, carbon dioxide (about 0.03 percent), and traces of other elements.

The *stratosphere* extends from 10 to 30 km of altitude. Its composition is almost the same as that of the troposphere, with the difference that water vapor is present in even lower percentages, and that some ozone is present in the upper regions (the importance of ozone will be evident when we speak of the problem of life on earth).

Above the stratosphere are other zones. The most important of these, because of its effect on the propagation of radio waves, is the ionosphere, in the neighborhood of 100 km. Here solar radiation frees electrons from atoms, making the atmosphere conductive.

Let us turn to the magnetic field. The use of a magnetic needle in determining direction was known as early as the twelfth century. But only around the sixteenth century was the earth itself recognized as the source of magnetism. The study of the earth's magnetic field has continued ever since. Even the modern compass is essentially made up of a magnetic needle which rotates freely in a horizontal plane and always points toward the north. This shows that there is a magnetic field in the space around the earth. Any magnetic field can be described by representing its lines of force, that is, those lines along whose tangents a magnetic needle aligns itself. The curious thing is that the earth's magnetic field can be compared to the field produced by a magnetized metal bar, oriented almost parallel to the earth's axis. The lines of force of this field are perpendicular to the earth's surface at two points: the magnetic poles (which do not coincide with the geographic poles). Their course is illustrated in Fig. 5.4.

One of the great problems of geophysics, which still remains un-

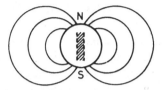

Figure 5.4

solved, is in understanding what this magnetic field is due to. We know for certain that it exists around the earth, as well as around other celestial bodies. It has been largely accepted that the earth's magnetic field is due to some electric current, for magnetic fields are usually generated by electric currents. We might think that some currents circulate in the earth's core, which is made of conducting materials. But we do not know for certain why these currents arise or how they circulate.

The magnetic field has important properties and, in particular, important repercussions for the earth and for our lives, especially from the point of view of the planet's history and evolution. Some rocks that are found on the earth's surface are magnetized, that is, they more or less possess the properties of a magnet. Some rocks possess these properties in a macroscopic way (magnetite, known since antiquity, is a natural magnet). Others possess them in a very small way, so that very sensitive instruments are needed to detect them. From many clues we can deduce that the magnetization of the latter rocks is due to the fact that they were formed in the earth's magnetic field. The magnetic field has induced the magnetization, which is oriented along the lines of force. We would then logically suppose that magnetism in all the rocks on the surface is oriented in a certain way, and that it has always been so. However, detailed analysis shows that the earth's magnetic field has not remained stable in time, so that the most ancient rocks present an orientation that is different from more modern rocks. Furthermore, in certain ages, the direction of the magnetic field has even undergone a *reversal*. The reason for these changes presents a problem that is difficult to solve, and so far we have only hypotheses.

5.4. The cosmogonic problem

As mentioned, the cosmogonic problem arises naturally and inevitably from the observation of what surrounds us. It can concern either the origin of the total system – the universe – or the origin of one of its subsystems. The earth offers an excellent example of such a subsystem.

What does a physicist mean when he speaks of the *origin* of the universe or of one of its subsystems? As already discussed, it cannot be a biblical meaning by which a material system emerges from a preexistent *nothingness* as a result of a fiat. This cannot be the *scientific* meaning because according to what we know today, matter does not seem to come from nothingness by physical law, therefore we must shift the argument and define it in a scientifically acceptable way – by reducing it to a hypothesis plus a definition.

The hypothesis is that the physical laws that we observe and study now have always been the same in the past and are the same everywhere (invariance principle).[10] The definition is the following.

Let us suppose that we consider a certain system S, which today (instant t) has the structure $S(t)$. If it is possible to find a system S_0 (instant $t_0 \ll t$) whose structure or composition is totally different from S, and is such that, evolving according to the laws of physics, it turns into S at the instant t, we shall say that S_0 represents the *origin* of S and that $t - t_0$ is the *age* of S.

This problem is certainly more limited and much less ambitious than the one we would have had to solve, had we adopted the biblical meaning. Nevertheless, it is still a formidable problem.

Furthermore, we must immediately note a fact that gives rise to many questions, including philosophical and epistemological ones. We might possibly find a solution to this type of problem, but it might not be a *unique* one. That is, we can find a system S' (t') with a structure completely different from $S(t)$, but also different from $S_0(t_0)$ which, left to itself, would also give rise to $S(t)$. And so there may be many others. If what we call *strong determinism* were in force and the application of physical laws to the present structure of the universe permitted us not only to *predict* its future structure with absolute certainty, but also to *retrodict* its structure, as it was millions or billions of years ago, the solution to the problem would have to be unique. The real trouble is that this point of view is untenable. Strong determinism is obviously excluded for two reasons, one from classical and one from quantum physics.

Let us for a moment put quantum considerations aside and assume the hypothesis (which in reality is very doubtful) that quantum indeterminism does not have macroscopic consequences. In classical physics the objection to a strong determinism comes from the second principle of thermodynamics. Let us consider a simple example. We pour two liquids – say, wine and water, in this order – into a container. After a certain time we find nearly a homogeneous liquid. This we could have *predicted*. But it is not possible to *retrodict* from which situation we started. In fact, if we had initially poured the liquids in inverted order, we would have reached the same final state! The fact that the two liquids have mixed is a result of the second principle of thermodynamics, by which, in an isolated system, entropy (i.e., disorder) is increased. To classical indeterminacy, one has then to add quantum indeterminacy.

The cosmogonic problem is therefore hard to solve, but it is not completely desperate. It is helpful to envisage different *classes* of solutions, characterized by a certain degree of reasonableness and plausibility. Then we try to reduce the number of possible classes gradually by means of ad hoc observations and the widening of general knowledge.

Leaving the problem of the origin of the universe for later consideration, let us begin with the earth, taken as the system S. There are many clues on the planet that suggest both an origin and a *history*. For example, an important idea emerges from the comparison of the chemical composition of the earth and its atmosphere with that of the sun and the other stars. The stars and interstellar matter are characterized by an abundance of hydrogen and helium relative to the other elements. We could say that the universe is composed roughly of 90 percent hydrogen and 10 percent helium, whereas the other elements practically represent mere impurities. But in the terrestrial atmosphere, hydrogen and helium are almost totally absent. Why such a difference?

Perhaps the various celestial bodies are made differently, as a result of unknown physical laws or, if we prefer, because of an unfathomable divine plan. But it is strange to note that if we do not consider hydrogen and helium, which are the lightest elements existing in nature, the chemical composition of the earth is not very different from that of the other celestial bodies. The percentages of the heavy elements are about the same. This fact strongly suggests that the earth may *originally* have had a composition perfectly analogous to that of all other celestial bodies, and that afterwards, for some reason, it may have lost the lightest elements – that is, almost all its hydrogen and a large part of its helium. This hypothesis is not only plausible but also corresponds to a very simple scientific explanation.

Today our knowledge of the distribution of temperature in the atmosphere is quite good. This distribution is characterized by a rather curious trend. In the troposphere it diminishes from about 20°C at ground level to about − 55°C at a level of 12 km; in this zone, the *tropopause*, there is an inversion and the temperature rises to about 0°C at 50 km. Here enters the *stratopause*, where a new inversion occurs: The temperature decreases to minimal values (− 70°C), measured in the *mesopause*, the zone at around 80 km. From this point on, a continuous temperature increase occurs and we even reach 1000°C in the highest region of the ionosphere, at about 400 km. All these data have been obtained through direct measurements carried out with rocket probes and artificial satellites.

On the other hand, we know that temperature is linked to the speed of thermal agitation of the molecules. The molecules of a gas are in continuous chaotic agitation, moving about, and colliding with and bouncing off one another. Every molecule has a certain kinetic energy $E_k = (1/2)mv^2$, and we know that the temperature of a gas is proportional to the mean kinetic energy of its molecules.

If we mix two different gases, we know that after some time they both have the same temperature: We have reached thermal equilibrium. In the atmosphere, which is a mixture of many elementary gases, all the components are in thermal equilibrium, at the various levels. Thus if we consider two molecules 1,2 of two different gases, as a result of the fact that the gases have the same temperature, we can write $(1/2)$ $m_1 v_1^2 = (1/2)m_2 v_2^2$ or $m_1/m_2 = v_2^2/v_1^2$. Let us say that molecule 2 is one hundred times heavier than the other, that is, $m_1/m_2 = 1/100$. In this case $v_2^2 = v_1^2/100$, that is, $v_1 = 10v_2$. Therefore we see that lighter molecules move faster in the atmosphere. The hydrogen molecules (molecular weight (m.w.) 2) are certainly the fastest, followed by those of helium (m.w. 4), and further along, by those of nitrogen (m.w. 28), oxygen (m.w. 32), and so on. The earth, though, exerts an attraction on these molecules, as on all other bodies. This force of attraction decreases in a way inversely proportional to the square of the distance from the center.

If we throw a body into the air, it reaches a certain height and falls back. But this is true as long as the body does not reach a certain critical velocity. In fact, if on one hand the terrestrial attraction slows down the movement, on the other hand the attraction decreases gradually as the body moves away. It is therefore possible that the attraction is unable to reduce the velocity to zero. The minimum velocity that a body must acquire in order to escape the gravitational action is called the *escape velocity*. For a probe or a spacecraft to escape terrestrial attraction and to travel toward another planet, it must reach the escape velocity. For the earth this is approximately 11 km/s.

If we calculate the mean velocity of the hydrogen molecules in the upper strata of the atmosphere, we see that it is precisely on the order of the escape velocity. At this point, we should say that the earth's atmosphere was at one time composed mainly of hydrogen – the most abundant element of the universe – and that this gradually escaped because of the high velocity of thermal agitation. Helium is four times heavier than hydrogen, and thus escaped in lesser quantities, leaving traces in the atmosphere. Even some heavier elements can escape terrestrial gravitation, for as a result of statistical fluctuations, some molecules sometimes reach a high velocity, exceeding the escape velocity. But these losses are so small that they can even be compensated by the gases emitted by volcanoes, for instance.

This same reasoning is naturally valid for the other planets and bodies of the solar system as well. First among them is the moon. We easily calculate that not only hydrogen and helium, but also nitrogen, oxygen, and water vapor have a velocity of thermal agitation equal to or slightly less than the escape velocity ($\sim$2.4 km/sec) from the moon. Thus it is impossible for the moon to maintain a gaseous covering that is not made up of very heavy elements; and in fact, the absence of an atmosphere has been known for some time, and has now been demonstrated by direct observation.

Let us consider for a moment the terrestrial crust. For a long time we thought that the earth's crust was fixed and immutable and that the earth had been created with the present seas, continents, and mountains. Geologists had long disproven that the situation was completely static. Nevertheless, scientists continued to think that changes occurred on the surface, in a *local* way, due to the erosion of atmospheric agents, to natural cataclysms, and so on.

In 1912 A. Wegener formulated the theory of *continental drift*, according to which the continents in their present form originated from the breakup of a much larger continental block, and continue to move on the earth's surface. This theory, which was criticized and rejected in its time, has by now overcome all opposition.

Wegener's idea was suggested as a result of a number of observations, the simplest and most evident of which concerned the correspondence between the outlines of the east coast of South America and the west coast of Africa[11] (Fig. 5.5). If we can imagine bringing these two continents together, we can easily see that the protrusions of the one correspond to the indentations of the other.

Naturally, other types of evidence are needed to support the theory; this evidence has been obtained in the last few years, mainly through studies of the earth's magnetic field. To complete the studies of the magnetization of superficial rocks of which we spoke in the preceding section, rock samples have also been taken from the ocean's bottom. Geological analyses of the ocean floor and magnetic analyses of the rock samples have led to extremely interesting results.

To begin with, geologists discovered the existence of peculiar mountain ranges, the midocean ridges on the ocean floor, each one roughly halfway between two continents. On each side of a ridge there is a depression, followed by a smaller mountain range, and so on. By analyzing the rocks drawn from elevations parallel to the midocean ridge,

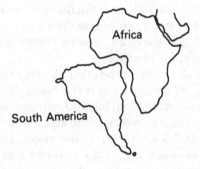

Figure 5.5

we discover that these rocks are alternately magnetized in opposite directions (Fig. 5.6). This is true for many stripes parallel to the ridge, except for some morphological details that we can overlook. The distance apart between two reversal lines of this type can vary between 10 and 100 km, depending on the location.

These facts, at first sight surprising, have been explained by a theory (*plate tectonics*), now supported by many other data and thus fully accepted. The two fundamental points of the theory state that (1) the earth's magnetic field is subjected to periodical reversals and (2) the ocean floor spreads outward from the central ridge.

There is not much to say about the first point. Field reversals surely exist, but we ignore their cause. Regarding the second point, we can easily believe that the asthenosphere, below the lithosphere, has convective currents, probably due to endogenous forces of thermodynamic origin. These currents are generated between the hotter and colder points of the plastic mass, similar to what occurs in liquids and gases. On the ocean floor, where the crust is thinner, an upwelling magma arrives which tends to move upward. Molten rock breaks through at a rift in the crest of the ridge and gradually solidifies, filling the gap that is left as the ocean floor on each side of the rift moves outward. The rock remains for some time in a plastic state and is therefore susceptible to magnetization. Naturally, magnetization occurs parallel to the existing magnetic field. Thus successive stripes, moving outward from the midocean ridge preserve the record of the reversals of the earth's magnetic field. There is very good agreement between the measurements taken in the Atlantic Ocean and those taken in the Pacific or Indian oceans.

Today in view of this and other evidence, the theory of continental drift is firmly established. Wegener spoke about a single supercontinent, the *Pangaea*, which existed once upon a time and which broke

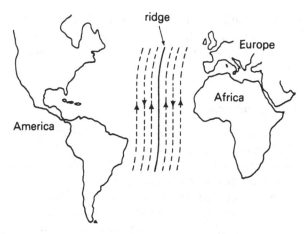

Figure 5.6

up into the continents as we know them today. The situation is probably more complex; continents can break up and drift apart, but they can also collide and join in the course of history. Thus America drifted apart from Europe and Africa, whereas India, moving northward, collided with Eurasia. At present, we can verify that the continents continue to move apart at a speed of a few centimeters per year. This shows that it took on the order of 100 million years to cover the present distance between Africa and America. Presumably, at the beginning of the mesozoic era the two continents were still united.

It is interesting to note what happens when two parts of the lithosphere meet while moving in opposite directions. This occurs, for instance, where the oceanic crust crashes with a continent that moves toward it. The crust is forced to go below the continent at a *subduction zone* and this causes the formation of a trench (e.g., the Chile–Peru trench, on the order of 8,000 meters deep, off the west coast of South America). It also gives rise to remarkable seismic phenomena such as earthquakes. It is sufficient to recall the famous San Francisco earthquake of 1906, and those which repeatedly occur on the west coast of America. The pushing effect can also generate mountain ranges in the continent itself, as is exemplified by the Andes.

Still more dramatic events are caused by the collision of two continents. For example, the collision of India with Eurasia gave rise to Tibet and the Himalayas.

At this point it is clear that the earth is anything but a static system, unchanged since the day of creation. A very picturesque though rough interpretation, is to think that we are living on the surface of a pot of boiling water; we do not realize this because our life is much shorter than that of the bubbles.[12] But it is also true that the succession of innumerable generations of living beings cannot be considered brief even on a geological scale; this makes it possible for life to influence the evolution of the environment.

5.5. The environment and the biosphere

Surprisingly, we may feel the need to speak of living organisms[13] when we speak of the physical environment in which we live. But today there are many good reasons for not strictly adhering to the traditional distinction between living and nonliving environment. For one thing, we can observe that efforts to define the concept of environment along traditional lines usually fail as soon as we try to reach a certain level of exactness and critical appraisal. Many scholars today are inclined to assume that when we study a certain object O, the environment must be simply defined as U-O, where U represents the universe. Indeed, sometimes there are good reasons for including O in the definition of its own environment!

Moving on now to discuss the living and nonliving in particular, we know, first, that the line of demarcation between the two realms is quite arbitrary; one can go almost continuously from the large molecules of organic chemistry to more complex structures, from these to others that are still more complex, and finally to viruses, bacteria, and so on. Second, it is not possible to overlook the fact that the *biosphere* definitely influences even the nonliving terrestrial environment. It is absolutely impossible to imagine a *geogony*, that is, a history of the earth's origin and development, which does not take into account the phenomenon of life. The earth is not merely a planet of the solar system, but it is also a planet on which life has developed.

As an obvious example, consider the fact that geological strata exist which are thousands of meters deep and made up of limestone, a sedimentary rock formed mainly by the shells of marine animals, and fossilized microorganisms. Sometimes these are deeply metamorphosized rocks, but their origin cannot be understood in any other way. This information is generally known; however, some facts concerning the atmosphere are not so widely known.

From many factors we have been able to deduce that the earth's primitive atmosphere did not contain oxygen, except in minimal percentages. Today we also know that most other planets have atmospheres that contain carbon dioxide, methane, and ammonia, but not oxygen. Therefore it is logical to ask from where this element came, or what generated it on earth.

Surprisingly, we are virtually sure that oxygen was initially produced by living organisms, then reused by life to support itself. This cycle has been made possible by the fact that the first living organisms in the ocean were probably *anaerobic* microorganisms, that is, microorganisms (largely *blue-green algae*) that do not need oxygen. These started synthesizing organic matter, fixing carbon from the atmosphere (composed in part of carbon dioxide, CO_2) and thus freeing oxygen. When the percentages of oxygen in the atmosphere was no longer negligible, beings that were better prepared for the new situation began to appear. These organisms could use the oxygen in the atmosphere to derive energy by an inverted process – by burning carbon or, in other words, by recombining carbon with oxygen. At this point the still evolving living organisms left the sea. Part of them, the plants, continued to fix carbon and free oxygen, until the present situation was reached. Regarding plants, we must say that they also consume oxygen for life, but during the day this process is considered negligible compared to the opposite one, photosynthesis due to *chlorophyll.*

Observing the degree of oxidation of the oldest rocks and taking into account other factors as well, it has been concluded that until 2,200 million years ago, the quantity of oxygen in the atmosphere was negligible. But 1,800 million years ago oxygen had already become an important component.

We feel naturally compelled to say that life has created its own suitable environment. This statement is further justified if we consider that because of the bombardment of ultraviolet radiation, life on the earth's surface would be impossible without the atmosphere as it exists today. We know that the sun emits radiations of all wavelengths. Among these, there is good radiation such as visible radiation which is necessary for photosynthesis in plants. But there is also *dangerous* radiation such as ultraviolet radiation (corresponding to wavelengths of about 2 to 3 $\times$ 10^{-5} cm) which is capable of destroying any form of life (in fact, it can be used to sterilize objects). On humans, ultraviolet radiation can cause skin cancer and other serious problems.

Life first developed in the sea. Here the upper layer of water completely absorbs ultraviolet radiation thus making the environment safe from this danger. Then, when oxygen began to appear in the atmosphere, a phenomenon of great importance occurred. When an O_2 molecule is struck by ultraviolet rays, it can break into its two atoms; these sometimes join other oxygen molecules, forming triatomic molecules O_3, called *ozone*. Ozone molecules absorb ultraviolet radiation; indeed, their absorption capacity is so high that relatively few ozone molecules are needed to constitute an efficient screen for the entire planet.[14] It is only after this phase that life can emerge from the sea.

A teleological interpretation to the presence of those few ozone molecules found in the atmosphere would be that they have been put there to protect our lives! Instead, we reverse the terms and say that if sea life had not started to develop oxygen in adequate quantities to form a sufficient ozone layer, evolution would have exploited other means (e.g., by remaining underwater). Any change in the environment would have given rise to new possibilities, although excluding others, and life would have taken advantage of the former. Moreover, the sun is a good disinfectant. This means that the atmosphere, even with its ozone, is not sufficient protection for a large number of microorganisms.

If ships could think they might think that seas, lakes, and rivers were made for them. We say instead that ships have been made as they are because seas, lakes, and rivers exist!

We can derive the following consideration from the preceding discussion: If life influences the environment, and can even make a hostile environment suitable to life, cannot the opposite happen as well? Can living organisms, particularly humans, not change the environment in such a way that it becomes less suitable to life? This is the problem of *environmental* science, which has acquired enormous importance today, and which sometimes goes under the rather inadequate name of *ecology*.[15] We might ask, why *precisely today*, and why is the problem so serious? After all, from the few concepts we have touched on here it could seem that the changes brought about by humans are only further contributions to the heating of this pot of boiling water, the earth.

The greatest problem now is that man does things very rapidly. Life has transformed the primitive environment, but is has done so at a very slow pace. Therefore it was possible for the new forms of organisms to adapt to the new environmental conditions; a modification dangerous for life as it was could be good for a new organism capable of exploiting it. Today instead we are witnessing an explosion of life, which on the basis of the data we possess, seems unprecedented. We must consider that man modifies the environment substantially, even on the scale of hundreds of years, whereas biological phenomena, those that have created and developed the biosphere, work in terms of millions or even tens of millions of years.

Things have changed radically since *intelligence* and *culture* were born, that is, when the moment came in which life no longer moved ahead in a certain direction through successive trials governed by chance, but was able to calculate its chances and to transmit the acquired knowledge to successive generations. The human race has been able to influence the environment through calculation and science. This has made the environment so much more favorable that our lives seem to have become extremely easy in respect to that of other animal species. These species have always defended themselves from extinction in a hostile environment with the only possible weapon: rapid reproduction, thus compensating for the losses. For humanity, which has practically succeeded in diminishing its losses by finding food for everyone, developing a defense against illness, and minimizing the effects even of natural cataclysms, the result has been a *population explosion*. But in this way the number of those who use and alter the environment, in order to live within it, increases more and more, thus creating a chain reaction.

Sometimes our intervention can seem too small to involve drawbacks. But this may merely be a matter of ignorance. A possible danger, for instance, is constituted by the supersonic planes of the last generation, which fly at high altitudes corresponding to the ozone layer. Their exhaust products could perhaps destroy the ozone.[16] As mentioned, there are few ozone molecules in the atmosphere; if they were all brought to sea level at normal pressure, they would form a layer roughly 3 mm thick. It is legitimate to ask how much ozone would be left the day in which fleets of supersonic planes flew. Recent research does not seem to lead to too pessimistic results. Nevertheless, it is peculiar that humans project and fly supersonic aircraft of this type before clearly understanding if the ozone is or is not substantially destroyed by doing so.

A similar cry of alarm went up regarding *spray* cans, whose propellent could also place the ozone in considerable danger because of its chlorine content ($ClO + O_3 \rightarrow ClO_2 + O_2$). Here the threat seems to be more serious. Some authors attribute the skin cancer rise of the last years to this cause.

Humanity's *massive* interventions in the environment are well known. We recall, for instance, the problem of combustion. We said that plants fix carbon and free oxygen. This oxygen can be used to support a living organism, which burns carbon by combining it with oxygen and releases a certain quantity of CO_2 into the atmosphere. Even plants follow this mechanism, but they do not burn all the carbon that they gradually fix. They store a part, which can be deposited in the earth and, in a process that requires millions of years, becomes coal, methane, and petroleum. We have been digging and burning all the fossil fuels that we find for nearly a century. This means that we are releasing into the atmosphere all the CO_2 that had been stored up over millions of years. Now CO_2 is a minor component of the atmosphere, but toward the end of the nineteenth century it began to increase, and it is still increasing. To understand the seriousness of this situation fully, we must bear in mind the particular function of CO_2 in the atmosphere: that of preserving the earth's heat. Visible solar radiation crosses the atmosphere and heats the earth. The earth releases part of this energy in the form of radiation of greater wavelength, in the infrared range. But this radiation is absorbed by the CO_2 in the atmosphere, so that the heat is stored for a certain time before dispersing in interplanetary space (the *greenhouse* effect). If the percentage of CO_2 were to increase at rates that seem predictable today, we would risk a nonnegligible mean temperature increase, a climatic change, and a variation of the thermal equilibrium. These changes are very small, but their effects can be macroscopic; it is sufficient to think that to increase or decrease the *mean* temperature of the earth by half a degree signifies passing from a tropical to a glacial era. At this rate we risk one day reaching such an antiglacial era that the polar ice caps will melt, provoking an increase in sea level of some meters, and therefore submersing an enormous number of square kilometers of land which is presently above sea level. It is necessary for humanity to realize that it is playing with fire and that certain problems must be confronted *before* the modifications we are introducing into the environment become irreversible.

From a theoretical point of view, man attempts to create *order* around himself, through his activity. But as we explained, he cannot contradict the second principle of thermodynamics expressed, for example, by equation (2.96). If he succeeds in diminishing the entropy of the systems, in which he is interested, he cannot avoid increasing it somewhere else, perhaps even in a much more conspicuous way.

One way of increasing entropy can also include diffusing substances that in nature exist in well-delimited areas (lead, mercury, etc.). Generally, this *pollution* is harmful. And, one should not necessarily think of material substances only. Even *noise* is a form of pollution and is due more or less to the same cause.

5.6. The origin and evolution of life

The problem of the origin of life, that is, knowing how the different forms of living organisms have appeared on earth, has not been completely solved, even today, although fully plausible hypotheses do exist.[17] We shall begin by examining the situation as it is currently understood.

Let us consider a living organism and examine its structure and fundamental components. Any living being is composed largely of certain molecules in long chains, which are called *proteins*. These constitute the most complex organic compounds that exist in nature, and are formed by the combination of simpler elements, the *amino acids*. There are twenty different amino acids that generally make up proteins. They can arrange themselves in long chains in different orders to give rise to an enormous number of different proteins. Proteins have structural functions of support of the organism, as well as of control and guidance of the internal activities of the organism itself. A very important class of proteins is that of the *enzymes*. These are particular proteins with catalytic properties that favor the development of the chemical reactions needed inside the organism.

The cellular construction of living organisms is necessarily achieved through the manufacture of proteins. If a builder were to undertake such a job, he would need a plan and a communication system to indicate to the person who materially carries out the construction, which *bricks* he must use, that is, which amino acids he must choose and how he must combine them. Nature has adapted to this necessity and has created a *code* with which it is possible to transmit this message, which is essential for the reproduction of the cell and of the entire organism.

In a living cell, we can distinguish between the *cytoplasm* and the *nucleus*, the latter containing those complex structures called *chromosomes*. These in turn contain *genes*, substructures that carry hereditary characteristics. Only in 1944 did O. T. Avery and co-workers clarify the chemical nature of genes. In 1959, thanks to D. Watson and F. Crick, it was possible to identify their structure and thus to understand the mechanism of genetic transmission. A gene is composed of a long, linear molecule of desoxyribonucleic acid (DNA), made up of a chain of elementary units. Actually, DNA chains always appear in pairs, in the sense that a DNA chain is always flanked by another (which we can call *complementary* to the first), and the two chains are located in space in such a way that they form the well-known *double helix* structure. Every unit of the chain contains a *character* of a code, equal for all organisms. The characters, composed of the DNA *bases*, are *guanine*, *adenine*, *cytosine*, and *thymine*. The order of the bases, that is, the way in which they follow one another in the chain, determines genetic information. This allows us to interpret the DNA chain as a form of writing based on a four-character alphabet.[18] The *orders* given by this written code are then carried out by other large molecular structures, known as *ribonucleic acid* (RNA), which are made up of four bases – guanine, adenine, cytosine, and uracil.

Let us briefly note how the message is read and how its prescriptions are carried out. A DNA filament wrapped in a helix gives rise to an RNA filament, called *messenger* RNA in this case, in which the information contained in DNA is reproduced. This RNA leaves the chromosome from which it has obtained the information, passes into the cytoplasm, and comes to rest on one of those bodies called *ribosomes*, made up of protein and *ribosomal* RNA, where the information passed from the messenger RNA is read.

There exists a third type of RNA, called *transfer* RNA, whose specific function is to join up with the amino acids required for the construction and to transport them from the cytoplasm to the ribosomes, where they will be ordered and tied together to form proteins, in the way specified by the messenger RNA. In brief, the information carried by the DNA-base sequence is transcribed into an RNA-base sequence, which in turn is *translated* into a sequence of amino acids in the proteins.

This mechanism of writing and reading a certain message plays an essential part, not only in the development of the organism but also in its reproduction. In sexual reproduction, for instance, it is known that the new organism inherits part of its genes from the paternal cell and part from the maternal one. The message, which naturally agrees with the general characteristics of the species, determines what its individual characteristics will be. The DNA chains of the new individual are reproduced exactly in all its cells. This greatly reduces the possibility of a disturbance of the organism's structure. In fact, a sufficiently energetic disturbance can provoke a genetic change, that is, a *mutation* in the elementary DNA units. It can thus modify the message. But if this happens in a tissue cell (skin, muscle) the phenomenon is normally limited to that cell and its possible daughter cells, and does not bear consequences. The results are instead remarkable when the changes occur in one of the germ cells. In this case the mutation can result in an organism that differs more or less noticeably from the other organisms of its own species. All this has been firmly established in general terms, although, as we said, many details (some of which are extremely important) still have to be clarified.

The essential problem for those who study the origin of life is that of knowing if these molecular chains and this code have been put together according to a preestablished plan, or if the process could have developed by itself under suitable environmental conditions.

This is a point of much discussion, but today a considerable number of scholars are convinced that in a certain environment (like that of the earth billions of years ago), the phenomenon could have started by itself. An essential condition is that processes that free energy sufficient to break molecular bonds exist in nature. Such energy must be at least on the order of 1 eV. We can certainly consider this condition satisfied, for we know certain processes on the earth that can set free even much higher energies. Such an example is cosmic *rays*, composed of particles that arrive in all directions from the cosmos, sometimes possessing extremely high energy (even billions of billions of eV).

When they enter the atmosphere they can break many molecular bonds. Radioactive substances in turn also emit particles that are capable of shattering molecules. But even lower energy processes, such as electric discharges in the atmosphere or volcanic eruptions (with very high temperatures in play), can produce the same effect.

On the other hand, we have a fairly good idea of how the earth's primitive atmosphere must have looked.[19] In all probability, the principal components of the atmosphere were hydrogen (with its simpler compounds such as ammonia and methane) and carbon dioxide. We can make a mixture of these compounds and subject it to processes that arise naturally in an atmosphere of this kind (electrical discharges, X rays, ultraviolet rays, and radioactive radiation).

Experiments of this kind have been carried out several times, after pioneering work by S. L. Miller and C. Ponnamperuma. It has been ascertained that in these conditions some more complex molecules can naturally be formed from simple initial molecules. If an electrical discharge passes through an atmosphere composed mainly of molecules of CO_2, NH_3, and H_2, some molecules break up and in their place we find atoms that tend to regroup, often reforming the same simple molecules, but sometimes forming more complex structures. At times, even amino acids or DNA bases can be formed. The case is very rare, of course, but considering the availability of an enormous number of molecules in the atmosphere and taking into account the many millions of years that may have passed before the appearance of life, it is not surprising that this phenomenon may have occurred often.

All we know today is that an atmosphere of primeval character, subjected to certain natural processes, can give rise to complex molecules, among them the DNA bases. After a long interval, of which we have very little if any record, we find life already evolved; that is, we find complexes of organic molecules that are already highly structured. What lies in the middle is rather obscure. We can only guess that the initial molecular structure gradually grew more complicated until it became a molecular aggregate capable of reproducing itself – of giving its imprint to the surrounding chemical elements. In this way, new similar aggregates were built, and so on.

This is, in an extremely simplified way, what biochemistry can tell us today about the possible origins of life. But to say that we know for certain that life had precisely this origin would be rather a daring statement.

Nevertheless, extremely important indications can be found in the *documents* left by the various phases through which life has passed on earth. By examining these documents, we can derive enough elements to work out a promising theory. But the elements are insufficient in number to explain everything, and in many issues there is ample room for doubt.

The record left by life on earth can consist of either cumulative signs, such as sedimentary strata left by the remains of living organisms (shells, coral, etc.), in which the structure of the single individual is no longer recognizable, or individual traces of living organisms. But the latter are mainly limited to comparatively recent periods, when living organisms had already developed substantial solid parts, that is, either outer skeletons (insects) or inner skeletons (vertebrates). This is what we see in *fossils*.

For many centuries, fossils had been considered mysterious objects, or even classified as *jokes of nature*, for it was difficult to imagine that those *things* had lived some day. We must in fact consider that fossil remains sometimes were found even at remarkable depths, and it was difficult to understand how living beings could have ended up down there. Furthermore, we also find forms among fossils that are completely different from present forms, and these have acquired a reasonable meaning only by using the theory of evolution.

On the other hand, if we were to think that the "jokes of nature" are due to natural fluctuations of the environment (as we sometimes find a stone or a piece of wood that reminds us of an animal), we would face great difficulties. The high degree of structural complexity and perfection of many fossils, together with the circumstance in which we find them and, moreover, some of them are perfectly similar, points to such an improbable fluctuation that we must rule it out, on the basis of equation (3.93). We must therefore be dealing with true signs, that is, with the past interaction of the environment with a conspicuous source of negentropy, such as might be presented by a living being.

The fact that fossils are datable has allowed us to derive extremely important information from them. For the entire nineteenth century and the first half of the twentieth, the dating was very crude, because of the imprecise procedures. The predominant procedure consisted of measuring the depth of the strata, evaluating the rate of sedimentation, and hence assessing the time necessary for the formation of a certain thickness. Then the dating of the strata could be correlated in one zone with that of the corresponding strata in a different zone – an enormous job with insufficient objective data. Precise dating has been made possible by radioactivity, a phenomenon that in many ways could be called the *natural clock*.

A radioactive element can be characterized by its *mean life*, that is, the average lifetime of its atoms; alternatively, one can use the *half-life*, the time necessary for half of the atoms to disintegrate or for the intensity of radiation to be reduced by half. This is possible because no external physical conditions are capable of altering the pace of radioactive decay, except for extremely energetic processes.

The final product of a chain of radioactive transformations produced by a parent element is a stable isotope of an element different from the initial one. For instance, the final product of uranium decay is a lead isotope. Moreover, since uranium 238 first emits an α particle and decays into thorium 234, which in turn decays by emission of an α particle, by measuring the present ratio between uranium and lead or uranium and helium (the α particles emitted) in a mineral, it is possible to determine the age of the rock or the time passed since its solidification. The half-life of uranium 238 is 4.5 billions of years, too long for the dating of comparatively recent rocks. The radioactive isotope carbon 14, which is present in all living matter and whose half-life is 5,700 years, is extremely useful to establish the age of relatively recent fossils or man-made articles with greater precision. But it becomes increasingly unreliable when the sample is older than about 30,000 years. Other methods, such as measuring the relative abundance of radioactive potassium and its argon daughter isotope or of radioactive rubidium and its strontium daughter isotope, are applied in different circumstances and in different age ranges.

A new and revolutionary method, called fission-track dating, has been recently introduced. It entails counting the number of tracks left in a rock by the heavy-ion fission products of uranium 238. It can be usefully applied over an enormous span of time, from billions of years to a few decades ago.

For a very long time it was thought that the earliest fossil records went back to the beginning of the *Cambrian* age, about 600 million years ago, when the *Phanerozoic* phase begins. Before, there was total darkness, and this constituted an enigma even for Darwin. The formerly accepted explanation was that organisms developed solid parts that could be conserved (even if petrified) only in that era.

In reality, the situation is different. Pre-Cambrian fossil documents are not in short supply, but must be sought exclusively in the microscopic, rather than in the macroscopic, realm, where classical paleontology had sought them. We can say that pre-Cambrian paleontology has originated in the last two decades. In a very short time it has produced results of extreme importance and is now in an explosive phase.

The solid earth was formed about four and a half billion years ago and the first sedimentary rocks are about 3,750 million years old. There are good reasons to think that microscopic life originated shortly thereafter, with an *Archaean* phase, of which we do have some microfossils (difficult to interpret), but mainly we have indirect documentation. Among these, we must chiefly mention *stromatolites*, microlamellar formations with characteristic alternations of calcareous and organic layers, produced by certain *anaerobic* microorganisms, even today. About 2,300 million years ago the *Proterozoic* phase took over (which extends until the Phanerozoic phase) with very diffuse stromatolites, some well characterized types of microorganisms and plentiful oxygen producers, such as blue–green algae.

About 1,500 million years ago cells with nuclei (*eucaoriotes*) appeared, which reproduce by division (*mitosis*). In this type of reproduction the daughter cells are identical to the parent; the variability is very limited and, consequently, the evolution is still very slow.

A formidable stroke was the invention (about 1,000 million years ago) of sexual reproduction (*meiosis*). The daughter cell inherits genes from both parents. The possibility appears of rapid variability, selection and adaptation which we usually associate with biological evolution. Toward the end of the Proterozoic phase, one-cell organisms (*protozoa*) were joined by pluricellular organisms (*metazoa*) and macroscopic organisms appeared. Thus we arrive at the Cambrian era with life already developed and differentiated in an imposing way.

Then begin the classical eras: the *Palaeozoic* (Cambrian, Silurian, Devonian, Carboniferous and Permian); the *Mesozoic* (Triassic, Jurassic, Cretaceous); the *Caenozoic* (Eocene, Oligocene, Miocene, Pliocene). Thus we arrive at about one million years ago, the epoch in which the *Neozoic* or *Quaternary* era begins (Pleistocene, Holocene). In the Quaternary, or perhaps in the Pliocene, humans appear on earth.

These eras constitute a classification of essentially empirical character, based largely on their characteristic fossils. For example, in the Cambrian era the characteristic fossils are the trilobites, crustaceans whose body appears longitudinally divided in three parts: whereas in the Carboniferous era the characteristic fossils are cephalopodal molluscs. Amphibians also begin to develop in the Carboniferous era, and the first reptiles appear as well. The Mesozoic era can be considered the era of reptiles, because in that period they developed in an imposing way, even reaching gigantic dimensions. Toward the end of the Mesozoic era, great reptiles start disappearing and some forms of birds are present instead. The following era, the Caenozoic, marks the development of mammals, and toward its end (we are still uncertain where to draw the limit), even hominids appear among the primates. All this has been fairly well established, thanks to the finding of numerous fossils.

The period during which life on earth has left documents reveals a succession of organisms whose complexity (or the number of *bits* necessary to describe them) gradually increases. With this documentation, the fundamental idea of evolution appears almost obvious. We must then make two considerations. The first is that in Darwin's time (1809–1882) the dating of fossils was still extremely uncertain, and it was difficult to place them in temporal succession. The second and most important is that the fundamental idea at the basis of Darwinian evolution is not so much that species are transformed into one another (in fact, this idea was already present in preceding writers), but that the fixation and affirmation of certain hereditary characteristics is due to natural selection. Roughly speaking, Darwin thought that the individuals of the species presented genetic variations in some way and that the natural selection chose those who possessed a better adaptation to the environment. The result was an elimination of the unfit and a transformation of the species. But what Darwin lacked was the possibility of understanding how this mechanism could be explained at the microstructural level.

Today we have a scientific basis for saying that things may have happened in a way extremely similar to that of Darwin's descriptions. This basis is corroborated by our knowledge of the structure and properties of DNA, which bears the blueprint code of the living organism. DNA is in fact subject to *mutations*, which can change the blueprint, even radically, and thus the structure of the organism. For this to occur we need an energetic process. To those already mentioned (X rays, cosmic rays, etc.) we can add the processes produced by certain chemical substances such as hyprite and nitrous acid, compounds structurally analogous to the DNA bases, and so on. Moreover, today we know that in the genetic patrimony, even of humans, mutations can occur that are often quite recognizable: For instance, we know that mongolism and hemophilia are caused by alterations of certain chromosomes. We also know how to induce mutations in an enormous quantity of plants and animals. In certain cases we are even able to induce mutations that turn out to be useful, especially in agriculture.

This alterability of the genetic matrix composed of DNA through processes, which are found in nature, could seem harmful at first sight. Instead, this factor is at the basis of the evolutionary process. Let us think of a species of organisms characterized by a certain type of DNA, which reproduce normally. If an individual undergoes a genetic mutation, it can transmit the mutation to its descendants; thus a new type of organism, different from its predecessors, is born. This new type naturally is compared to the *normal* type[20] and, if the mutation is harmful, that is, it has made the individual less suited than the other to the environment and to the struggle for life, it will be eliminated by selection.

When we think that a living species is generally formed by an enormous number of individuals and that the species lives a vast number

of years, it is by no means unbelievable that a favorable mutation sometimes takes place, by which the *new* individuals are better suited for
survival – finding food, seeking a sexual mate, and reproducing – than
the others. Thus a new variety of the species, or even a new species,
appears.

This kind of interpretation does justice to what Darwin had guessed
more than a century ago, and also explains why other theories of evolution, existing as alternatives to Darwin's, are far from having the
same scientific dignity. No one can swear to the validity of the Darwinian view. On the other hand, it is difficult to deny that his theory
represents a most rational and scientifically sound approach.

Naturally, there are many difficulties! For example, Weizsäcker
writes:

> A great mathematician, van der Waerden from Zürich, once discussed this
> matter with me at length. Van der Waerden claims to be able, by calculating
> the probability of the steps necessary to make the evolution of the eye of a
> vertebrate conceivable, to prove that the combination of these steps is so
> improbable that the five billion (or perhaps only 2 billion) years available as
> a geological scale, are not at all sufficient. The value of expectation of the time
> required for the appearance of this structure according to the Darwinian model
> is too great, according to van der Waerden's calculations. I have discussed
> it with him, but he did not convince me (Weizsäcker, 1970, p. 286).

I have the impression that Weizsäcker discusses the problem too easily.
Van der Waerden's objection, like a number of other objections raised
by anti-Darwinists, is very serious. This merely means that we lack
some essential knowledge (as Darwin himself did), not that Darwin's
view is necessarily wrong. To dismiss a scientific theory, simply because it does not yet explain *everything* and because there are difficulties, is an error repeated again and again by many scholars in the
course of history.

At this point it may seem that we have digressed from the cosmological problem. However, as noted, it would certainly be a mistake
to exclude the biosphere from cosmology: (1) because in dealing with
cosmology, it is impossible to neglect such an important fact as the
biological evolution that has occurred on this planet (even if we do not
presently know if this type of evolution has occurred only on the earth)
and (2) because other types of evolution, totally unknown and unthinkable for us, might have occurred somewhere else. Clearly, then, those
who restrict themselves to a purely physical concept of the universe
may have a limited view: The universe may be much more than a simple
set of particles that move according to certain laws.[21]

To conclude, we can briefly mention another problem, still completely open to philosophical speculation and scientific research. The

biosphere has given rise to a new phenomenon, which perhaps a biosphere per se would not necessarily imply: the birth of the *noosphere*, the birth of intelligence. Is this merely a matter of chance? The only certainty we have to proceed on is the fact that on earth the noosphere has been the last to arrive, after billions of years. Aside from this, we can only ask questions such as: Does the noosphere constitute a point of arrival, or should we expect a further evolution? If so, in what direction? These are questions open to any answer.

5.7. Windows on the universe

Cosmology requires the accumulation of considerable information from the observation of the universe. Before presenting the comparatively brief description of the solar system, our galaxy, and the more distant regions of the universe, let us examine the means of physical investigation that have allowed us to accumulate these data.

Let us reexamine that zone of the earth's atmosphere situated at an altitude of about 100 km, called the *ionosphere*. In reality, some ionospheric layers can already be found at 50 km of altitude, whereas the most distant are found at around 400 km. For some of these, the position is stable, whereas for others it varies even with the hour of day. It is characteristic of these layers that ionization processes occur there. The electron bonds of some atmospheric molecules are broken, which leads to the formation of pairs of particles with opposite charge. In the simplest case, ionization can be considered as the passage from an initial state, composed of a neutral molecule, to a final state, composed of a positive ion and a free electron. The energy necessary for a process of this kind to occur is furnished by solar radiation, mainly ultraviolet rays, which are almost completely absorbed.

Thanks to the presence of free electrons, the ionospheric layers thus become conductors of electricity. This fact determines a result of the utmost importance for the development of communications on our planet: The ionosphere is capable of reflecting radio waves. We shall explain.

When we apply an electric field to a region of space (or to a material body) in which there are free electrons, these are set in motion, producing an electric current. The electromagnetic waves of which we have spoken at length (§2.12), consist of oscillating electric and magnetic fields that propagate in space. When an electromagnetic wave reaches a conducting body, the electrons are made to move back and forth, following the oscillations, and in this way give rise to an alternating electric current.

On the other hand, we know (§2.11) that an alternating current radiates an electric and magnetic field; that is, it generates electromagnetic waves. Radio waves, for instance, are generated by sending an alternating current of very high frequency through an *antenna*. As a result, when the alternating electric field of the wave (overlooking the magnetic field in the interest of simplicity) reaches the surface of a conducting body and makes the electrons oscillate, these reemit a wave of the same frequency as that of the impinging wave. In this way the *reflected* wave is born. In the case of the ionosphere the electrons are relatively few and dispersed. They cannot follow the oscillations of the electric field if they are too fast. This explains why the ionosphere does not reflect light waves (of very high frequency) but does reflect radio waves (of lower frequency).

The usefulness of the ionosphere for radio communications became apparent even before its structure and properties were known. Much of the merit goes to G. Marconi. Marconi is an interesting and, in a sense, problematic figure. He was perhaps one of the last representatives of that category of people brilliant enough to obtain concrete results of notable scientific import, although lacking a basic academic background. Indeed, his kind has often been opposed by the academic world.

We know that the transmission of electromagnetic waves had already been demonstrated by H. Hertz, crowning in this way and completing, the work of J. Maxwell. Hertz built the first electric oscillators and was able to detect the waves they generated, but the distance between oscillator and detector could not be greater than a few meters. At that point this seemed an interesting laboratory experiment. Marconi had the merit of building an antenna with which waves could actually be sent at great distances. Later he conceived the daring idea of attempting the radio transmission between Europe and America, notwithstanding the adverse opinion of scientists on the possibility of success. Their doubt was based on knowledge of the behavior of electromagnetic waves, a behavior that, even for radio frequencies, is analogous to what we can observe at light frequencies: straight line propagation, reflection, and refraction. Diffraction is also possible and becomes quantitatively more notable as the wavelength increases. For this reason, radio waves can go around even a fairly large mountain. But even taking this into consideration, there seemed to be no way to overcome the enormous obstacle represented by the earth's curvature.

In spite of these predictions, Marconi did not give up, and in 1901 he succeeded in transmitting radio waves across the Atlantic Ocean. The success of his experiment called for an explanation, and O. Heaviside and A. F. Kennelly proposed one, hypothesizing a reflecting

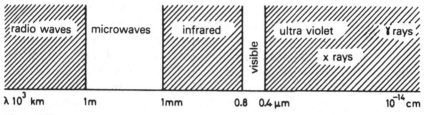

Figure 5.7

layer. The existence of this layer was demonstrated by E. V. Appleton, who identified the structure and characteristics of the ionosphere.

The ionosphere also presents a negative side: it not only reflects the low frequency waves coming from the earth's surface but also the electromagnetic radiation that arrives from the cosmos. This fact leads to an important question. What can we say about our place of observation in regard to the cosmos? Are we able, from earth, to obtain much or little information about the rest of the universe?

As long as we knew only of light rays we could have believed that this was an excellent place of observation, because it was sufficient to raise our eyes to the sky to see everything. Perhaps certain objects appeared extremely small and certain details were lost; but this was due to their distance, and telescopes had enormously improved the situation. But when electromagnetic waves coming from space were discovered, it was understood that an enormous quantity of signals arrives from the cosmos, of which we are capable of intercepting only a small part. Today we can think of being inside a closed box provided only with a few narrow cracks; and this certainly does not constitute an ideal place of observation!

Let us in fact consider the entire spectrum of electromagnetic waves which extends from very long radio waves ($\lambda \sim 10^3$km or more) to high energy γ rays ($\lambda \sim 10^{-14}$cm). We know that radio waves are in the most part reflected by the ionosphere; infrared radiation is absorbed by molecules of CO_2 and H_2O present in the atmosphere; ultraviolet radiation is absorbed by the ozone, and so on. The *windows* left free for us are essentially two: that of the *visible* range and that of *microwaves* (Fig. 5.7).

To get an idea of the width of these windows, let us take as indicative limits of the electromagnetic spectrum $\lambda = 10^{-14}$ cm on one side and $\lambda = 10^8$ cm on the other; thus there is a factor of 10^{22} from end to end. Recalling that to go to a higher octave in music means to double the frequency, we can say that the electromagnetic spectrum contains about 73 octaves (in fact, $10^{22} \simeq 2^{73}$). Of these 73 possible octaves, the visible band covers *only one*, whereas microwaves cover about ten. It is therefore evident that we normally live in the darkness and have only a narrow crack through which our eyes can observe the rest of the universe. Only by resorting to artificial instruments, can we use the second window of the box in which we are contained.

But today space travel has rendered us capable of punching a hole in the box and going out beyond the atmosphere, revealing the other radiations as well. This has represented a major turning point and has given rise to surprising developments in astronomy. Astronomy in the last decades has found itself in an extremely privileged position with respect to other sciences (some of which are nearly stationary) and has made such progress that even concepts that were once considered firmly established, are being discussed again. We have suddenly found ourselves in possession of an enormous amount of information which needs to be digested, so much so that astronomers were dismayed. Some of this information has already been inserted in a reasonable theory, but much of it is still waiting to be adequately used.

Two interesting considerations emerge. First, nature has given us a window too narrow to provide a complete vision of the universe from the earth's surface, but wide enough for humans to understand that there is a marvelous world beyond it, and therefore want to investigate it. The window was sufficiently large to permit the explosion of a Copernican revolution, the appearance of a Galileo, a Newton, and so on, up to space travel.

The second consideration is more like a question and concerns the fact that our eyes are sensitive only to luminous waves. Would it not have been more logical for nature to provide animals with sensitivity for both windows, including that of microwaves?

Let us see if this second solution would have been more functional. From thermodynamics we know that all bodies emit electromagnetic waves. The emission varies with the temperature of the body: An increase in temperature is accompanied by increased emission, and the maximum of emission moves gradually toward the shorter wavelengths. In other words, although a body emits noticeable radiation in the visible range when its temperature is higher than about 400°C, as it gradually cools down, it begins to radiate mostly in the infrared range, until at ordinary temperature, the maximum corresponds to the far infrared or to microwaves of short λ. This is why the sensitivity to microwaves would not represent a viable solution. In fact, taking into account the intensity of the microwaves emitted by the sun, we conclude that the microwaves emitted by bodies at ordinary temperature create a noise which covers the radiation scattered by the particular object we would like to see. In an environment in which this radiation propagates in all directions, it would be very difficult to distinguish one object from another.

As stated, a finalistic explanation of the universe and of life is untenable today; but it is *as though* things were conceived in a finalistic way. Nature has not wasted anything!

There is a curious fact concerning the sensitivity of the eye. The human eye has maximum sensitivity in the yellow-green range ($\lambda = 5.5 \times 10^{-5}$ cm); the maximum of the fish eye instead is shifted toward shorter wavelengths ($\lambda = 5.0 \times 10^{-5}$ cm). On the basis of a finalistic explanation we could say that the reason for the shift is that red light does not pass through a great thickness of water. As for the choice of the green–yellow band for the maximum sensitivity in humans, analysis of the solar radiation arriving through the visible light window has shown that the maximum intensity is actually at wavelength 5.5×10^{-5} cm. The human eye therefore is most sensitive to precisely that solar radiation that reaches it with maximum intensity.

Once again, we are forced to reflect on the absolute contingency of the present structure of humans. Humans are made this way because their environment is made this way. This has a remarkable significance also in answering the question, if and where life can exist in the universe, and if and where beings similar to us can exist. Before answering this question, we should ascertain the specific characteristics of the environment and investigate how humans or living organisms in general can be made to match them.

Returning to our closed box, where the window of the visible band had always been used, we must note that with the advent of microwave techniques, we have begun to exploit fully the second window as well. This marks the birth of *radio astronomy*, which gathers and processes the signals that reach us at those frequencies. In more recent times the advent of artificial satellites and rockets has allowed us to enlarge our field of observation. Instruments capable of detecting and resolving infrared, ultraviolet, X, γ radiation, have transmitted a great mass of new information to the earth.

In the present state of things, almost all the information that we have concerning the universe is carried by electromagnetic waves. But in addition to these waves, *cosmic rays* (high energy particles whose origin has not been totally clarified) reach us from space. These rays, encountering nuclei of the atmosphere, give rise to nuclear disintegrations with production of new particles. A chain of processes takes place, whose effects can be detected at sea level. But only outside of the atmosphere is it possible to examine the *primary* cosmic rays, the greater majority of which turns out to consist of protons and of very few light nuclei. Recently, useful data on the (extremely rare) heavy nuclei of cosmic rays, which provide precious information for cosmogony, have also been gathered.

Even *neutrinos* reach us from space. However, it is extremely difficult to detect them, for they interact very little with matter. A neutrino arriving from space has a very high probability (almost certainty) of crossing the atmosphere and the earth without having any interaction with matter, that is, without being detected. In all probability, even *gravitational waves* due to rapid shifts of enormous masses reach us. Today we are attempting to develop a neutrino astronomy. We are also trying to detect gravitational waves. But we are certainly very far from replacing electromagnetic waves as the most efficient tool of investigation.

5.8. The solar system

The sun is a star and, as far as we know today, a very common one. But earth, at least within the solar system, seems for several reasons a peculiar object. The sun occupies an intermediate position among all the stars we know of because of various characteristics. It is a gas globe, largely hydrogen, with a diameter about 100 times that of the earth. Its volume is thus 10^6 times that of the earth, but its mass is only 3×10^5 times greater, hence it is less dense. At the center, at about 700,000 km from the surface, the pressure is enormous and the temperature is valued at 15 million degrees. Away from the center, the temperature gradually decreases to the 6000 K which can be measured on the surface. It is in the center of the sun that enormous amount of energy is produced by the nuclear reaction of fusion,[22] similar to what we know how to use today in building H-bombs, but which we do not know how to control as a terrestrial energy source. This reaction can consist of many steps, but its end result is that hydrogen is burned and transformed into helium; at the same time, an enormous quantity of energy is released (§4.23). This was shown by H. Bethe back in 1938.

The energy produced at the sun's center must cross hundreds of thousands of kilometers of gas in order to escape. However, because of the high degree of ionization, the gases that constitute solar matter are rather opaque, and the more external layers steadily absorb the radiation and reemit it in all directions, both toward the interior and the exterior. Thus only a small percentage of the energy produced is able to propagate to outer space. The energy included in the innermost layer produces a certain radiation pressure which, in reality, is not very important in the sun, but can play a major role in more massive stars. The radiant energy that reaches us comes almost entirely from an outer layer, whose thickness is on the order of 100 km. We call this layer the *photoshere*. Around this brilliant surface are two other gas layers, which become more tenuous with distance. These are the *chromosphere* and the *corona*.

In the photosphere some peculiar details, called *sunspots*, can be distinguished. They constituted one of the most important discoveries of Galileo (1611). It seems odd that the Western world had to wait for Galileo to observe what can be seen with the naked eye.[23] These spots are darker zones on the surface, whose dimensions change in a regular cycle. They become numerous and readily visible to the unaided eye when there is a maximum of activity. They can be detected by looking at the sun through a smoked glass or simply at sunset. Because Aristotelian philosophy presupposed a sky and sun perfect and incorruptible in all their parts, there is a strong suspicion that scholars did not see what they did not want to see. Galileo himself had to fight for the acceptance of his observations. He noted that the spots are moving across the sun's surface, and from the direction of this motion deduced that the sun's axis of rotation was inclined with respect to the line of conjunction between the earth and sun.

All solar activity undergoes great variations every eleven years; the spots appear at a latitude of about 40°, move toward the equator and disappear, only to reappear at medium latitude at the beginning of the following cycle.[24] The sun's cycle seems to affect the earth's climate. It is certain that it affects vegetation, as the strengthening or weakening of the tree rings every eleven years show. Besides, the sun continuously emits a flow of particles (mainly protons and electrons), called the *solar wind*. During the frequent explosions that occur on the star's surface, the solar wind becomes more intense and causes irregularities of the earth's magnetic field (*magnetic storms*), auroras and other phenomena. The sun rotates once every 27 days, but the speed of rotation is different at different latitudes. Some workers suspect that it can vary greatly even with depth. Certainly, the dynamics of the sun's interior are very complicated, and we still know very little about it.

From time immemorial, humans have noted five bright bodies that seemed to move among the stars, following paths more complicated than those of the sun and moon. These bodies–Mercury, Venus, Mars, Jupiter, and Saturn–were called *planets*, that is, wandering stars.

The first known astronomers, largely Greek and Egyptian, proposed various theories to explain the structure of the universe and the movement of the planets, but the theory of the *geocentric* system is linked to the name of Ptolemy. According to the Ptolemaic conception the heavenly vault revolves about the earth and all planets move with uniform motion in circular orbits whose centers also move with uniform motion in almost concentric circles about the earth.

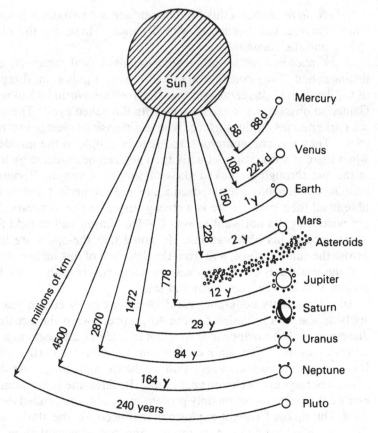

Figure 5.8

This system, which placed the earth at the center of the universe, remained almost unchallenged until the sixteenth century. In fact, a decisive turn occurred only in 1543, the year in which Nicholas Copernicus published his work, *De revolutionibus orbium coelestium*. The Copernican system, which placed the planets in orbits around the sun, was definitively adopted only after a severe struggle, with the help of decisive contributions by Galileo, Kepler, and, finally, Newton. The astronomical observations made by Galileo through his telescope, Kepler's enunciation of the three laws that govern the movement of the planets, and Newton's formulation of the law of universal gravitation were able to demolish the Ptolemaic and Aristotelian conception, laying the foundation of classical astronomy.[25]

Uranus, Neptune, and Pluto (discovered, respectively, in 1781, 1846, and 1930) were successively added to the six known planets. Figure

5.8 shows a schematic representation of the solar system, including the distances of the various planets from the sun and their periods of revolution.

Apart from Pluto, which is rather anomalous, the planets are normally classified in two distinct groups: the group of small planets of the *terrestrial* type, called *inner planets* – Mercury, Venus, Earth, and Mars – and the group of *giant* or *outer planets* – Jupiter, Saturn, Uranus, and Neptune. Let us now look briefly at the most interesting characteristics of each planet.

Mercury holds many "firsts" among the eight inner and outer planets: It is the closest to the sun, the smallest, and has the most eccentric orbit (i.e., the least circular). Its equatorial radius is about 2,400 km, that is, about 0.4 times that of the earth, and its mass is approximately 0.06 times that of earth. Its mean density, almost equal to the earth's, is about 5.4 g/cm^3, which makes one think of a central core rich in heavy metals. Mercury completes its revolution around the sun in 88 days and rotates on its axis once every 59 days. As G. Colombo noted, the period of rotation is two-thirds that of revolution, which is due to a resonance phenomenon. The distance from the sun varies from 46 million kilometers at the perihelion to 70 million kilometers at the aphelion, forming a rather marked ellipse. The *precession* of the perihelion is an interesting phenomenon. A displacement of the perihelion along the orbit can occur when the orbit of a planet is disturbed by the gravitational fields of nearby planets, but in the case of Mercury the effect is quantitatively greater than its predicted value. This discrepancy was explained by Einstein's gravitational theory, as a result of general relativity. Even the most recent theoretical studies and the most precise measurements have substantially confirmed the predictions of Einstein's theory. Concerning environmental conditions, the nearness of the sun gives rise to temperatures on the order of 350°C on the lighted hemisphere, which go down to −170°C on the nocturnal hemisphere. Observations seem to establish the absence of an atmosphere, or at least of an appreciable layer of atmosphere. This was predictable, because on the lighted side the gas molecules easily attain the escape velocity (about 4 km/s). These conditions make the presence of any form of life extremely improbable. Space probes have revealed that Mercury also has a crater structure like that of the moon.

Venus has a size almost equal to that of earth, because the diameters differ only by about 600 km. Venus's mass is about eight-tenths that of the earth. This makes the planet very similar to our own, but the environmental conditions seem notably different. First of all, the planet rotates much more slowly (243 days), in the direction opposite to that of the earth's rotation, and has a very long day. Venus is large enough to hold many gases, and in fact has an atmosphere. Indeed, the atmosphere that surrounds the whole planet is so dense that it completely hides the surface. As a result, it is not possible for us on earth to make out any details on the surface in visible light. But some structures (such as craters) have been revealed using radar techniques. Additional information has been furnished by space probes, which have landed on the planet's surface.

The atmosphere is almost exclusively made of CO_2. We have already mentioned that the primeval atmosphere of earth very probably contained large quantities of CO_2 and that the first living organisms actually absorbed CO_2 and stored carbon, releasing oxygen. We are therefore led to surmise that Venus, although similar to our planet, does not bear life. This conclusion may be too drastic, but it is certain that life on Venus, if it exists, cannot be as developed as that on earth. Several facts contribute to this certainty. The pressure of Venus's atmosphere at ground level is almost 100 times that on the earth's surface. The distance from the sun is roughly two-thirds of the earth–sun distance; since the sun's radiation is inversely proportional to the square of the distance, Venus receives more than twice the amount of solar radiation received by the earth. Moreover, the CO_2 atmosphere does not permit the planet's heat to disperse. For this reason, temperatures of hundreds of degrees Celsius are reached. Thus it appears very doubtful that Venus can bear life, at least similar to life as we know it. Nevertheless, Venus can still hold surprises for us. For example, one can surmise that the heating of the surface and the consequent freeing of carbon from rocks are cause and effect of one another and correspond to a comparatively recent phenomenon. In the past there could have been on Venus conditions closer to those on earth.

The main features of our own planet have already been discussed; but we should say something about the earth's satellite. The moon is no longer the mysterious object it once was, but it is still very interesting. The moon's diameter is 3,470 km, little more than one-fourth that of earth, and its mass is slightly more than one one-hundredth that of earth, which means that the mean density of the moon is small (precisely, 3.3 g/cm^3). The ratio of the size of the moon to the size of our planet is very large, if compared with the corresponding ratios for the giant planets and their satellites. For this reason, the earth–moon system constitutes in a way a double system; it is possible to think of two planets rotating around one another rather than of a planet and its satellite. The mean earth–moon distance is about 384,000 km, but it does not remain constant. In fact, the moon gives rise to tides in the terrestrial oceans whose effect is a torque that tends to slow down the earth's rotation. Correspondingly, there arises a torque applied to the moon, tending to accelerate its orbital motion. The result is a gradual increase in the length of the terrestrial day and an increase in the distance from earth to moon. We can thus assume that the two bodies were very close to one another 4 to 5 billion years ago, which seems to be the approximate age of the double system. The moon has no atmosphere, as was predictable from its small mass and low escape velocity (about 2.4 km/s), and as has been confirmed both by astronomical observations and by the measurements made *in loco* by astronauts. Thus there is no atmospheric erosion and weathering on the moon's surface. On the other hand, the characteristic crater structure is due to the impact of large fragments of solid matter, which were prevalent in the early history of the solar system (about 4 billion years ago) but which today have practically disappeared. Only small fragments (meteorites) remain, which continue to generate minor craters every now and then. Volcanic activity may also have been intense in the past (it is observed that lunar *seas* might be seas of solidified lava), but it stopped a long time ago. Today the moon appears quiescent and represents a static, or nearly static, world. For this reason, its surface has retained features very similar to those of more remote times and has preserved a precious historical documentation. It is not impossible that sometime we may find on the moon traces of the life that developed on earth in very ancient, pre-Cambrian times. For the microorganisms of the Archean phase must not have been greatly different from large molecules (even a virus for that matter, is a large molecule) and as such, could have been found even in suspension in the atmosphere. Some of these molecules might have reached the escape velocity (by way of fluctuation) and could have reached the moon. But so far, this is only a daring speculation.

Mars has always excited human minds for many reasons. For a long time it has also represented the most likely candidate for having a life somewhat similar to that on earth. They hypothesis is anything but excluded, even today. But it was certainly more plausible some years ago before space probes were able to approach the planet and provide new, more precise data. It has been confirmed that the Martian atmosphere is quite thin; its pressure on the planet's surface is valued at less than one one-hundredth of earth's atmospheric pressure, whereas its thickness is limited to a few dozen kilometers. The important component is CO_2, (about 96 percent) whereas the rest appears to be composed of nitrogen (2.5 percent) and argon (1.5 percent). The presence of oxygen is almost negligible (0.1 percent). The length of the day and the tipping of the rotational axis are practically equal to those of the earth. But the mean surface temperature is much lower (about $-50°C$). When we observe Mars with a telescope, we see the two white polar caps that extend and contract with the seasons. It seems that they are mostly composed of water ice, although some frozen CO_2 may be present. In the atmosphere there are often clouds of water ice, finely divided into tiny icicles. Spacecraft exploration has revealed that Mars's surface presents a surprisingly detailed structure, with volcanoes, dunes, canyons, riverbeds (now dry, but which in the past certainly transported some liquid, probably water). Violent winds that cause frequent and extended dust storms are characteristic features of the Martian world. Regarding life on Mars, space landers have recently cast much doubt on its existence, although the question is not definitively settled. In any case it is clear that life on Mars would be quantitatively much less developed than that on earth. To complete the description of Mars, let us state that it has two satellites, Deimos and Phobos, both very small, only a few kilometers in diameter.

Jupiter is the first of the series of giant planets that are all much less dense than the earth. It has a volume 1,316 times greater than that of the earth, whereas its mass is barely 318 times greater. Thus its density is less than one-fourth that of earth. This has led astronomers to surmise that the planet is an enormous ball of liquid hydrogen, surrounded by an atmosphere of hydrogen and helium. There may also be a small solid core, made of heavier elements. As far as the possibility of life is concerned, Jupiter is very far from our capacity of intuition, for its conditions are extremely different from those on earth. The great distance from the sun is responsible for the very low temperature (about $-150°C$) in the outer layers. We can know nothing of the inside, but the fact remains that the planet radiates more heat than it receives. Thus there must be a heat source somewhere inside. Jupiter's atmosphere seems to be very active. Radiostronomical observations also indicate the presence of electrical discharges, which could be remarkably important for the synthesis of organic substances, as discussed in regard to the origin of life on earth. By observing Jupiter through the telescope, we note a series of light and dark stripes parallel to the equator and the famous *red spot*, situated on the southern hemisphere, about 50,000 km long and 16,000 km wide, which was observed for the first time in 1665 by G. D. Cassini. The light and dark bands are interpreted as currents at different altitudes in the atmosphere, connected with the fact that the period of rotation is different according to the latitude. The red spot, since 1830 has changed form, color, and visibility, and has also moved longitudinally, but it has never disappeared and has always remained more or less in the same area. No one knows exactly what it is. Some authors now tend to think that the red spot is a particularly stable cyclone. This hypothesis is confirmed by the fact that in 1972 a small spot of the same type was discovered, which lasted about two years. Jupiter has fourteen known satellites, the first four of which were discovered in 1610 by Galileo, who called them the Medici satellites. Jupiter's satellites had great importance even afterward, when in 1674 they allowed O. Römer to make a first estimate of the speed of light. The space probes that have recently approached the Galilean satellites have sent us very interesting information about these bodies. It is most striking that their surfaces appear widely different from one another. Io's surface is particularly surprising because it shows a number of very active volcanoes. A tenuous ring circling Jupiter, analogous to Saturn's rings, has also been observed.

Saturn is almost as big as Jupiter, but even less dense (its volume is about sixth-tenths and its mass about three-tenths that of Jupiter). It too is surrounded by an impenetrable atmosphere in which we observe light and dark bands, but its appearance was thought for a long time to be unique among all the planets because of the presence of the famous rings. We generally distinguish three rings, of which the central one is brighter. They are extremely thin, because their width is comparable to the diameter of the planet, whereas their thickness is barely 16 km. This makes their total mass quite small. For a long time this appendix to Saturn was a mystery. Galileo himself, who was the first to notice it, did not succeed in observing its structure because the resolving power of his telescope was not sufficient, and thought of a system of three bodies attached to one another (§3.13). Later it was possible to detect that the rings are made up of an enormous number of disjoined small bodies (a few centimeters in diameter, probably composed of ice). Indeed, the rings are not completely opaque, but show the stars behind them. Thus there must be some opening through which light passes. Moreover, we observe that the rings do not rotate with the same angular velocity, but that the innermost ring rotates faster than the others. By contrast, in the case of a rigid body, the velocity of the outer parts should be greater than that of the inner ones. We conclude that each of the particles that make up the rings behaves more or less like a single satellite of Saturn. It is difficult to escape the temptation of giving a genetic interpretation to an appearance, which for a long time has represented a very strange anomaly among the known celestial bodies. A satellite can be considered to be made up of many parts, subject both to their own overall gravitation (i.e., they stay together because they attract one another) and to the gravitational pull of the planet. The latter is exerted differently on the various parts that comprise the satellite. If they were independent of one another, the nearest particles would revolve about the planet at a greater angular velocity. Normally the gravitational force of the satellite is greater than the difference between the forces exerted by the planet on the nearest and farthest parts. But if the distance apart of the two bodies becomes small enough – the *Roche distance* – the gravitational difference becomes greater than the cohesive force, and thus tends to break up the satellite. Some scholars have thought that Saturn's rings had precisely this origin; the countless particles of which they are comprised could have been the fragments of a satellite that had come too close. But many authors think instead that the rings may have originated together with the planet.

Another remarkable peculiarity of Saturn is represented by the fact that the largest of its ten satellites, Titan, has an atmosphere. Because it is much smaller than Mercury and has low density, its escape velocity is comparatively small, and at first sight it may seem strange that Titan can hold an atmosphere. But it is necessary to bear in mind the much greater distance from the sun, which leads to temperatures on the order of $-150°$ and $-200°C$, and thus to quite small thermal velocities for the gas molecules, less than the escape velocity. Titan's atmosphere seems to contain great quantities of methane. This last fact might confirm the hypothesis that originally the same type of atmosphere was present on all bodies of the solar system.

Uranus is the first of the planets discovered in comparatively recent times. It was observed by chance for the first time in 1781 by W. Herschel. It is large in respect to the earth (diameter about 4 times, volume about 67 times and mass about 15 times greater), but it is small in respect to Jupiter and Saturn. Its atmosphere contains hydrogen, helium, and methane. Uranus is characterized by the fact that its axis of rotation is nearly parallel (8°) to its orbital plane, whereas all the other planets present axes at least 60° away from their orbital planes. It is difficult to explain this peculiarity. Uranus has five satellites, the largest of which, Titania, has a diameter about one-half that of the moon. Very recently, it has been discovered that Uranus is surrounded by rings, in much the same way as Saturn. But the rings are so thin as to escape direct observation.

Neptune, discovered in 1846, is the first planet whose presence had been hypothesized on the basis of calculations (made to explain the perturbations of the motion of Uranus), before it was observed. The first calculations were completed by J. C. Adams, a young man still unknown in the scientific field. He concluded that the deviations in the orbit of Uranus could be explained by the presence of a rather large planet, which would have to be found in a certain area of the sky. No one paid much attention to young Adams. After several years the mathematician U. J. J. Leverrier in a completely independent way, reached the same results and asked J. G. Galle to perform an experimental check; half an hour after the search began, the telescope confirmed the presence of the new planet. Neptune's features are very similar to those of Uranus, in respect to which it is slightly larger. But it does not present the anomaly of the very small inclination of the axis of rotation on the orbital plane. We know of two satellites.

Pluto is the last planet of the solar system to have been discovered (1930). It too was discovered following mathematical predictions based on the perturbed orbits of Uranus and Neptune.[26] Pluto, in fact, is a small planet, less than the size of the moon, located at an enormous distance. We know very little about its physical characteristics, except that its temperature must be extremely low. Peculiar features are the large tilt of the orbital plane with respect to the ecliptic and the large eccentricity, even larger than that of Mercury, which already represents an exception among the almost circular orbits of the other planets. In 1978 a satellite was discovered.

A few other characters of the solar system cannot be neglected: *asteroids, comets, meteors,* and *meteorites.* But we shall be very brief, even if the importance of these bodies to cosmology has been rapidly increasing in the last few years.

The distances of the planets from the sun tend to show a certain regularity, as we can also see by observing a scale map of the solar system. This regularity can be translated into an empirical mathematical formula, called *Bode's law.* According to this law, the distance d of the various planets from the sun, measured in astronomical units (an astronomical unit (AU) equals the mean distance from the earth to the sun), can be calculated with the formula $d = 0.4 + 0.3 \times 2^n$, where the exponent n is a whole number equal to -1 for Mercury and increases by one unit for each more distant planet.

Table 5.1 shows the values obtained with Bode's law and the values of the distances that result from astronomical observations. The figures would agree fairly well, were it not for the fact that no planet corresponds to the value $n = 3$ and for the discrepancy existing in the case of Pluto. Bode's law dates back to the end of the 1700s, a century and a half before the discovery of Pluto, and in those times the only anomaly to explain was that of the vacant place. Incidentally, the discrepancy between the computed and measured values for the distance of Pluto could be explained today by the hypothesis, corroborated by other facts, that Pluto was originally a satellite of Neptune.

G. Piazzi, convinced of the soundness of the law, devoted himself to the search for the missing planet until, in 1801, he found, at the predicted distance, a celestial body which he called Ceres. We can speak of it as an asteroid rather than a planet, for its diameter is only 955 km. But with the passage of time, more and more new asteroids have been discovered. They form a belt, more or less at the distance

Table 5.1. *Bode's law*

Planet	Distance given by Bode's law	Observed distance
Mercury	$0.4 + (0.3 \times 2^{-1}) = 0.5$	0.4
Venus	$0.4 + (0.3 \times 2^0) = 0.7$	0.7
Earth	$0.4 + (0.3 \times 2^1) = 1.0$	1.0
Mars	$0.4 + (0.3 \times 2^2) = 1.6$	1.5
?	$0.4 + (0.3 \times 2^3) = 2.8$	—
Jupiter	$0.4 + (0.3 \times 2^4) = 5.2$	5.2
Saturn	$0.4 + (0.3 \times 2^5) = 10.0$	9.5
Uranus	$0.4 + (0.3 \times 2^6) = 19.6$	19.2
Neptune	$0.4 + (0.3 \times 2^7) = 38.8$	30.1
Pluto	$0.4 + (0.3 \times 2^8) = 77.2$	39.5

from the sun predicted by Table 5.1, and today we can count about 10,000 of them. The orbit has been determined only for fewer than 2,000 asteroids; the others have been discovered, but have not been followed with continuity because they are too small. Ceres remains the largest of a dozen asteroids with diameters greater than 150 km, whereas all the others have diameters that range from a few dozen kilometers down to several kilometers, or even to several hundred meters. The presence of such a large number of tiny bodies (system S) in the same region in which we would have expected to find a planet, leads us to think that they are fragments of a primordial planet (system S_0). But the inverse interpretation probably comes closer to the truth. Ceres may represent a body that had begun to approach planetary dimensions by aggregation of smaller bodies, just as the formation of the solar system came to an end. Some workers today think instead of the original existence of a small number of large asteroids placed roughly in the same orbit. The perturbations produced by Jupiter would then have led them to collide with one another and to split into the present asteroids. Some smaller fragments, thrown far out of the original orbit, would have constituted meteorites.

Among the components of the solar system, there are some that have a negligible dynamic importance but a spectacular appearance: *comets*. Generally, comets are made up of a tiny fragment of solid matter, the head, surrounded by a nearly circular envelope of gas, and the tail, by an enormous cloud of extremely rarefied gas. The tail develops fully only when the comet is close enough to the sun; the gas evaporates out of the head and produces that splendid effect that has always fascinated humans. The tail is always turned in a direction opposite to the sun, whether the comet is nearing the sun or moving away from it.

This happens because the pressure exerted by solar radiation and the solar wind is sufficient to repel the extremely rarefied matter of the tail.

Curiously, all comets have an extremely elongated orbit, so that for a long time scientists wondered whether their orbits were elliptical, parabolic, or hyperbolic; or in other words, whether comets were objects belonging to the solar system or crossed it coming from interstellar space. Today we are fairly sure that they rightly belong to the solar system. To be exact, we are dealing with countless (probably billions) nuclei of condensation of the original nebula, which circulate at an enormous distance from the sun (up to 100,000 astronomical units). Sometimes the perturbations caused by a nearby star sends one of these bodies toward the sun. The successive perturbations that comets undergo by passing very close to a planet, can sometimes change the elliptical orbit, sending them away forever. When such perturbations do not occur, comets follow their orbits regularly, and thus reappear periodically. The most famous comet is that of Halley, which has a period of 76 years, and whose next passage should occur in 1986.

We have already spoken about *meteorites*. We add that they are normally classified in two categories according to whether they are mainly made of metal or stone. The stony meteorites known as *carbonaceous chondrites* are of the utmost importance for cosmology, for they are believed to consist of materials that were formed 4.6 billion years ago when the nebula that gave rise to the solar system began to condense. If this is true, they were formed well before the most ancient rocks of the earth and the moon had crystallized, and preserve important clues about the origin of the solar system. As to metallic meteorites, they might be the remnants of the metallic cores of larger bodies that underwent catastrophic collisions.

When a meteorite crosses the earth's path, it enters the atmosphere, undergoes an enormous heating due to friction, and melts superficially. What remains of the meteorite can reach the ground with a very violent impact which forms a crater. On earth we know of a dozen meteorite craters. Among the most famous are that of Arizona, which presents a diameter of 1,300 meters and a depth of 170 meters, and the series of small craters (the largest has a diameter of about 50 meters) caused by the meteorite that fell in Siberia in 1908. We must observe that the meteorites falling on the earth are certainly not less in number than those falling on the moon or Mars, whose surfaces present an enormous number of craters. Calculations show that to bring about the present crater structure of the moon, taking into account the absolute lack of weathering agents on its surface, the fall of a large meteorite every 20–50,000 years is sufficient.[27] This time is extremely brief in cosmological terms, but is, however, very long in terms of human life. The waters on the earth's surface, atmospheric precipitation, and winds tend to erase the traces of such events quite rapidly. For this reason, we can observe only the craters produced in recent times on earth, which are necessarily in a very limited number. In this respect, entropy on the earth's surface increases after the interaction with an outer system, much more rapidly than that on the moon or on Mars.

Meteors are much more frequent than meteorites. They are very small fragments, almost dust grains that, with the heat of friction, evaporate completely, giving rise to a wake of ionized incandescent matter, which emits light. The falling stars are nothing but meteors. Meteors can come from any direction and any region of the sky: In this case they are isolated fragments, also called sporadic meteors. In some periods of the year there is an increase in the number of meteors coming from a well-determined region of the sky: we then speak of a *swarm*. Meteor swarms seem to follow well-determined orbits, and thus return periodically to the same point. The Perseides swarm, which returns every year, reaching a maximum around the tenth of August, is particularly interesting. Today we tend to interpret the swarms as remnants of comets, continuing to rotate in the orbit of the original body.

Micrometeorites are constituted by even smaller fragments of matter, so small and light that they can cross the atmosphere slowly and land intact on the terrestrial surface. The diameter of these microscopic fragments is on the order of one-thousandth of a millimeter. And yet they are of a considerable importance, much greater than that of bigger meteorites. In fact, several million tons of micrometeorites fall on earth every year. The thickness of the layer they produce on the ocean floor can give an idea of the age of the ocean.

5.9. The origin of the solar system

Based on many facts we assume that all components of the solar system have a common origin. We have already discussed some of them. All planets, with the exception of Uranus (which it is difficult to define precisely, because its axis lies almost in the orbital plane) and Venus (which rotates very slowly), as well as most satellites, rotate about their axes in the same direction. Also, they all revolve without exception in the same direction around the sun. The orbits described by the planets are ellipses of very little eccentricity (i.e., very nearly circular) and lie on planes that are inclined slightly with respect to one another. All, except Uranus, have axes of rotation that are nearly perpendicular to the plane of the orbit. The distances of the planets from the sun vary regularly, according to a mathematical law. In addition, the planets with larger masses have lower mean density.

These analogies, agreements, and regularities cannot be attributed to a pure fluctuation of the system. As a result, several cosmogonic hypotheses have been formulated with a view to explaining the origin of the solar system. Before discussing some of them, we shall mention two facts that any theory should consider – the distributions of mass and angular momentum. The mass of the system is almost all concentrated in the sun. The planets do not exceed altogether one one-thousandth of the solar mass. By contrast, the angular momentum is almost all carried by the planets. We recall that angular momentum is given by the product of the mass of a rotating body times velocity and times the distance from the axis of rotation. In an isolated system, such as the solar system, the sum of the angular momenta of all the components remains constant. Observation and calculation lead to the conclusion that Jupiter alone possesses 60 percent and that all the planets together reach 98.5 percent of the total angular momentum. The sun rotates slowly on its own axis, and thus in spite of its enormous mass, it possesses only 1.5 percent of the total angular momentum.

One of the first hypotheses on the origin of the solar system was originally formulated by Descartes in 1644, taken up again by Kant, and then by Laplace at the beginning of the nineteenth century. Laplace assumed the existence of an incandescent nebula formed by gases that moved in a vortex. A progressive cooling and gravitational action made this matter condense toward the center, thereby causing the velocity of rotation to increase. The progressive increase of the centrifugal force caused the successive detachment of rings rotating in the same direction. The central mass that remained behind gave rise to the sun, whereas each ring, after breaking up and then recondensing, became a planet. But unfortunately, this hypothesis is not easily reconciled with the condition concerning angular momentum. In fact, the central mass that has gradually contracted and of which the sun was comprised afterward, should have maintained a high velocity of rotation and a high angular momentum greater than that of the planets. The original Kant–Laplace theory has not surmounted these and other objections; however, by and large, the hypothesis of the primordial nebula, taken up again and developed in a thousand different ways, is still the best one.

Today, on the bases of considerable evidence, we believe that the solar system was formed 4.6 billion years ago out of a cold cloud of interstellar gas and dust. The cloud, having a mass of a few solar masses, began to contract gradually, under the pull of its own gravitational forces and to spin faster and faster until it reached the form of a disk. Under these conditions, the temperature gradually rose at the center, until all the solid grains existing in the cloud evaporated. More and more material moved toward the center of the disk, giving rise to the sun. What remained behind, partly condensed, giving rise to the planets, and partly dispersed into space. Probably the cloud first condensed into a myriad of very small fragments (*planetesimals*), and some small protoplanets (and satellites) gradually grew by capturing those fragments. The high angular momentum of the sun possibly was channeled into a very high solar wind at the initial stage. It was also thought that the sun may still have a core spinning much more quickly than the surface layers.

Earlier this century J. Jeans had formulated a completely different hypothesis, using an old idea of Buffon (1745). He surmised that a large star passing very near the sun had caused, with its gravitational pull, the ejection of a quantity of solar matter. This gigantic jet, subjected to the simultaneous attraction of the sun and the star, thinned at both ends and then broke into a series of fragments. As the star followed its path and disappeared into space, some of the fragments gave rise to the planets. This hypothesis may be qualitatively plausible, but not quantitatively. Accurate analyses carried out later with more highly evolved mathematical apparatus have shown that it is untenable. If two large spheres like two stars are made to interact in the described way, we do not reach the creation of fragments of notable angular momentum but, rather, the creation of a filamentary structure destined to disappear into space.

From our point of view, a very important feature distinguishes Kant–Laplace's hypothesis from that of Jeans's. This is the question of whether the solar system represents an extremely rare object in the universe or an object that we can encounter, perhaps in slightly different forms, frequently. In a Kant–Laplace hypothesis there is nothing that can make us think the solar system unusual. On the contrary, the phenomenon of the condensation from a cloud of dust and gas may have been at the origin of all the stars we see. It is therefore possible to think that planetary systems might exist in the vicinity of many stars. By contrast, because the stars are very far apart with respect to their diameters, the possibility that two of them might meet or pass very close to one another is extremely rare. We might compare the stars of our galaxy to a few dozen tennis balls distributed in a volume equal to that of the earth and moving at a speed of a few centimeters per hour. The probability that two will meet, even in a period of time of cosmological order, is almost zero. Therefore, Jeans's hypothesis would necessarily imply that the solar system is unique, or almost unique.

Today instead it is widely believed that our solar system could be one of many planetary systems existing in the universe. Some readers may question why we have dealt with this particular subject – the origin of the solar system – independently of the more general one – the origin of the universe. But as we shall show later, the solar system is considered today to be, somewhat, an object of the second or third generation. The universe as a whole originated much sooner. Therefore the formation of the solar system represents an event quite distinct from and, in a sense, independent of the birth of the universe.

5.10. The stars

With the exception of the sun, all the stars are very far away from the earth, so far away, in fact, that we are in the same position today in regard to the stars as we were two centuries ago in regard to the moon and the planets: There is no imaginable way of reaching the stars. Whereas everything concerning the solar system today can be considered within our reach, simply requiring adequate technical and economic commitments (as large as it may be), the situation is completely different for the exploration of the stars and their possible planetary systems.

The nearest star, Alpha Centauri, is 4.3 light years away.[28] In terms of astronomical units (AU), the most distant planets of the solar system are 40–50 AU from the sun, whereas the closest star is more than 200,000 AU away! It is easy to understand the difficulties of overcoming these distances, and discussing some of these problems from specifically a technological view can be interesting. In order to reach a star within a reasonable period of time, that is, at least during a human lifetime, we must attain a speed approaching that of light. Naturally, if a spaceship could depart immediately with speed very close to c, it would take only 4.3 years to reach Alpha Centauri. But this, even with a propulsion system that would supply the needed thrust, is impossible – not only for a human crew, which certainly cannot bear accelerations far greater than g, but also for a crewless spaceship. The instrumentation, unavoidably complex and delicate, would not be able to withstand such an acceleration without damage.

One can try to calculate how much time would be needed to reach the speed of light maintaining a low acceleration, equal to g. Obviously, the exact calculation, requiring the use of relativity, leads to the conclusion that the time required would be infinite, because c represents an unattainable speed limit. But we are interested only in knowing the order of magnitude of the time needed to reach a speed close to c; therefore we can still use the classical formulas. From $v = at$ we derive in our case $t = c/g$, that is, about 3.10^7 s, or a little less than a year. Accepting this approximation, we can conclude that it would be possible to reach Alpha Centauri in six to seven years, taking into account the year required for deceleration, as well.

So far, so good. We are still in the sphere of human dimensions.[29] But what about the problem of a propulsion system capable of supplying an acceleration g for a minimum of two years? No chemical propellant, neither existing nor conceivable, is able to provide such power. Just think of the enormous quantity of propellant that would be needed on board at the time of departure and of the consequent increase in the weight of the spacecraft in order to get an idea of the difficulties involved! Naturally, we can think of nuclear propellants, but calculations show that even these are unsatisfactory. As a last resort, let us then imagine that we are able to derive energy from the process of annihilation which occurs when matter and antimatter, stored in two separate containers, come into contact.[30] In this case it would in fact be possible to launch the spacecraft with the necessary tonnage of cargo and to maintain the desired acceleration, but the energy developed at the departure would be accompanied by such a shower of γ rays, that life on an entire continent would be destroyed!

These considerations, I believe, are sufficient to understand that here we are beyond both our technological and theoretical capabilities. Thus we can assume two different attitudes! We can hypothesize that discovery of a new physical law allowing us to overcome the obstacle will never happen, leading to the conclusion that the stars are, and always will be, *unreachable*. However, we can be much more cautious and cease making hypotheses about the future, leaving the door open to solutions that are even today unimaginable. The choice between these two attitudes is purely psychological. Personally, I prefer the second, but there is no concrete fact that can shift the balance of one toward the other.

In conclusion, it is by no means necessary to verify personally how things exist in the universe.[31] What is important is the ability to derive information, in one way or another, about distant objects; and in this sense, the results already obtained are more than satisfactory. The situation is not very different from that of microphysics. Who would think of personally going inside an atom or a nucleus?

The apparent *luminosity* of a star depends on two factors. One is the total radiation that it emits, say, in the visible spectrum. The other is the star's distance from the earth. Obviously, a star of great intrinsic

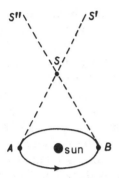

Figure 5.9

brightness but very far away can appear less brilliant than another, intrinsically less luminous but closer. In classifying the stars, we usually refer to the absolute brightness, or to the luminosity that results if each star were brought to one and the same standard distance from the earth. Clearly, if we can measure the apparent luminosity of a star, it is sufficient to know its distance in order to calculate its absolute luminosity, and vice versa.

Although the stars are extremely far away, some of them fortunately, are not so far away that we cannot measure their distance. This is possible by making use of the parallax effect. The situation is that shown in Figure 5.9. An observer on earth, located at A, sees the star S projected in the direction of a star S' against the background of the very distant stars. The earth travels around the sun, and after six months it is found at B, from which S is no longer seen in the direction of S', but in the direction of another very distant star, S''. If we can measure angle ASB and if we know the size of the earth's orbit, it is quite simple to derive the distance AS. This is made possible by the fact that we have a sky of very distant, or fixed, stars, that never seem to move relative to us. The nearest stars (from 4 to about 500 light years) appear instead to oscillate in the heavenly vault with a one-year period. Parallax measurements can reach even a bit farther, taking advantage of the sun's motion through the galaxy. But using this method we cannot go beyond about 1,500 light years.

The universe has many secrets and aspects that are extremely difficult to clarify. But it almost seems that for any enigma there is a key left somewhere that leads to the solution. If the nearest star were located beyond the 1,500 light years within which the parallax is measurable, or even if the earth did not follow its orbit around the sun, it would not be possible to deduce information concerning the distance of the stars. But having succeeded in measuring the distance to the nearer stars, we have been able (taking advantage of various phenomena of other kinds) to derive the distance to the farther stars as well. Facts of this kind suggest that the universe is *cognitively connected* in the sense that no part of it appears to exist, which is not knowable from a chain of observations carried out in another part; for example, from the part that is nearest to us. If this is true, there also follows that every part of the universe must somehow *influence* all the others.

The sun, like all the other stars, emits an extreme amount of energy. Until a few decades ago (up to the nuclear age) this constituted a great mystery. In fact, we were completely unable to understand what processes could produce such energy. Calculations show that if this energy were produced in a conventional way, by chemical processes, or even if it were attributed to gravitational forces (the sun contracts, and in so doing loses gravitational energy), the life span of the sun would be very brief, much briefer than the 5 billion years that constitute its real age. Nuclear physics has solved this problem. As already noted, the nuclear fusion of hydrogen into helium occurs inside the sun, and the temperature and pressure near the star's center are sufficient to ensure the continuation of the reaction.

Before our century it was thought that the microscopic composition of matter did not have great cosmological importance. But today it would be unthinkable to understand something of the birth and evolution of the stars (and therefore of the universe) without having some knowledge of the atomic elements and their nuclei. Even in this case, everything in the universe seems to be connected: The very small – the atomic nucleus – is important in knowing how the very large – the universe – is made. We know a great deal about the present structure of the stars. But, oddly enough, we also know how to trace their evolution in the course of time or their *life*, to use a term that is normally applied.

It is surprising that an observation that for us is necessarily *synchronic* (i.e., carried out at a well-determined time – that in which we live) should enable us to trace the *diachronic* aspect of the stars through periods of time much longer than our lives, than the period in which science has developed, or even than the period that encompasses the existence of humans on earth. Think of an observer who comes to earth from another planet with no knowledge of the human race. He will notice a number of beings, some of whom are small and do not speak, some who are larger but speak poorly, and others who are still larger and speak correctly, perform work, and so on. He would probably be able to infer that those beings are a population (humankind) in evolution. Similarly, we observe stars of different sizes, different brightnesses, different colors, and even different behaviors. By gradually

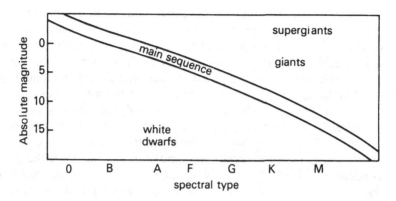

Figure 5.10

combining new observations, we have reached the realization that the stars are in some ways analogous to a human (or animal) population. There is evidence that they did not all originate at the same moment, that hour X which conventionally can be called the origin of the universe. There is also considerable evidence that the stars continue to be born and die, and therefore that older and younger stars exist among them. Naturally, their life spans are extremely long, ranging from a few million to a few billion years.

Observation has shown that the absolute luminosity of most (but not all) stars is linked to color, in the sense that the absolute brightness increases, going from red to blue. Usually the color of a star is characterized by the difference of its brightness in red and blue light. To obtain this difference, we compare two photographs taken with a blue and red filter. Conventionally, a star's color or *spectral type*, to use a more technical term, is indicated by letters of the alphabet, going from O for blue–white stars to M for the reddest ones (Fig. 5.10). There is a precise correspondence between the color and the surface temperature of a star. On the surface of white stars the temperature is around 40,000 K. It gradually decreases to about 3,000 K for the colder red stars. Plotting the absolute luminosity (or better yet, its logarithm, which is proportional to the absolute magnitude) against the spectral type, we obtain the diagram in Figure 5.10, known as a Hertzsprung–Russell diagram. Most stars belong to the *main sequence*. The sun is a yellow star, situated toward the center of the main sequence.

It has been observed that the bluest, most luminous stars are usually seen in a rather definite area of the sky, known as the central plane of the galaxy. Here we can find clouds of gas and of interstellar dust, often very extensive. The present-day hypothesis is that various condensation nuclei originate in these enormous nebulas, due to gravitational forces. As the gases are gradually attracted to and condensed in the nucleus, the temperature rises. When it reaches a certain level, it is capable of setting the process of nuclear fusion in motion. The energy released in this way starts to counterbalance the tendency toward collapse, that is, the concentration of the whole mass of gas in the center. When equilibrium is established, the mass of gas takes a spherical shape, and thus a very large, very luminous star is born!

The equilibrium condition occurring in the main sequence can be of short or long duration, depending on the original mass of the gas that condenses to form the star. If this mass is much greater than that of the sun, the energy developed in the gravitational collapse is so high that it causes the internal temperature to reach extremely high values, thus making the fusion process ever more rapid. As a result, the star burns its fuel rapidly, with a lifetime valued at about 50 million years. But if the original mass is on the order of that of the sun, or less, the fusion proceeds with greater regularity and less energy is radiated. The star has a yellow or reddish color and a life span of several billion years. All these statements can be derived by rather reliable calculation. On a few occasions, comparisons of recent photographs with photographs taken a few decades ago, in an area of the sky containing clouds of interstellar gas, have led to the discovery of a new, very luminous star. It is amazing that we are able to observe the birth of a star directly, considering the enormous length of their lives in respect to those of human beings. But we should take into account that there are 150–200 billion stars in our galaxy.

Let us now see what happens at the end of the life of a star when almost all its hydrogen has been burned. At that point the equilibrium between internal pressure and gravitation is broken. The outer parts fall onto the inner ones and the star undergoes a collapse. As a result, the internal temperature keeps rising and a new reaction may begin. This time the energy is so high that it may favor the fusion of helium into carbon, and a new equilibrium is established. Carbon in turn can burn, and there may occur a succession of collapses and equilibria, more and more heavy elements being formed at the core. One of the collapses can become explosive. As the core continues to contract, the more external parts of the star are thrown outward, expanding extremely rapidly. In this way the radiant surface of the star increases enormously and the star becomes very luminous. After undergoing a phenomenon of this kind, a star that is very weak because of its distance and which may be too far away to be seen with the naked eye, suddenly becomes extremely brilliant. Events of this type, in which the stars tend to appear for the first time in the sky, have been noticed since antiquity. The way in which the process develops depends mainly on the initial mass of the star. In an ordinary explosion of a *nova* the star brightens by a factor of about 10,000 and after some weeks or months disappears or turns again into a weakly luminous star. In the case of a *supernova* the star can become ten billion times more luminous than an ordinary star.

Supernovas are believed to appear about twice a century in the galaxy, but not all of them are visible from the earth. Tycho Brahe observed a supernova in 1572, J. Kepler observed another in 1604. The supernova which appeared brightly in the sky in 1054, and which was described by Chinese astronomers, is particularly interesting. We are absolutely certain that precisely this supernova gave rise to what is known today as the Crab nebula. In the last thousand years the expansion of the gases has continued, so that the star has given rise to a nebula. It is still quite luminous, but it continues to expand and the destiny of the exploded gases is naturally that of dispersing and cooling until they become absolutely nonluminous. The study of supernovas is developing very rapidly, thanks to the observation made of them in other galaxies. It is particularly important to observe their behavior in the early stages of the explosion. Many features of their life cycle are beginning to be fairly well understood, as well as their role in the universe. They are probably responsible for the production of cosmic rays and the nucleosynthesis of heavy elements.

If we begin by assuming that stellar matter was initially composed only of hydrogen and helium, as seems most likely, the first generation stars obtained their energy by the fusion of hydrogen and became enriched with helium. After burning almost all their hydrogen and following successive collapses, these stars produced quantities of other heavy elements, up to iron. At this point the star explodes, and fragments of matter that contain very heavy elements are thrown out into space. The second generation stars then, contain, besides the overwhelming predominance of hydrogen, very small quantities of heavy elements, with the possibility of producing elements even heavier than iron, thus following the same cycle. By proceeding in this way, we surmise that the stars of the successive generations contain ever greater percentages of heavy elements, which seems to be confirmed by spectroscopic analysis. As far as the sun is concerned, its chemical composition indicates that it is a second or third generation star.

If the star that has exhausted all its nuclear fuel is not sufficiently massive, it can end its life without explosion as a *white dwarf*, that is, as a body of stellar mass and planetary size. These surprising features, which were the source of notable perplexity toward the turn of the century, can now be theoretically understood. The gravitational force has supercompressed stellar matter to the point that atoms have partially lost their electron clouds and matter consists of an ion lattice embedded in a *degenerate* electron gas. Such a gas cannot be further compressed because the electrons – obeying the exclusion principle and the Fermi-Dirac statistics – occupy all the quantum states of lower energy and cannot go to still lower states. Density in these conditions can be millions of times that of water. A star of this type, which is still luminous only because it is still warm, might perhaps be called a warm "star corpse."

White dwarfs have weak luminosities and appear at the bottom of the Figure 5.10. The same diagram also shows the group of the *giant* stars – not too frequent in the galaxy – that are ten or a hundred times brighter than the sun and that have only a mass two to four times the solar mass. It is believed that giant stars represent a later stage in stellar evolution. By contrast, the *supergiant* stars that are 10^4 to 10^6 times brighter than the sun and have twenty to forty times the solar mass, are believed to be very young stars in rapid evolution.

5.11. Neutron stars, pulsars, blackholes

When a supernova explodes it can leave a very dense core at its center. This can be a white dwarf but when the mass is larger than the limit of 1.44 solar masses indicated by S. Chandrasekhar, not even the degenerate matter of a white dwarf can be stable. However, it may take still a higher density if all the electrons are *absorbed* by the nuclei. Precisely, each proton combines with an electron giving birth to a neutron plus a neutrino that flies away. In this way a *neutron star* is formed – a kind of object predicted by R. Oppenheimer in the thirties.

To understand what the density of a neutron star might be, let us recall that an atom has a diameter on the order of 10^{-8} cm, whereas a neutron has a diameter on the order of 10^{-13} cm, that is, 100,000 times smaller. With this ratio between the linear dimensions, the volume of a neutron star is 10^{15} times smaller than that of a normal star, having the same mass. Although one cubic centimeter of water weighs one gram, one cubic centimeter of neutrons would weigh a billion tons!

All this appeared quite disconcerting for some time, also because it was believed that neutron stars did not emit any kind of radiation and therefore that the possibility of revealing their presence was almost nonexistent. But the situation changed drastically when in 1967 a completely new kind of celestial body, one that emits radio waves, was discovered. These bodies were called *pulsars* (an abbreviation of *pulsating star*). The radio waves they emit consist of pulses repeated at precise and constant intervals on the order of a second or a fraction of a second. No other natural source was known to produce signals of this kind.

As a matter of fact, variable stars, whose luminosity increases and decreases periodically, exist even among less exotic stars. We know of various types: the most prominent being the Cepheids, whose periods range from one day to a few dozen days. We also believe that we know the mechanism that causes the variations in their brightness: They are probably very young stars in typical phases of nonequilibrium, which alternately increase in volume and then collapse, in an attempt to reach the equilibrium that characterizes stars in the main sequence. An experimental verification of this hypothesis has been made possible by the fact that in the successive phases of contraction and expansion, the radiating surface layers facing us move back and forth. This motion can be revealed thanks to the Doppler effect. Given a source that emits waves at a certain frequency, an observer can receive waves of lower or higher frequency respectively as the source moves away from, or toward, him. In the case of variable stars the emission spectrum in the visible region is periodically shifted toward red (lower frequencies) and toward violet (higher frequencies). A phenomenon of this kind, that is, the succession of collapses and expansions, requires periods of time on the order of one day, and certainly cannot happen in periods of fractions of a second. Besides, the regularity presented by the Cepheids is fairly good, but not comparable to that of pulsars.

The investigation of the origin and mechanism of pulsars led initially to various theories. Some hypotheses even attributed the emission of these signals to intelligent beings. We must recognize that when the first pulsar was discovered, the temptation of attributing the chronometric regularity of the emission to a nonnatural process was strong. But the regularity per se does not at all rule out a physical origin, and the subsequent discovery of other pulsars, located very far from one another, has contributed to the rapid withdrawl of hypotheses of messages from other inhabitants of the galaxy.

Even today the problem of pulsars cannot be thought of as completely clarified. Everyone agrees to consider them according to the original proposal of T. Gold – rapidly rotating neutron stars. Some doubts remain concerning the mechanism of energy emission. We said that a neutron star is a very small object, having a diameter on the order of 10 km. If we think of the angular momentum of a star larger than the sun, and if we recall the never disproven physical law of the conservation of angular momentum, it is easy to understand that when the star contracts, its angular velocity must increase. If it is reduced to the size of a neutron star, its rotation must be very rapid. The perfect regularity of rotation should not surprise us too much. Recall the fairly consistent regularity shown by the earth's rotation (although disturbed by various factors such as the internal movements of the magma and the tides caused by the moon), to understand how a neutron star (with a perfectly homogeneous interior and free from interactions with other bodies) can maintain its period with precision. We are less sure, though, of the mechanism by which pulsars emit radiation. It is probably tied to the presence of a very intense magnetic field.

The idea that pulsars can derive from cataclysmic events, such as those that lead to a neutron star, was confirmed by the discovery of a pulsar that has one of the briefest known periods, in the Crab nebula. It is now well known that the explosion that gives rise to a supernova does not involve the whole star – only the outer layers move away at great velocity – whereas the central ones collapse again, and in a sufficiently large star they can give rise to a neutron star. Thus a neutron star is a part of the ashes left over from a supernova explosion. The discovery of the Crab nebula pulsar has also finally suggested a plausible explanation for the high intensity of the radiation, both in the visible and microwave band emitted by the nebula. Scientists were unable to understand, almost ten thousand years after the explosion, how it could still radiate so much energy without a continuous energy supply. Today we think that the pulsar itself transmits energy to the nebula through some (still unknown) interaction. If this explanation is correct, a gradual decrease of the rotational energy, hence of the angular velocity of the pulsar, should occur. Recent measurements have in fact demonstrated this trend. The effect is very small, but present technology permits extremely precise measurements of the frequency of the impulses; it turns out that the period of this particular pulsar, which is on the order of 0.01 s, increases every year by 3×10^{-9} s. Similar results have been obtained for other pulsars.

Neutron stars constitute an extremely important stage in the history of astrophysics, representing the first sensational case of a new type of celestial body, first theoretically foreseen and then observed. It is also true that the existence of Neptune and Pluto had been predicted before their discovery. But they were planets, that is, commonplace objects which even the average person considered familiar. Here instead the prediction concerned objects of a completely new and exotic kind. As a result, the fact that the prediction has been confirmed by observation, provides a new and formidable argument in favor of the uniformity of the universe and physical laws that govern it. In the wake of this success, other theoretical predictions have received more and more credit. One of these is extremely exciting: *blackholes*.

The theory of neutron stars is still incomplete, but everyone agrees on several points. One is that neutron stars can exist only within a limited range of masses; roughly not less than one-tenth that of the sun nor greater than three solar masses. What happens to a collapsing object whose mass is greater than this critical value? To solve this problem one must resort to the theory of general relativity. Space–time is not a rigidly fixed and absolute entity in which masses move. Rather, its structure is directly linked to the presence of matter. It can be represented as a curved, four-dimensional continuum whose curvature depends on the distribution of matter. The gravitational motion of matter in turn depends on the space–time curvature. As we know, a result of general relativity is that a beam of light that passes near a body of great mass does not proceed in a straight line, but is deflected by the body's gravitational field. For an ordinary star, the deflection is so small that it is barely discernible. However, when we deal with a collapsed star, which reaches incredible densities and has a mass greater than that of a neutron star, calculations lead to the prediction that a beam of light is not merely deflected, but can even be trapped. If light is emitted parallel to the surface of the star, the rays are bent to the point that they behave like satellites and can only rotate around the enormous mass of the body. Furthermore, no material substance can emerge from it, for the escape velocity is greater than c.[32] We can therefore speak of an invisible world, incommunicable with the rest of the universe. This fully justifies the name blackholes, which has been attributed to such bodies. What happens inside a blackhole and near its surface is currently the object of much theoretical work and speculation. Not everything is perfectly understood in this fascinating branch of general relativity. It may seem impossible to reveal the existence of a blackhole. At first sight, blackholes appear to represent parts of the universe that are not cognitively connected with us and which must remain in eternal isolation. But this is not so.

To begin with, it was shown by S. W. Hawking in 1974 that there must be a quantum effect by which the tenet that a blackhole does not emit any radiation at all is incorrect. For we recall that *empty* space is swarming with pairs of virtual particles and antiparticles that are continuously created and annihilated. In the presence of a blackhole one member of the pair – say, the antiparticle – may be captured and fall into the blackhole, leaving the particle free to escape to infinity. Of course, this *materialization* of the virtual pair occurs at the expense of the energy of the blackhole. Alternatively, one can view the process in the following way. The antiparticle traveling *forward* in time is equivalent to a particle traveling *backward* in time (see §4.22). Thus the particle starts from inside the blackhole, *tunnels* (see §4.10) through the potential barrier existing at the surface of the body, and once outside is scattered by the gravitational field and rebounces forward in time. On performing precise calculations, we find that the blackhole emits radiation with Planck's distribution. For a mass of stellar order the temperature turns out on the order of 10^{-7} K, consequently, the amount of radiation is absolutely negligible. But Hawking has found that the temperature rapidly increases as the mass decreases and is not at all negligible for a *microscopic* blackhole. For example, a blackhole having the mass of a mountain would have the size of a nucleus and would send out considerable radiation until – having exhausted most of its energy – it would burst into γ rays. Hawking suggests that such tiny blackholes might have formed in the early stages of the universe and that it should even be possible some day to observe one of these bursts.

Further we note that a blackhole can be observed *indirectly* in a number of ways. As we have discussed several times, the requirement to see an object *directly* before acknowledging its existence is somewhat naïve. There is always a long chain of mediations between the object and our mind.

Various spotlike X ray sources with surprising properties have recently been discovered using space rockets and artificial satellites. Each of them may be ten thousand times more powerful than the sun and undergoes rapid fluctuations in intensity, which resemble those of pulsars, as well as slow fluctuations lasting several days. The only plausible explanation for this behavior is that the source is a binary system composed of a normal star and a companion collapsed star (a neutron star or blackhole) which rotates around it. The gaseous matter emitted by the normal star is attracted by the compact star and rushes toward it and forms an *accretion disk* with such enormous acceleration that it emits X rays. Naturally, these X rays can leave the system and radiate in space only if they are emitted when the matter is still farther away from the center of the compact star than the *Scharzschild radius*, at which the escape velocity equals c. The rapid fluctuations are due to the rotation of the compact star around its axis, and the slow ones are due to the revolution of the compact star around the normal one, the former sometimes hiding behind the latter. Perhaps the most notable fact is that in some cases, the mass of the compact star is so great that we can exclude, in the light of our present knowledge, that we are dealing with a neutron star. It must be a blackhole. If this is so, we have found a definite link with those almost isolated worlds and a means with which to investigate them.

In ending these notes on the more recently discovered stellar objects, we should also mention a most puzzling star called SS 433. It is a star of very feeble luminosity, probably a supernova remnant, emitting radiowaves, visible light and X rays. It shows very intense hydrogen and helium lines. Each one of these lines appears at its normal wavelength but is accompanied by two lines shifted toward the red and the violet, respectively. This shift can only be interpreted as a Doppler shift (see §5.13). Hence there must be matter moving toward us and matter fleeing from us. But the speed of these movements is incredible, up to 50,000 km/s, that is, one 1/6 the velocity of light. Moreover, both lines travel in opposite directions, first away from the normal line then toward it in an overall period of 164 days.

It is difficult to avoid the conclusion that SS 433 is a rotating object. Someone has proposed that it is a neutron star that emits two gas jets at fantastic velocities along the lines of its magnetic field. The magnetic field rotates (precesses) about an axis not coinciding with the jets, hence each jet describes a cone. The gas of each jet alternatively comes toward us and flees from us. Other models such as that of an accretion disk have been suggested, based on the fact that the star appears to be a member of a double system. However, the theory meets with several difficulties and the object still remains a mystery.

This discussion leads us to believe that the star zoo is far from complete. We surmise that as our experimental tools become increasingly powerful and refined, we shall be able to discover many new and surprising objects and shall encounter new and very difficult problems.

5.12. The galaxies

All the stars visible to the naked eye, as well as most of those observable with the most powerful telescopes, are part of a well-determined system: the galaxy. The term, galaxy, or Milky Way, dates back to the ancient Greeks. Its primitive meaning indicated the white band that crosses the celestial vault like an arch, from one extreme of the horizon to the other. Galileo, observing the Milky Way through his telescope, was the first to realize that the stars which comprise the galaxy were so densely packed and so far away from earth, that they could not be perceived individually. Since Galileo's time, further studies have ascertained that the Milky Way merely gives the appearance of the whole star system (which also includes the solar system and which we continue to call our galaxy) when seen from our point of view.

The structure of our galaxy is quite well known today. This knowledge has been made possible not only by observations of its interior but also by the discovery of a great number of other galaxies outside our own. By combining the information obtained directly with that derived from examination of external galaxies, we conclude that our galaxy is shaped somewhat like a flattened disk with a central bulge and that the sun occupies a position away from the center. The main concentration of stars lies in the central plane of the disk (the equatorial plane). The diameter of the disk is about 80 to 100 thousand light years, in contrast to its thickness, which is barely 1,000 light years.

From our observations of external galaxies, it was possible to deduce that all stars and interstellar matter rotate around the center of the galaxy with a period of about 250 million years at an angular velocity that decreases as the distance from the center increases, as occurs with the planets that rotate around the sun. Only the galaxy's central nucleus seems to rotate like a rigid body. If it were possible to view the galaxy from the outside, we would see that it has various spiral arms. The sun, located in one of the arms of the spiral, is about 30,000 light years away from the center.

The spiral structure of the galaxy was partly deduced by observations of external galaxies, but it was mainly verified directly by radio astronomy. Observations in the visible range had already permitted us to obtain certain results by turning our telescopes toward objects that certainly lie in the arms of the spiral, like the hot supergiant stars, and proceeding to accurate measurements of their distance. Radio astronomy has permitted a more precise investigation, thanks particularly to the presence of interstellar hydrogen, characterized by an emission line with a wavelength of 21 cm. Other observations had led us to the nearly certain conclusion that hydrogen existed only in the arms and rarely outside of them. Furthermore, by examining the shifts of the hydrogen line due to the Doppler effect, one was able to calculate the velocity of rotation of the other arms in respect to us. In this way we have reached a satisfactory description of the galaxy's structure.

The youngest stars and almost all the clouds of hydrogen, dark cosmic matter, and interstellar dust lie on the equatorial plane, in the arms of the spiral. This fact is confirmed by observations of those external galaxies that are seen from a side (rather than full face). We can clearly see some dark areas composed of clouds of opaque interstellar matter, distributed only along the equatorial plane.

The oldest stars are generally found outside of the equatorial plane, in masses that take the name *globular clusters*. Globular clusters generally have a diameter of a few hundred light years and can contain tens of thousands of stars. It seems certain that they contain very little if any interstellar matter. Globular clusters are distributed in a nearly spherical volume centered around the galactic nucleus and having a diameter roughly equal to the diameter of the disk.

A cosmogonic interpretation of these facts is fairly easily imagined. The present form of the galaxy is quite probably the result of the rotation of a gigantic cloud of prestellar matter (protogalaxy) essentially made up of hydrogen and helium. This cloud gradually began to condense, forming the nuclei from which the first stars originated. Continuing to rotate, the cloud became increasingly flatter, due to centrifugal force, leaving behind the globular clusters of old stars. Interstellar matter, the basis for the formation of new stars, was distributed on the galactic plane. It is precisely in this area that we see the brightest, youngest stars and where new stars can be born.

As early as the eighteenth century, scientists and philosophers (among them Kant) had expressed the opinion that our galaxy was not the only example of its kind existing in the universe, on the basis of purely speculative reasons. In the twentieth century, scientists had a fairly clear idea of the existence of our galaxy, but they continued to argue about whether nebulas were galactic or extragalactic objects. Thus the rather vague term *nebula* was (and had been for some time) attributed to all the more or less luminous objects that appeared through the telescope as fixed, extended, and diffused. For all practical purposes, nebulas were zones of the sky distinguished from their backgrounds by a fairly continuous distribution of luminosity. This luminosity could also be negative, in the sense that the nebula could appear as a dark cloud that made these zones stand out against a background of luminous stars.

Among the luminous nebulas, some types characterized by an elliptical or spiral form attracted a great deal of interest. But no one could say whether or not these objects were inside the galaxy. Only in 1923 did E. Hubble, using a telescope more powerful than those of his predecessors, distinguish some stars, globular clusters and clouds of matter in the Andromeda nebula, the nebula closest to the earth. He thus concluded that it was an outer galaxy.

Let us now open a parenthesis in order to complete our discussion of the measurement of distance. We stated that the method of triangulation based on the parallax or the sun's movement, is only applicable to stars within a certain limit, beyond which geometric methods no longer work because the parallax becomes smaller than the resolving power of the instruments. Spectroscopic analysis has permitted us to overcome this obstacle. Spectral lines indicate the presence of certain elements and their intensity can provide other information, for example, the temperature of the body. By surveying the spectra of closer stars whose distance can be directly measured, astronomers discovered a surprising correlation between the type of spectrum and luminosity. This permitted them to deduce the absolute brightness of a star from an examination of its spectral lines. Then it is sufficient to measure the apparent brightness to deduce the distance. This estimate was not possible for external galaxies before the individual stars to which the method is applicable could be discerned. But once individual stars were resolved in the Andromeda nebula, it became an easy matter to calculate the distance of this system. This distance was on the order of a million light years and this is why scientists deduced that it was a galaxy outside of our own.

Today we know that an enormous number of other galaxies (certainly more than a million) exist. We also know that galaxies tend to group together. There are systems formed by galaxies close to one another, which are probably gravitationally interdependent. Even our galaxy is part of a small group, the *local group*, which contains about 20 galaxies whose mean distance is about two million light years. The galaxies farthest away from the earth are located billions of light years away.

Galaxies can be grouped into three principal types, represented in Figure 5.11. There are *elliptical* galaxies, indicated by the letter E, which can vary from a nearly spherical form to a more or less flat ellipse; normal *spiral galaxies* (Sa, Sb, Sc), which have a nearly circular symmetry of the nucleus and arms; and finally, *barred spirals* (SBa, SBb, SBc), characterized by a luminous bar across the nucleus. As already pointed out, observations of spiral galaxies show that clouds of gas and dust are generally found in the arms. Therefore we believe that a continuous *siderogenesis*, or formation of stars, takes place there.

Today we have reason to believe that the arms of the spiral represent something analogous to waves, which do not move together with the matter of which they are composed. Each star, in the course of its rotation around the center of the galaxy, enters an arm, crosses it in a time period of about a hundred million years, then leaves it and moves toward the next arm. Gas clouds do the same. In entering an arm, gas clouds undergo a compression that can give rise to siderogenesis. The solar system may have been born in this way. Elliptical galaxies, however, have no gas clouds, They contain many red stars and therefore give the impression of being the oldest galaxies.

We have reached the limits of the observable universe, let us say roughly 10 billion light years away. Our technical means – radio telescopes – are not powerful enough to go further. Is this the real size of the universe? For now, only cosmological theories can answer this question. And the answers they provide are hypothetical.

Perhaps it is useful to note that there is an area of our galaxy that has recently attracted a great deal of attention. This is the central zone or *nucleus* of the galaxy. Until a few years ago, observation of the nucleus was inhibited because of the high absorption of light by the dust clouds distributed on the galactic plane. But these clouds exert

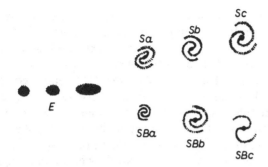

Figure 5.11

a lesser disturbance on infrared, radio waves, X rays and γ rays. And it is precisely through these regions of the spectrum that we have begun today to gather important information on the nucleus. We are dealing with an extremely complex system and disconcerting properties. The stellar density reaches a level several tens of millions of times greater than that around the sun. Gaseous clouds of 10^5 and 10^6 solar masses are ejected at speeds of several hundreds of kilometers per second. Presumably, the nucleus emits also gravitational waves.

Today many workers believe that some galactic nuclei may contain giant blackholes, up to 10^5 solar masses. A blackhole of this kind may have originated from a collapsed star (some ten solar masses) that started to increase by gradually swallowing the neighboring stars. These stars are attracted by the gigantic gravitational pull of the object and come close to the blackhole until they reach the Roche distance; at this point they are split into many fragments and form an accretion disk around the blackhole. Here matter in rapid rotation and acceleration emits very intense radiation, until it loses most of its energy and falls into the blackhole.

A theory of this kind can help to clarify the nature of those fantastic objects called *quasars* (*quasi-stars*) that have been discovered in the early sixties and today number about two thousand. Quasars represent the remotest luminous sources known – at the boundary of the observable universe, as witnessed by their Doppler red shift. The amount of radiation emitted by a quasar – ranging over all the spectrum from radiowaves to γ rays – may be 100 or 1000 times that of a normal galaxy, although the size does not exceed a few tens of light years!

Today many astrophysicists believe that a quasar is a galaxy whose nucleus includes a supergiant blackhole – on the order of a few million solar masses. The galaxy about the nucleus is invisible because it is too far away and too faint in respect to the nucleus. This theory is corroborated by the existence of the *Seyfert galaxies* – a connecting link between normal galaxies and quasars – which appear as very luminous objects surrounded by comparatively faint envelopes.

Why are quasars observed uniquely at those extreme distances? To answer this question we must recall that when an object is 10^{10} light years away from us, we observe it as it was 10^{10} years ago, much younger and much closer to the origin of the universe. At that epoch the size of the blackhole was presumably less than the Roche radius and, consequently, the approaching stars could be fragmented and captured in the accretion disk. In later stages when the blackhole grows beyond the Roche limit, whole stars are swallowed before fragmentation can occur, hence the quasar ceases to send out light. Thus the quasars closer to us might be *extinct*, because they are too old.

Finally, another class of puzzling and fascinating objects must be mentioned, the *radio galaxies*, which emit enormous amounts of radiowaves. Some of them represent the largest bodies seen in the sky, up to several million years in size. Analysis of the radiation sent by giant radio galaxies reveals that it must be generated by high energy electrons moving in a magnetic field (*synchrotron radiation*). But no one knows from where the energy that feeds this process may come. Perhaps some mechanism analogous to that explained for the quasars is at work.

5.13. Cosmological hypotheses

In developing cosmological or cosmogonic hypotheses and theories, we make use today of a guiding principle suggested by many observations but certainly not strictly proven. This is the *cosmological principle*. It states that the universe, considered in its large structures, must appear the same when observed from any one of its parts. This is the maximum extension that the nonanthropocentric attitude, maintained since Copernicus and Galileo, can assume.

Note that this is not exactly the same as the interpretation of the uniformity of nature which we have expressed so far as the space–time invariance of laws. The latter invariance is compatible with a nonuniform structure of the universe, contrary to what the cosmological principle postulates.

Observation of the external galaxies has provided an extremely important bit of information that dominates all the present cosmological and cosmogonic theories. This is the *recession of the galaxies*. To clarify this concept, we should add a few words concerning the Doppler effect, which we have previously mentioned.

Let us consider a phenomenon of wave propagation – sound. We may think of a certain periodical configuration that shifts and which, in this case, consists of a series of compressions and rarefactions of the medium in which the sound is propagated. The pitch of a note perceived by a listener corresponds to the frequency of the sound signal, that is, to the number of waves that reach him in a unit of time. If the source emits sound at a constant pitch, an observer in motion notes that the pitch increases as he approaches the source and decreases as he moves away from it. This is readily explained, for the observer who is moving closer to the source receives more sound waves per second than if he were standing still. Conversely, the observer who is moving away from the source receives fewer sound waves, and thus hears a lower sound. A similar phenomenon takes place if the observer stands still and the sound source moves. The example of the train whose whistle becomes more acute as the train approaches and deeper as the train moves away, is part of everyday experience. This effect, linked to the name of C. Doppler, is easy to note and measure in the case of sound waves, whose speed of propagation (about 340 m/s) is small, but is ordinarily negligible for light. However, when the relative velocity of the source and the observer is very high, the effect can be detected even for light waves, using the spectroscope. The results of the classical theory are not altered substantially by relativity, and the phenomenon remains virtually the same. Precisely, we see the light of a source that is moving toward us, shifted toward the violet end of the spectrum, and that of a source moving away from us, toward the red end.

Examination of the spectral lines of isolated stars in external galaxies shows that they are shifted toward the red. In 1930 Hubble formulated the law that bears his name, according to which the red shift increases with the distance d of the galaxy observed: precisely, the recession velocity V is proportional to d. Thus Hubble's law is written as

$$V = Hd \tag{5.9}$$

The values of *Hubble's constant* H corresponds to about 20 km/s for every million light years. Therefore a galaxy at a distance d of one billion light years should recede at a speed of 20,000 km/s. Note, however, that in this case the light emitted by this galaxy takes a billion years to reach us and so we observe the galaxy in the state it was in a billion years ago. This fact leads to a correction that we must always take into account in cosmological speculations.

Before the discovery of the recession of the galaxies, when we thought that the universe could be infinite and, on the average, homogeneous and static, a paradox indicated by P. L. Chéseaux (1744) and H. W. M. Olbers (1826) and which generally goes under the name of the latter, carried great importance. It was argued that the sky should appear uniformly luminous, with a brightness equal to the mean brightness of the stars, because, every ray extended from our eye would have to encounter a star sooner or later. There are many ways to show the fallacy of this conclusion. However, the progressive shift toward the lower frequencies is enough to obviate the paradox.

Hubble's law, equation (5.9), can easily lead us to assume that scientists have returned to a kind of anthropocentric view, because d actually represents the distance from us. But this is not so. A little reflection leads to the conviction that the situation is seen in a similar way by any observer whatsoever; for him too, the others recede at a velocity proportional to their distance. The phenomenon is perfectly similar to an *explosion* in which each fragment sees the others recede at a velocity proportional to their distance.

When did the explosion occur? The length of time that has passed since then can easily be found with equation (5.10). At the time of the explosion the reciprocal distances were practically nil. A body that has crossed a distance d at a velocity V has taken the time

$$t = \frac{d}{V} = \frac{1}{H}$$ (5.10)

Thus we find a time equal to about 15 billion years. But recent studies have demonstrated that this *expansion of the universe* undergoes a certain deceleration. Therefore the time that has passed since the *big bang* may be somewhat less than that calculated on the basis of equation (5.10).

The big bang hypothesis, initially formulated by G. Gamov and collaborators, has had its ups and downs but, essentially, it has steadily gained ground, particularly in the last few years.

A rival theory, which for some time enjoyed much success, is the *steady-state theory* proposed by H. Bondi, T. Gold, and F. Hoyle. It was formed on the basis of the *perfect cosmological principle*, according to which the universe (which is infinite) is (on the average) equal in appearance and structure in all its parts, and is immutable in time.[33] Because the universe is expanding, we must resort to the hypothesis of a *continuous creation of matter*. This may seem bewildering, but when the calculations are done, we see that the rate of production of new matter is so slow that it could not possibly be revealed in the laboratory.

In spite of its aesthetic merit, the steady-state theory has been practically abandoned, mainly because of the discovery of universal *background radiation*, discovered in 1964. This is a type of electromagnetic radiation, observed in the microwave and infrared bands, which comes to us from the universe in an isotropic way, that is, equally from all directions. From what we know it represents (on the basis of Planck's formula) radiation in thermal equilibrium at a temperature of about 3 K. Everything points to the idea that this radiation is what remains of the initial *photon gas* that existed at extremely high temperatures in an epoch soon after the big bang and which cooled as a result of the expansion.

We think, then, that the system S_0, the distant origin of the universe S which we observe, was composed of a ball of fire in which the temperature reached hundreds of billions of degrees. Calculations based on all we know of modern physics, lead to the conclusion that the initial evolution of the system must have been extremely rapid. The temperature at the various stages determines the ratio between the density of the hadron, lepton, and photon components of the universal fluid.

As the universe expands and temperature decreases, successive epochs are reached where the average energy available per degree of freedom is no longer enough to provoke certain reactions or to create certain particles. Consequently, the abundance of those particles remains *frozen* at the thermodynamic equilibrium reigning at that temperature and their further evolution is virtually separated from the rest of the universe. The number of protons and neutrons becomes frozen at a very early stage. After a few seconds the temperature has decreased to under 6 billion degrees. The photons have an energy less than 0.51 MeV (the mass of the electron) and are no longer able to create pairs. The positrons disappear and only the few electrons that counterbalance the electric charge of the protons remain. The dominant component consists of photons; it is the fire ball.

After some three minutes the nucleons present start a first nucleosynthesis, forming deuterium, helium, and a few light nuclei. About 100,000 years later the atoms are already formed and the photons cease to have enough energy to excite or ionize them. Electromagnetic radiation decouples from matter and starts its independent cooling until the present 3K are attained. A universe of the type we know begins, dominated by matter – that is, by baryons. The formation of the galaxies begins, as does that of the stars, and in the latter the nucleosynthesis of gradually heavier elements, as we have explained.

The reader will have easily guessed that this simplified description summarizes a complex theory, an enormous job of synthesis and comparison of thousands of different observations. Particle physics, nuclear physics, atomic physics, classical fluid mechanics, thermodynamics, electromagnetism, and above all, general relativity, contribute to this synthesis.

Einstein's gravitational equations actually accept as a solution the expanding universe and the *singularity* represented by a zero radius at the time of the big bang. Nevertheless, many facts remain to be clarified. For instance, the present measurements of the mean density of the universe do not permit us to decide with certainty if the universe is totally Euclidean or positively or negatively curved, if it will reach a maximum radius and then recontract or expand forever.[34]

We have already explained in what conventional way we must use the word *origin* when discussing cosmological systems. What can still strike us is the impossibility of directly verifying the relevant hypotheses and theories as is usually done in physics. But in the case when the system with which we are dealing is the universe in its totality, we encounter a very singular circumstance: The idea of verifying the past may not be so crazy. To this end, let us recall what we have observed concerning the very distant galaxies. We do not see them as they are today, but as they were when they emitted the light that reaches us. Because of this, objects such as the quasars and the radio galaxies, at the limit of what we are capable of observing, carry us several billion years back in time, that is, to a time not very far from the cosmogenesis. And it is quite meaningful that many elements deduced from these observations seem to agree fairly well with the type of evolution described by the big bang theory.

5.14. Life in the universe

On earth, the phenomenon of life is very conspicuous and important. We have seen that it is not possible to construct a reasonable geology or geogony without taking the biosphere into consideration. On the other hand, life so far has not been observed elsewhere outside the earth. This naturally leads us to ask: Is it a unique phenomenon?

The cosmological principle would suggest a negative response. If nature has the same general appearance anywhere in the universe, how can we think that such a conspicuous and decisive phenomenon occurred only in a particular, infinitesimal area of it? A skeptical attitude concerning the negative response is, in a sense, quite reasonable; nevertheless, we should not forget that the cosmological principle, although it is a useful heuristic and interpretative guide, is not a dogma. It would be antiscientific to assert something with certainty on the basis of that principle alone. We must instead analyze, without prejudices, what reasons of fact favor the existence of biological phenomena even outside of the earth.

It is perhaps wise to begin by making a general consideration. We are acquainted with a certain type of life: what developed on earth. Given the semicasual way in which we think evolution has taken place, we would be naïve to think that the same animal species, and perhaps people equal to ourselves, who speak the same languages, must exist elsewhere in the universe! Everyone is ready to assume that the species found on another planet should be very different.

Instead it is more difficult to come to grips with another question. We are acquainted with a life based on carbon compounds, on those carbon compounds such as amino acids and nucleic acids. Are other forms of reproductive organization of matter possible? Would life based on silicon be possible? But this is not all. In the almost degenerate system, near absolute zero, in which we find ourselves, only life based on extremely weak bonds such as molecular (chemical) ones, is conceivable. But inside the stars, would not life on the nuclear level, with much stronger bonds and much shorter lifetimes be possible? Clearly, we have reached the boundaries of science fiction and really cannot say anything. I have mentioned this possibility only to emphasize the need to be free from any prejudice and open to any possibility.

Our working hypothesis must therefore be quite restrictive. Let us suppose that elsewhere life has developed in a way similar to our own, on the same chemical basis. To support or deny this hypothesis, we should first verify if an environment analogous to our own exists in some part of the universe. If we were to give credit to a cosmogonic hypothesis, say, like that of Jeans, by which the solar system was formed by the passage of two stars at grazing distance from one another, we would immediately be able to say that in all probability there is only one solar system. But Jeans's hypothesis has been abandoned today in favor of other hypotheses that correspond to much more frequent phenomena. Direct observation allows us neither to affirm nor to deny the presence of planets such as the earth around the other stars. We must think that the star nearest to us is so far away that even a planet of Jupiter's size would be impossible to see. It is true that a favorable indication is given by observations of the movement of some stars whose perturbations must be due to large dark bodies that gravitate around them. But the quantitative explanation of the observed anomalies requires that these planets be much larger than Jupiter, and we are not sure that life is possible on bodies of that size.

Nevertheless, even the existence of a planetary system analogous to ours is not sufficient in itself. It is also required that the star at the center should have certain properties. We must rule out all the stars at the beginning of the main sequence, which only live a few million years, that is, too short a time to allow the development of life. On the other hand, the stars near the end of the main sequence, which live several billion years, are too cold and do not have the possibility of supporting life phenomena similar to photosynthesis. Therefore only the stars such as our sun, in the middle of the main sequence, must be taken into consideration. But this is certainly not a great limitation. As a matter of fact, it is estimated that there are roughly 200 billion stars in the galaxy, several tens of billions of which respond to our needs. From this number we must take away another substantial slice, that of double or multiple stars, which constitute the majority.[35] This is because, in such systems, the orbits of possible planets are more complicated than in systems with one star only, and the successive movements toward and away from each of the stars of the system would lead to jumps in temperature too large for life to bear.

After this reduction, we estimate that a billion stars would still be available. At this point we must say that it is very probable that planetary systems exist, with planets of the terrestrial kind, unless a precise cause, today totally unknown to us, has prevented their formation. Indeed, today it is believed that isolated stars, without companions or planets, are exceptional.

As a conclusion, the present view of many scientists is more or less the following one. It is very probable that life has developed somewhere else in the universe, including higher forms of organisms and perhaps even thinking beings. If all this happened only on earth, we must be able to understand why; this *why* today escapes us. It is a cautious attitude. But it is the only attitude that the information in our possession permits. Attempts to obtain new data on the problem have been made, but so far without results. We have not yet found anything in our solar system suggesting that there is or there was even some lower form of life, let alone the other systems, so far removed from us that we know nothing of their evolution.

We often ask: If thinking beings more evolved than we had been born somewhere else, would they not have made their presence known by now, would they not have come to earth? This is a curious problem. It is difficult to give an affirmative response to the second part of the question, for we have no idea whatsoever of how they could have made an interstellar trip over distances of hundreds of thousands of light years in a period of time comparable to a human life span.[36]

But we know that it is possible to send messages and it is certain that at the level of technology we have reached, we are capable of receiving messages if someone on the relatively nearer stars were to transmit. Without doubt it would be possible to distinguish an intelligent transmission from a purely natural phenomenon, and even to decipher the code (see, e.g., Bracewell, 1974).

Electromagnetic radiation with the 21 cm wavelength appears very suitable for transmission, as was suggested in 1959 by G. Cocconi and P. Morrison. In fact, since it is the wavelength widely emitted by hydrogen throughout the universe, it would be a point of reference known to everyone. Furthermore, it would be very suitable where there is an atmosphere similar to our own, which at that frequency is transparent and has a minimum of background noise.

Listening has been tried as well, for example, by F. Drake, a pioneer in this field, and by several others, both in the United States and in the Soviet Union. So far nothing has been obtained. But this proves very little, because the listening has been directed toward one star in a thousand out of those that are believed to present favorable conditions (see Sagan and Drake, 1975). The search must continue.

In concluding our remarks on the universe and our place within it, I would like to propose a subject for reflection. Everything today makes us assume that humans are a part of the universe and that other parts of the same type may exist elsewhere as well. But these parts of the universe have a very singular property: they are able to make the universe become part of themselves! In fact, *knowledge* does exactly this.

Now, if we examine facts in a scientific spirit, considering the universe as a set of physical elements with a given structure, we ask: How is it possible that this set can be represented on one of its subsets, humans? Naturally, if we were speaking of infinite sets, it would be possible. But if, as we usually assume, humans are a finite set of atoms, with a structure that is definable in a finite way, is it not incredible that this representation can occur?

This is a problem that, in various forms, has challenged philosophers and poets for centuries and which has often found mystic, fantastic, or purely rhetorical answers. I certainly do not claim to solve the problem. I will only say that a key for understanding this representation is the *redundance* of the world, which we have already discussed (§4.19). Physical laws, starting from the elementay ones known to everyone, even at a prescientific level, allow us to bring together infinite information about the universe in a finite sequence of symbols. This has made possible the birth of intelligence, of language, and of culture. However this book ends here and another should begin; I shall refrain from starting it.

Notes

Chapter 1. The method of physics

1 There are a number of suggested readings that might be helpful. The following list is by no means exhaustive and is restricted to modern authors: Ackermann, 1970; Agassi, 1975; Agazzi, 1969; Bachelard, 1965; Braithwaite, 1953; Buchdahl, 1969; Bunge, 1967, 1973; Carnap, 1961, 1966; Geymonat, 1960; Hanson, 1969; Hempel, 1965; Hutten, 1956; Nagel, 1961; Pap, 1962; Popper, 1959, 1968, 1972; Reichenbach, 1951, 1970; Sneed, 1971; Stegmüller, 1974, 1975; Suppes, 1969; Tondl, 1973; Zinov'ev, 1973.

2 See, e.g., Bohm, 1957; Bohr, 1963; Bondi, 1967; Born, 1943; Einstein and Infeld, 1942; Einstein, 1949, 1950; Feynman, 1965; Heisenberg, 1958, 1971; Heitler, 1962, 1970; Margenau, 1950; Pauli, 1958; Schrödinger, 1951, 1957; Weizsäcker, 1971.

3 In this context one may mention the journal, *Foundations of Physics*, H. Margenau and W. Yourgrau, eds. (London: Plenum Press), specifically devoted to the foundations of physics, viewed from within. Of course, its articles are not always easily understood by the layperson.

4 Although I am not a follower of B. Croce, in this case I cannot help mentioning his distinction between *history* and *chronicle*. A history of science is something very different from a chronicle of science.

5 A number of specific historical works will be cited wherever appropriate. An excellent general reference is Geymonat, 1972. See also, Butterfield, 1949; Dreyer, 1959; Koyré, 1959, 1966, 1961.

6 A few references are: Barber and Hirsch, 1962, Casimir, 1971, Haberer, 1969; Salomon, 1970; Ziman, 1968, 1976.

7 However, one should remember the important experiment of the neopositivist school, consisting of an encyclopedia of unified science, published in Chicago at the end of the 1930s (cf. Neurath, Carnap, and Morris, 1970).

8 Many physicists prefer this *ostensive* definition, saying that physics is what physicists are dealing with. Obviously, this definition is not entirely satisfactory; for instance, a physicist is not dealing with physics when dealing with his income tax.

9 For example, one should not forget J. Kepler, the great contemporary of Galileo; see, e.g., Holton, 1973.

10 The object of our study will therefore be post-Galilean physics. One should not conclude, however, that science did not exist before Galileo. On science prior to and during the scientific revolution see, for example, Boas, 1962; Butterfield, 1949; Crombie, 1959; Dijksterhuis, 1960; Dreyer, 1959; Enriques and Santillana, 1973; Hall, 1956; Koyré, 1957, 1966; Kuhn, 1957; McMullin, 1963; Neugebauer, 1952; Rossi, 1962, 1971; Singer, 1959.

11 If this happens, biology will be *reduced* to physics. *Reductionism* should be thought of as a methodological precept which has had a great success up to now. But it would be contrary to the scientific attitude to make it become *faith* and to believe a priori that it will never find the road barred. Today there is no evidence to support that belief. However, the success obtained so far by reductionism renders the opposite faith – that of *antireductionism*, even more absurd. In dealing with this problem, one should be careful not to be conditioned by emotional factors. We shall discuss the subject in §4.28.

12 In addition to the works already mentioned, see Bernardini and Fermi, 1965; Geymonat, 1965; Hall, 1963; Koyré, 1939; de Santillana, 1955; Shea, 1972. And, of course, one should read Galileo's works!

422

13 The most scrupulous science historians sometimes warn against writing the history of science using our present image of science. This attitude reflects, I am sure, a good deal of wisdom. However, as is often the case, if one insists on the point too much, one risks being void of wisdom. With what eyes can we see if not with our own?

14 Someone might say that we are outlining the internal history of science, or a rational reconstruction of it, as distinct from the external history that describes facts in their actual sequence (see Lakatos, 1971, p. 91). It is not necessary to come to the conclusion (which perhaps indulges too much in humor) that "an approach which is purely historical is not enough. It was said by Mark Twain that: 'in real life the right thing never happens at the right place at the right time: – it is the business of the historian to remedy this mistake.' It is the business of the philosopher to remedy it also." (Temple, 1959).

15 This is probably a fortuitous – but fortunate – coincidence of terms. In fact, some historians would remark that by *sensible experience* Galileo meant an experience of the *senses*. Okay. But it is difficult to think that Galileo did not imply that the experience should also be *reasonable*.

16 As Nagel says: "since facts do not proclaim themselves to be relevant or irrelevant for a given problem, the scientist must adopt at least some preliminary hypothesis as to what sorts of facts are pertinent to his problem." (Nagel, 1967, p. 10).

17 As Bunge says: "At least the set concept will occur in any law statement, so that at bottom there are no nonmathematical laws." (Bunge, 1967, p. 343).

18 For information on the formal criteria with which this correspondence must comply, see, for example, Carnap, 1961; Hempel, 1965; Bunge, 1967. We shall forgo such discussion, whose results are fairly intuitive.

19 The assertion that the numbers of a set are proportional to those of another means that the numbers of the latter are obtained by multiplying the numbers of the former by one and the same factor, called the *proportionality constant*.

20 One cannot do without measurements by simply writing a proportion such as, say, $s_2/s_1 = t_2^2/t_1^2$. This is equivalent to taking s_1 and t_1 as units of measure of distance and time, respectively.

21 Sometimes physicists are accused of not being *interested* in such problems. This is not so. Suppose, for instance, that someone, worried about the economic crisis, sees a doctor during visiting hours, 2 P.M. to 4 P.M., and discusses his problems. If the doctor dismisses him abruptly, this does not mean that doctors, in general, are not interested in economic problems. Instead, the patient must recognize that although medical science can cure a liver ailment, it cannot solve an economic crisis.

22 This is only true in the physics of everyday objects or of very large objects. In microphysics the situation is very different. Therefore we are not yet able to discuss the process of inductive inference fully. However, once the necessary physical knowledge is set forth we shall be able to do so.

23 Naturally this holds true at the level of *intuitive* arithmetic. In a formal theory (which might even assume that $2 + 2 \neq 4$) the problem shifts from the level of content to the level of *syntax*, but remains essentially the same. The experiment no longer concerns the *theory* but the *metatheory*, that is, the context in which the theory itself represents the object to be studied. For example, in order to convince myself that the formal system adopted is not contradictory, I must perform an experiment. The sameness of the result is guaranteed only by *the postulate of the invariance of human reason*. P. W. Bridgman goes even further and says: "Mathematics thus appears to be ultimately just as truly an empirical science as physics or chemistry, and the feeling that it is something essentially different arises only when we do not carry out our analysis far enough" (Bridgman, 1936, p. 52); see also, Putnam, 1969, p. 216). I would perhaps be more cautious in generalizing. It is enough for me to emphasize that the dichotomy, mathematics – empirical science, the former being the realm of absolute certainty and the latter the realm of doubt – is the result of a misunderstanding.

24 Ordinarily, however (as W. V. Quine reminds us), philosophers who talk about possible worlds take for granted that their thoughts are anchored to *this* world.

25 See Bridgman, 1927. In order to realize in what sense many physicists understand the operational construction of their science, see Ageno, 1970.

26 Note that Bridgman has stated:

I believe that I myself have never talked of "operationalism" or "operationism," but I have a distaste for these grandiloquent words which imply something more philosophic and esoteric than the simple thing that I see. What we are here concerned with is an observation and description of methods which at least some physicists had already, perhaps unconsciously, adopted and found successful – the practise of the methods already existed. What I have attempted is to analyse these successful methods, not to set up a philosophical system and a theory of the properties that any method *must* have if it hopes to be successful (Bridgman, 1955, p. 1).

27 More precisely, as we are considering a subjective impression of color, we should say that the *trichromatic coordinates* of the light coming from the object are confined to certain values. But this is merely a technical detail, which is added for the sake of precision.

28 It is essentially a matter of language. As Carnap says:

Some philosophers maintain that modern science, because it restricts its attention more and more to quantitative features neglects the qualitative aspects of nature and so gives an entirely distorted picture of the world. This view is entirely wrong, and we can see that it is wrong if we introduce the distinction at the proper place. When we look at nature, we cannot ask: Are these phenomena that I see here qualitative phenomena or quantitative? That is not the right question. If someone describes these phenomena in certain terms, defining those terms and giving us rules for their use, then we can ask: 'Are these the terms of a quantitative language, or are they the terms of a prequantitative, qualitative language?' (Carnap, 1966, p. 59).

29 In technical terms one might remark that any dictionary presupposes a *metalanguage*, known to the reader, by means of which the terms of the language, namely, the words, are defined. The dictionary is circular, because any one term of the language can in turn become a part of the metalanguage when it is used in the definition of another term. The metalanguage of physics contains the names of the elementary quantities in their intuitive sense.

30 In this connection, see (Bunge, 1967, Chapter II).

31 This term, which was introduced by logical positivists, is not found in any ordinary physics textbook. But the underlying concept is currently applied in all sciences.

32 Nor is it a question of the *scientific explanation* of a fact or a law by means of more general facts or laws. In the latter case it is customary, rather, to speak of *explanandum* and *explanans*. Not all authors do, however, (Popper seems to retain the verb *explicare* also in this case). See also, Bunge, 1967, p. 3. I shall not argue with this kind of explanation, because in my opinion it does not have great epistemological importance. Carnap has said: "We do not have to say: 'Don't ask why,' because now, when someone asks, why, we assume that he means it in a scientific, non-metaphysical sense. He is simply asking us to explain something by placing it in a framework of empirical laws." (Carnap, 1966, p. 12).

33 This distinction (partially known to Leibniz or even to the stoics) was brought to the attention of scientists by G. Frege in 1892. He spoke about *sense* and *meaning* (*sinn* and *bedeutung*), whereas others use *connotation* and *denotation* or *meaning* and *reference*, respectively. This unfortunate plurality of terms may be confusing.

34 See Hempel, 1970, p. 108. Note that the author gives this example only in view of an elementary exemplification. But it is precisely in this view that we shall refer to the same case, to show that it is unwise, to distinguish primary from derived quantities in an essential way.

35 The thesis that any observation is theory-laden has been supported by a number of

authors, in particular, very effectively and with a wealth of examples, Hanson, 1969, Chapter I. The same author says that the theory according to which "our sensational data registration and our intellectual constructions thereupon are cleft atwain, is an analytical stroke tantamount to logical butchery" (Hanson 1967, p. 90). Popper affirms that the language itself is always "theory impregnated" (Popper, 1972, p. 20, 146).

36 For example, Popper says: "the customary distinction between *observational terms* (or *non-theoretical terms*) and *theoretical terms* is mistaken, since all terms are theoretical to some degree, though some are more theoretical than others" (Popper, 1965, p. 119).

37 Every circle symbolically represents a class, and every point inside it represents an element of the class. The area common to two circles represents the *intersection* of the two classes, that is, the set of elements belonging to both.

38 A more detailed discussion of this problem can be found in Dalla Chiara and Toraldo di Francia, 1973.

39 Weizsäcker rightly says: "The real number is a free creation of the human mind, which perhaps is not adequate to reality" (Weizsäcker, 1971, p. 163). When discussing this, one sometimes has the impression that few are aware of how big infinity is, and how small zero is!

40 In this connection, and for bibliography, see Cole, 1973.

41 In the laboratory it is usual to carry out several measurements on the same quantity. As a result, not all the numbers of the interval turn out to be equivalent. A distribution of *probability* is associated with them. These technical details, however, discussed in the *theory of experimental errors*, are not very important for our discussion now. Rather, let us remark that the interval can also be infinite. This happens when the value is found to be, say, $\geq a$. In this case, the interval extends from a to ∞. We shall confine ourselves to the typical case where ϵ is finite.

42 In fact, this and similar statements that will be encountered as we go along, need some qualification. In principle, a new measurement can fall anywhere, even outside the interval. However, if ϵ is properly chosen, the probability that this may happen is so small that it can be absolutely neglected. This assertion is a result of the *law of great numbers* which will be described in probability theory (§3.10).

43 The convention thus accepted can appear fairly arbitrary. But it will be justified in the next section.

44 However, it will be in order to note that Aztecs and Mayas added the toes to the fingers, and had base 20; whereas Sumerians and Babylonians made use of base 60 (see, e.g., Guitel, 1975).

45 All this is oversimplified. In practice, as already remarked, several measurements are carried out, and statistical averages and relative errors are determined. But these technical details are not essential in understanding the fundamental concept.

46 Equation (1.4) asserts that if the measurements are repeated with precisions $\epsilon_a' \ll \epsilon_a$, $\epsilon_b' \ll \epsilon_b$, the two resulting intervals will of necessity fall within a total interval 2ϵ.

47 Of course, one has the right to attribute a *metaphysical* meaning to equation (1.3). But what results is no longer considered part of physics. And more important, it cannot have intersubjective value.

48 Popper says:

Yet whenever I used to write or to say, something about science as getting nearer to the truth, or as a kind of approach to the truth, I felt that I really ought to be writing "Truth" with a captial "T," in order to make quite clear that a vague and highly metaphysical notion was involved here, in contradistinction to Tarski's "truth" which we can with a clear conscience write in the ordinary way with small letters (Popper, 1963, p. 231). Okay. But Tarski's merit lies precisely in the fact of freeing us from that vague and metaphysical concept! A small letter, modest but certain, is better than a vague, nonverifiable (and rather useless) capital letter.

49 This means assigning a *correspondence*, within a certain interval, between the values

of b, and the values of f. Usually this is done by indicating a certain number of mathematical operations (*algorithm*) to carry out on b. For example, in the very simple case discussed immediately following equation (1.4), we have $f(b) = b - k$ and the mathematical operation to be carried out on b is the subtraction of the constant k. However, the set of all possible functions is much wider than that of the *computable* functions. This is a problem of the greatest interest. Among those who have contributed chiefly to its solution are: A. Church, S. C. Kleene, K. Gödel, J. Herbrand, A. H. Turing, E. L. Post. See, for example, Kleene, 1967, p. 36; Hermes, 1969. This subject will be discussed again in §1.12.

50 According to the *instrumental* point of view, a law such as equation (1.4) or (1.5) is only an instrument, so as to infer new data from a series of data. As Suppes keenly observes: "Thus, in the familiar syllogism 'all men are mortal; Socrates is a man; therefore, Socrates is mortal,' the major premise 'all men are mortal,' according to this instrumental viewpoint, is converted into a principle of inference. And the syllogism now has only the minor premise 'Socrates is a man.'" (Suppes, 1967, p. 64). I maintain that a physical law such as equation (1.4) or (1.5) is not a *rule of inference* but a *major premise*. We shall return to this subject (§2.29). For a discussion on the instrumental viewpoint, see, e.g., Popper, 1968, p. 107.

51 Similar to the case of one variable, $f_1(a_n)$ in practice is understood to indicate a number of mathematical operations to be carried out on the a_n. Equation (1.6) requires that the result be zero.

52 These concepts are only intuitive, and not completely rigorous. We shall discuss these in more detail when we talk about the general problem of induction (§4.21).

53 Just to give a very simple example, if $f_1(a_n)$ is acceptable, in most cases $f_2(a_n) = f_1(a_n) + \gamma$, with a sufficiently small γ, will also be acceptable.

54 The explication of the term *simple* is not an easy matter, if one requires a reasonable degree of objectivity. Referring to the simplicity of the *computation* represents a crude anthropomorphism; nature does not make computations. On the concept of simplicity, see, e.g., Bunge, 1963; Popper, 1959, Chapter 7; Walk, 1966, p. 66.

55 The strip can be visualized, as in Figure 1.5, in the two-dimensional case (two quantities a_1, a_2). In three dimensions (three quantities a_1, a_2, a_3), it would be a layer. In the general case of many dimensions, one should speak of a *hyperlayer*, but there should not be any difficulty in understanding the meaning.

56 In ordinary language, one says that it did depend on them, but only *undetectably*.

57 However, this is not always the case. Sometimes in theoretical physics it is expedient to choose the units in such a way as to have $c = h = 1$. In a sense, this is equivalent to identifying energy with both mass and frequency.

58 In modern language one should perhaps say that although nature presents its functions in *extensional* form, we tend to put them in an *intensional* form. Of course, in the set-theoretical foundation of mathematics, a function is specified by its extension. Physicists, however, are largely accustomed to think of a function as a computational prescription. Let us repeat simply: Nature does not make any computations!

59 The confinement to polynomials using only a few terms embodies a methodological criterion, which all physicists believed were sound. Sometimes it is worded as: one must choose the mathematical form having the least number of constants to be adjusted by experience. Be careful! One incurs a logical error if one fails to distinguish between *adjustable* and *adjusted* constants, because the number of adjusted constants merely depends on a notational convention. Consider, for instance, the polynomial $x - x^3/(1 \cdot 2 \cdot 3) + x^5/(1 \cdot 2 \cdot 3 \cdot 4 \cdot 5) - \cdots$, with an infinite number of terms and an infinite number of k coefficients. We know from mathematics that such a polynomial (or *series*) represents $\sin x$. When use of this notation is made, one usually thinks that the function does not contain any arbitrary constant!

60 Clearly, this relation, when written as $a' - a - f(\bar{x}, a_1, a_2, \ldots, a_h)(x' - x) = 0$, is of type such as equation (1.6).

61 By doing this we introduce a hypothesis of continuity, expressed by the old tenet: *natura non facit saltus* (nature does not jump). Function f does not change its value

abruptly in going from one point to another quite nearby. This hypothesis, of course, is only justified when none of its results are found to clash with experience.
62 Of course, such a law will only be verified up to a certain precision ϵ.
63 Strictly speaking, the notation da/dx should be used when a depends only on x, and not on the additional variables a_n, as well. In our case one should talk about a *partial derivative* and write $\partial a/\partial x$. This technicality need not interest us here.
64 Here superscripts denote indices, not exponents.
65 Some philosophers of science have argued that the great success of infinitesimal calculus in physics proves that *exact* measures, expressed by real numbers, have a physical meaning. The argument is wrong.
66 But there remain some open problems about the set-theoretical concept of *continuum*, on which calculus is usually based.
67 *Actual infinitesimals* were banned from mathematics, when L. A. Cauchy, C. Weierstrass, and other great mathematicians of the past century put mathemetical analysis on rigorous grounds. Only recently, at the beginning of the 1960s, have infinitesimals regained citizenship rights thanks to the *nonstandard analysis* developed by A. Robinson.
68 The rod, for one thing, must be rigid, undeformable, free from perturbations, and so on when it moves. The precise definition of each one of these requirements leads to tremendous difficulties or circularities. A great deal has been written on the subject. A good review will be found in Grünbaum, 1973, Chapter 1.
69 These are, precisely, the cases where the quantity to be measured shows the kind of *additivity* that enables one to determine how many units must be added in order to match it. Usually, such a quantity is called an *extensive* quantity, whereas other quantities are said to be *intensive*. Perhaps this somewhat old-fashioned distinction does not have great conceptual importance, however, it can have practical importance, in that it is expedient to choose only extensive quantities as fundamental quantities.
70 The meter was originally defined as one forty-millionth of the earth's meridian. As a result, every novel and more precise measurement of the meridian entailed a change of the meter. This procedure was abandoned in 1875; today one compares the meridian with the standard meter and not vice versa.
71 An obvious exception is represented by the laws of invariance, which express the quality of the values of *one and the same* quantity at two different places or times. Further, it will be noticed that in microphysics one can consider the ratios of two *homogeneous* and fixed quantities (as, e.g., the ratio of the mass of the proton to the mass of the electron). These ratios are constant and do not depend on the units of measurement. They are said to be *dimensionless* or to represent *pure numbers*. Unfortunately, today we are unable to derive such constants from a theory; all we can do is to try to memorize them.
72 It is sometimes said that, in this way, one obtains a *coherent* or *consistent* system of units.
73 G. Gamow, 1970, p. 203, jokingly calls them "the three kings of physics," in order to stress their pompous conventionality.
74 As already mentioned, nature does not carry out computations. Still less, it does not raise lengths to the second or third power (which, incidentally, does not have any meaning).
75 It may be of interest to remark that the ancients grappled with a similar problem when they investigated the *proportions* of the human body. These were also represented by pure numbers (ratios of lengths) or, what essentially amounts to the same, by geometrical relations. A celebrated example is represented by Leonardo da Vinci's *Homo ad circulum et ad quadratum* (Gallery of the Academy, Venice). This may arouse suspicion (but merely a suspicion) that the present investigations may be as futile as the old ones.
76 The axiomatization of physical theories was the sixth of the famous twenty-three problems proposed by Hilbert in 1900 at the International Mathematical Congress in Paris. The problem is still largely open, in spite of several important attempts by

D. Hilbert, C. Carathéodory, A. Reichenbach, J. von Neumann, J. C. C. McKinsey, A. S. Wightman (see Bunge, 1973, Chapter 7).

77 The reader should be warned that the term "model" has in modern logic a different meaning from (perhaps, diametrically opposed to) that it generally has in physics. Further, unlike the latter, the meaning, which was introduced chiefly by A. Tarski, is very precise. When reading a book or a paper on the philosophy of science, one should be alert when a model is mentioned. One must first establish whether the language is that of logic or that of ordinary physics.

78 H. Post (1975) describes, in this case, "floating models."

Chapter 2. The physics of the reversible

1 In fact, water vapor is invisible. What we actually see is the cloud of tiny water droplets, into which steam condenses when coming into contact with a cool atmosphere.

2 Obviously, we are referring only to what happens in the *physical medium* before the ear or the eye are attained. What happens in the sense organs or in the mind of the observer belongs to physiology or psychology, respectively. Nonetheless, a number of physicists and biophysicists are investigating these fascinating subjects.

3 From here on, in the expression of physical laws, the accuracy $\pm \epsilon$ will be understood, without being explicitly stated. But the reader should remember that an ϵ *is always present*, even if its precise specification may not be of interest.

4 It is nevertheless of interest to recall the diametrically opposed view expressed by B. Russell:

we must entirely reject the notion of a *state* of motion. Motion consists *merely* in the occupation of different places at different times, subject to continuity as explained in Part V. There is no transition from place to place, no consecutive moment or consecutive position, no such thing as velocity except in the sense of a real number which is the limit of a certain set of quotients. The rejection of velocity and acceleration as physical facts (*i.e.*, as properites belonging *at each instant* to a moving point, and not merely real numbers expressing limits of certain ratios) involves, as we shall see, some difficulties in the statement of the laws of motion; but the reform introduced by Weierstrass in the infinitesimal calculus has rendered this rejection imperative. (Russell, 1938, p. 473).

The opinion of such an authority, even if expressed originally a long time ago (1903), may serve to emphasize the diversity between the frame of mind of the physicist and that of the mathematician.

5 Gravity acceleration is not a universal constant, for its value varies with location and decreases when leaving the earth's surface.

6 He lacked even a clear understanding of the experimental possibility to produce a vacuum or of the nature of atmospheric pressure. The latter was detected and measured by Galileo's pupil E. Torricelli.

7 Usually, vectors are represented by boldface symbols, whereas their moduli are represented by normal characters.

8 Let the arc AB be described in time Δt. Then $AB = v\Delta t$, hence $\alpha = v\Delta t/r$ represents the angle under which AB is seen from the center, or what amounts to the same, the angle $C\hat{A}D$. If α is very small, $CD = v\alpha = v^2\Delta t/r$. By dividing this increment of velocity by Δt, one obtains the equation (2.8) of the acceleration.

9 For history and bibliography see Andrade, 1965; Hall, 1963, Hesse, 1961; Pala, 1969.

10 An excellent discussion of this concept from the historical point of view is given by Jammer, 1957.

11 These concepts were subjected to analysis by E. Mach at the end of the last century. A large number of authors have raised doubts about the physical *reality* of forces and masses. In my opinion, such problems do not make much sense. However, for a defense of this reality, see Hooker, 1974.

12 As is generally known, one must write $W = Fl \cos \alpha$.
13 For information about the manifold interactions between the socioeconomical world and science see, for example, Bernal, 1953.
14 Remember Heraclitus' saying: "It is impossible to plunge twice into the same river."
15 Here, of course, Heraclitus is opposed by Parmenides who asserted that what *is*, is immovable, is not engendered, and does not have an end.
16 This corresponds to the scholastic tenet: *nihil agit in distans nisi prius agit in medium* (nothing acts at a distance unless it first acts in the medium). For the history of the problem see, for example, Hesse, 1961.
17 This is a *universal constant* which cannot be done away with, because the units of force, mass, and distance have already been fixed.
18 Sometimes violating historical truth, Newton has been granted a methodological attitude that he really did not take. In optics, however he did make hypotheses! Nevertheless, the failure of all Newton's attempts to imagine a reasonable hypothesis that could account for universal gravitation was, in my opinion, definitely instrumental in leading to the modern idea that although hypotheses of that kind may somtimes be useful for heuristic purposes (but sometimes misleading), they are not in the least necessary in order to develop physical science. (For a critical analysis of how Newton arrived at the law of universal gravitation, see, e.g., Cohen, 1974, p. 299.)
19 I do not intend in any way to deny that knowing more about gravitation would be beneficial. All I say is that equation (2.24) represents per se an important piece of knowledge about the physical world. It would be ridiculous to say that a person having one billion dollars is not rich because he does not have two billion dollars.
20 In this connection it is of interest to quote A. A. Zinov'ev, an author not suspected of being idealist: "It is impossible to find a definition of existence and non-existence which would satisfy all sciences and all instances of knowledge. There are in different sciences and in different sections of the same science different notions of existence and non-existence. Instead of clear definitions one usually finds vague conventions" (Zinov'ev, 1973, p. 19).
21 As Quine says: "Reification is a theoretical move, distinct from the observational component of an observation sentence" (Quine, 1970, p. 95).
22 When one takes this attitude, the entire problem may become vaguely similar to that of Homer's existence, as was described by a high school pupil: "Homer's poems were not written by Homer, but by another man who had the same name."
23 It all depends on the degree of *ontological commitment* one wants to choose. In any case, a perfectly admissible *intension* does not commit one to rule out the possibility that the corresponding *extension* may be empty. As argued by Quine, in his beautiful essay on what there is: "We can use singular terms *significantly* (my script) in sentences without presupposing that there are the entities which those names purport to name" (Quine, 1953, p. 12).
24 Of course, the field extends to infinity. However, at a great distance from A and B, it becomes negligible.
25 The electric field varies from point to point in space. Consequently, it must be measured *at a point*. Those who think that measuring with infinite precision may make any sense say that the field cannot be operationally defined because any measuring apparatus (probe) has a certain size and is not a point! But it should be evident that one measures the field existing at $x \pm \epsilon_x$, $y \pm \epsilon_y$, $z \pm \epsilon_z$ at time $t \pm \epsilon_t$. Of course, compared with more accurate measurements this represents only an *average* field. This is what happens for *all* physical quantities (at any rate, with respect to t); one cannot help wondering why the electric field should have struck the mind of some philosophers in this way.
26 Of course, this pun arose because when the positive direction of the current was first established, electrons were unknown and physicists had not realized that electrons are responsible for current.

27 This expression merely means that the experiment is carried out at a distance from the wire, much smaller than the length of its straight portion.

28 This is merely a *classical* description, whereas at the atomic level one should use quantum mechanics. However, the qualitative result would be identical.

29 Of course, a surface of infinite extent will only mean a surface whose extent is much greater than the wavelength.

30 A similar concept was perfectly clear to Galileo, who in the *Assayer* says: "For that reason, I think that tastes, odours, colours and so forth are no more than mere names so far as pertains to the subject wherein they seem to reside, and that they only exist in the body that perceives them. Thus, if the living creature were removed, all these qualities would be removed and annihilated." This kind of consideration will lead to Locke's distinction of primary and secondary qualities later on.

31 We are talking about the *normal* eye, not of the *shortsighted* (myopia) or *longsighted* (hypermetropia) eye, where the image forms before or behind the retina, respectively.

32 This kind of consideration was brilliantly developed by E. de Condillac in 1754. In his *Treaty of Sensations* he imagined a statue, internally constituted as a human being, which in turn acquires our five senses, starting from smell. In this way, Condillac accounted for the rise of all intellectual activity purely on the basis of sensation (*sensism*).

33 Of course, the object must first be recognized as such, in spite of its apparent small-ness. We are used to performing this task, and therefore may be unaware of its intrinsic difficulty. Adults are often surprised that little children *cannot see at a distance*. If when facing a broad landscape, you tell a little child, "Look at that far away house," he may fail to see it. Only much later will he acquire the training needed to recognize distant objects.

34 Fairly similar considerations were developed by *form psychology* (or *Gestalt psychology*). This school of thought, started in the first decades of our century by M. Wertheimer, K. Koffka and W. Köhler, maintains that the perception of an object is not decomposable into elementary sensations. What we perceive is a totality, built according to well-determined laws of organization. The concept that, within certain limits, we cannot see but what we are in the habit of seeing, has been extended by some to a context as wide as that of the scientific view of the world (Kuhn, 1970).

35 Today the Huygens–Fresnel principle is no longer an additional and ad hoc hypothesis. It is a mathematical result of Maxwell's equations, as was shown much later by G. Kirchhoff, who was able to give it a definitive formulation.

36 During dt the quantity vt increases by $v\,dt$. Dividing by dt, one obtains v.

37 Among a few important exceptions was Bishop Berkeley, who subjected the notion of absolute space and time to sharp criticism. Although Berkeley had a determinant influence on many philosophers, he had scarce, if any, influence on physicists. But he was, in a sense, a precursor of Mach and Einstein (see Popper, 1964, p. 166).

38 Those who maintain that metaphysical principles can be of help to the scientist should meditate on the relative positions of Newton and Berkeley. Newton who asserts that God is present everywhere, and exists always and everywhere, *constitutes duration and space*, is in favor of absolute time and space. Berkeley instead is in favor of relative space, and says:

that dangerous dilemma, to which several who have employed their thoughts on that subject imagine themselves reduced, to wit, of thinking either that real space is God, or else that there is something beside God which is eternal, uncreated, infinite, indivisible, immutable. Both which may justly be thought pernicious and absurd notions (Jammer, 1954).

39 To see this in a very simple way, imagine, after Einstein, an observer traveling with an electromagnetic wave at speed c. He would see a *static* and sinuisodal field which is inconsistent with Maxwell's equations.

40 Oddly enough, this problem dates back to the seventeenth century, when Torricelli first produced vacuum. His contemporaries were startled by the fact that light ostensibly could travel through a vacuum. There was an argument – several people maintained that Torricelli had not produced a true vacuum because light could be transmitted through it.

41 As a matter of fact, some attempts were made to interpret the result in different ways, for instance, by the ad hoc hypothesis of Lorentz and Fitzgerald, to be discussed later, or by the hypothesis that the earth *drags* the ether. It had to be a *total* drag, not a partial one like that proposed earlier by Fresnel in another connection.

42 The notion of inertial frame was first introduced by L. Lange in 1885 (see Jammer, 1954).

43 Without saying, this also makes sense and can be verified, but only up to a certain precision. In absolute terms, the assertion does not make sense because the stars move relative to one another.

44 There are many elementary expositions of relativity, see, e.g., Einstein, 1950; Reichenbach 1965. There is a lot of literature about Einstein himself, see Schilpp, 1949; Whitrow, 1967; C. Lanczos, 1974.

45 For the historical record, one must recall that the general postulate had been set forth by H. Poincaré in 1904. Einstein was not aware of this; however, Poincaré had not been able to derive from his postulate the developments that are due to Einstein.

46 We omit the easy arguments by which these formulas are derived, and invite the reader to assume them. However, for the reader who needs to renew his acquaintance with elementary trigonometry: Given the angle α build a rectangular triangle ABC on it. Then define $\sin \alpha = BC/AB$, $\cos \alpha = AC/AB$, $\tan \alpha = BC/AC$. There results $\tan \alpha = \sin \alpha/\cos \alpha$, and by Pythagoras' theorem $\sin^2 \alpha + \cos^2 \alpha = 1$. Solving the system of the last two equations for $\sin \alpha$ and $\cos \alpha$, one gets $\sin \alpha = \tan \alpha/(1 + \tan^2 \alpha)$, $\cos \alpha = 1/(1 + \tan^2 \alpha)$.

47 This is usually convenient because *our* velocities are extremely small (with respect to light speed, of course). As J. A. Wheeler says, we do the same thing when we measure the width of a road in meters and the length in kilometers.

48 Strictly speaking, it is not necessary to have $x'^2 - c^2t'^2 = x^2 - c^2t^2$; it suffices to have, say, $x'^2 - c^2t'^2 = \alpha(x^2 - c^2t^2)$ with a constant α. But if $\alpha \neq 1$, there follows an asymmetry between K and K' which does not seem acceptable.

49 Note that the imaginary unit is not more imaginary than the real unity. They are both abstract mathematical entities that are treated according to some well-determined rules. What is important is that with the *complex numbers*, consisting of a real plus an imaginary part (the latter being a real number times i), one can set up a perfectly coherent algebra that contains, as a particular case, the algebra of real numbers.

50 Of course, one can develop the argument without leaving the field of real numbers. Actually, there is a tendency today in relativity theory to use only real quantities. If this is done, space–time takes a *pseudoeuclidean* metric, meaning that in Pythagoras' theorem there also appear *differences* of squares (as in $x^2 - c^2t^2$).

51 H. A. Lorentz had discovered this transformation about a year before Einstein, but

432 Notes to pp. 120–40

attributed merely a formal significance to it. Einstein, unaware of Lorentz's contribution, gave a clear physical derivation of the transformation.

52 Note that even when a vehicle moves at the high speed of 1000 km/h, its v/c ratio is still less than one millionth. One readily sees that in this case, revealing the difference between both transformations is almost a desperate undertaking.

53 The existence of a physical quantity with an upper unattainable limit, has aroused much amazement and speculation. Someone has been tempted to derive profound philosophical implications from this. But to some extent, the fact is only a result of the arbitrary and unnatural way in which velocity is defined as a derived quantity, or as the ratio between distance and time. If one adopts the more natural definition of velocity as that of a *primary* quantity (§2.2), the situation can entirely change. For those with a little more mathematical background than required for understanding this book, we add a few remarks. Let us set $\alpha = iV/c$ in equation (2.54), from which $v/c = \tanh(V/c)$. The quantity V thus defined is none other than the primitive velocity, measured with the device shown in Figure 2.1. When V is very small, it coincides with v. The (experimental) law of uniform motion must then be found to be $l = ct \tanh(V/c)$, instead of equation (2.2). Velocity V adds in the *natural* way, $V = V_1 + V_2$, and can grow to infinity. When $V = \infty$, one has $l = ct$, thus light has infinite velocity. Notice we are only showing that relativity can be expressed in a different *language*, not that its *content* could be different.

54 To show this one can multiply by the square root and square both sides, finding $1 = 1 - 3v^4/4c^4 - v^6/4c^6$. If v/c is, say, 10^{-6}, the error made is on the order of 10^{-24}!

55 As a matter of fact, the last statement is not fully correct because of the second law of thermodynamics, to be studied later (§3.6). However, it would be pointless to complicate the issue here.

56 As a point of history, Berkeley forestalled Mach in suggesting this principle.

57 To be exact, there should be in this case some additional effects that we are neglecting here for simplicity. For one thing, the weight in Figure (2.59b) is greater than in Figure (2.59a); further, one could use a gyroscope or a Foucault pendulum to reveal the rate of rotation.

58 In a spaceship, which at departure has, say, acceleration $3g$, everything happens as though three new planets, identical to the earth, were suddenly added (and superimposed) to the earth.

59 This involves, once again, the equality of the inertial and gravitational masses, which has been confirmed with enormous precision by R. Dicke and V. Braginski. But the principle of equivalence is also valid for massless particles such as photons, as has been ascertained by R. Pound, G. Rebka, and J. Snider. As of 1975, R. Colella, A. W. Overhauser, and S. Werner have verified the equivalence even for single particles with quantum behavior.

60 In treatises on relativity, one usually writes $2g_{12}$ in place of g_{12}. The same is true for the mixed product coefficients in equation (2.76). However, this need not interest us.

61 More precisely, one should not talk about Kepler's ellipses which are curves in three-dimensional space. One should discuss instead the world lines described by the planets.

62 Special relativity is indispensable, at least in particle-accelerator engineering.

63 As a point of history, the possibility that the gravitational field should travel at a finite speed had long been suggested by various authors, for instance, Faraday and Maxwell.

64 As a rule, a theory does not consist of a single law; however, a law can represent a theory in its own right. This is why we now have to repeat, although in a more general form, some facts about the laws of physics.

65 This notion will become clearer and more precise when we study the general problem of induction (§4.21).

66 In my opinion, the assertion that a theory has unlimited validity is downright nonsense.

67 It is often believed that this is a very unpleasant occurrence, only halfheartedly

accepted by the scientific community. That is also a mistake, perhaps caused by the wary attitude that scientists *must* take when confronted with new evidence. However, one of the best cherished goals of an experimental physicist is to find a new experiment that clashes with someone of the firmly established and time-honored laws of physics! And the theoretical physicist is anxiously waiting for such evidence, in order to be able to work out a new theory.

68 The circumstance that Lavoisier's law was not "falsified" by experience, but by theory, is immaterial. If one could measure masses with sufficient accuracy, one would obtain the result experimentally.

69 The reverse order can be found even in Galileo's work (see, e.g., Dijksterhuis, 1961).

70 This statement is perhaps too drastic. As a matter of fact, the "defendant" theory is often granted an appeal. But the procedure is essentially the same. After Bacon, we can call E a *crucial* experiment.

71 Popper, arguing with Keynes, expresses the opposite opinion (Popper, 1965, p. 247). I would rather agree with Keynes. And then, Popper himself says very well:

And although I believe that in the history of science it is always the theory and not the experiment, always the idea and not the observation, which opens up the way to new knowledge, I also believe that it is always the experiment which saves us from following a track that leads nowhere, which helps us out of the rut, and which challenges us to find a new way (Popper, 1959).

72 The fact that there exists a domain D_n, in which the assertions of T_{n+1} coincide with those of T_n, can be termed the *general correspondence principle*, by analogy with the correspondence principle of quantum mechanics (see Post, 1971; Koertge, 1973, p. 167).

73 Historians sometimes take the authors who do such *violence* to history to task. I believe that doing a strictly *external* history of science is a mere dream. External history can be done only in the light of an internal history, perhaps built unconsciously (see Lakatos, 1971).

74 As already remarked, the historian must take note of the fact that Michelson's experiment had little, if any, influence on Einstein.

75 In nineteenth-century Italy, when one wanted to do away with an inconvenient opponent, one used to say, "He spoke badly of Garibaldi!" (Giuseppe Garibaldi was the hero of Italian independence and it was inconceivable to speak badly of him.)

76 Sometimes one gets the impression that all these "refutations" are piled up with the only purpose of showing that science is impossible. Those who enjoy the exercise of perpetual confutation remind me of the well-known story of "the jaguar's friend." I am not one of the jaguar's friends. I know that science *is* possible and am very grateful to those who seek to explain how and within what limits it is possible.

77 I am referring, of course, only to the serious ways of confutation. But there are other ways, more difficult to tolerate. One has only to think of the childish game of asking why. Starting from a completely reasonable assertion, anyone is able to give rise, by means of a series of questions, to a *regressus ad infinitum*, or to circularity. Another bad habit of some opponents of a given *x*-ism, consists in affirming that *x*-ism is *dead and buried*. However, the animosity and the vehemence that accompany that statement lead one to suspect that *x*-ism is still alive, for nonnegligible reasons.

78 Remember Einstein's words:

For even if it should appear that the universe of ideas cannot be deduced from experience by logical means, but is, in a sense, a creation of the human mind, without which no science is possible, nevertheless this universe of ideas is just as little independent of the nature of our experiences as clothes are of the form of the human body. (Einstein, 1950, p. 2).

79 The facts need only be observable *in principle*. An assertion about the other side

of the moon was not metaphysical in Wittgenstein's times, as it is not metaphysical today.

80 As Neurath put it: "Sentences are to be compared with sentences, not with 'experience,' nor with a 'world,' nor with anything else" (Weinberg, 1948). Neurath, quoted by J. R. Weinberg in *An Examination of Logical Positivism*. Paterson, N.Y.: Littlefield, Adams, 1960, p. 277.

81 While sharing Popper's distrust toward what cannot be falsified and toward psychoanalysis in particular, I would like to remark that the demarcation line seems sometimes to have an *historical* value. As an example, one cannot rule out the possibility that tomorrow we shall be able to pinpoint a neurophysiological structure in the brain corresponding to the Oedipus complex. Alternatively, we might find for the *explicandum* "Oedipus' complex" such an *explicatum* that can enable us to decide from behavior whether or not a person has that complex. What is unobservable today can become observable tomorrow. By contrast, the neopositivist affirmation that, say, "the idealism-realism dilemma represents a pseudoproblem, because such terms have no meaning," is intended to be valid forever. For a critical appraisal of psychoanalysis as a pseudoscience see Bunge, 1967, p. 40.

82 If there are, they are recognized as such only decades later, with hindsight.

83 The opinion that programs are accepted or rejected only on the basis of a judgment about their progressivity can be confirmed by the fact that some workers (Jánossy, 1970–72) still believe it possible to build relativity, without ruling out the existence of the ether. As Jánossy rightly remarks, the impossibility to reveal motion with respect to something does not in the least imply that this something does not exist. For instance, it may not make sense to reveal motion with respect to a uniform field, still the field exists. In my opinion, the only important fact to ascertain is whether the ether is *necessary* in order to account for the physical world. If not, one may as well apply *Ockham's razor* (*entia non sunt multiplicanda praeter necessitatem*), and refrain from multiplying unnecessary entities.

84 See Kuhn, 1970. Analogous ideas, even though with different shades, have been expressed by a number of other authors; see, for example, Toulmin, 1961; Feyerabend, 1965, 1972. For criticism about these ideas see, for example, Kordig, 1971.

85 There are some figures deliberately drawn with the purpose to give rise to this phenomenon. However, the Gestalt switch is observed in everday life much more frequently than we realize.

86 All this is strictly connected with the question of the *continuity* or *discontinuity* of the history of science (see, e.g., Agassi, 1973). This question has strained the minds of quite a few scholars. In my opinion, it does not make much sense, because in science history one cannot give a precise definition of continuity, as is done in mathematics. There is no point in arguing about the use of an approximate expression, which should be left to the choice and taste of each writer.

Chapter 3. The physics of the irreversible

1 This is so much so that the observation that de facto all planets revolve in the same direction raises a question, which can only be answered by suitable cosmogonic hypotheses (see §5.9).

2 Which, incidentally, represented a contradiction because no ordinary fluid can be imponderable, or penetrate all bodies, as does heat. The issue is somewhat analogous to that of the ether.

3 Gases are, so to speak, the guinea pigs of thermodynamics. It is comparatively simple to perform experiments using them (especially imaginary experiments!). As will be seen in due course, many important results have *general* validity, that is, are independent of the material system investigated; in these cases it is expedient to refer to a gas.

4 If the container is very large, one must take into account the effect of gravity. The

upper layers of the gas weigh on the lower ones and increase the pressure there.

5 Assuming the law to be valid down to $t = -273$ would entail making a tremendous *extrapolation*. Before reaching $V = 0$, one would arrive at a very small volume, where the gas could no longer be considered rarefied. Thus the vanishing of the volume cannot represent much more than an *indication*.

6 Apart from a slight correction, which need not interest us here, the *molecular weight* M of a substance indicates how many times a molecule of that substance is heavier than a hydrogen atom. If m_0 denotes the mass of the hydrogen atom, the molecule concerned has the mass Mm_0. A *mole* of the substance is defined as M grams of that substance; it contains a number of molecules given by $N = M/Mm_0 = 1/m_0$, where m_0, of course, is expressed in grams. Thus N is a fixed number, independent of the substance. It is called *Avogadro's number* and has the value $N = 6.02 \times 10^{23}$. Mark the fantastic order of magnitude.

7 In 1811 Avogadro put forward the hypothesis that at equal temperature and pressure the molecules of any gas are at equal distance d from one another. Each molecule then corresponds to volume d^3, therefore the volume occupied by a mole is Nd^3, equal for all gases.

8 Obviously, the simple observation that work can be converted into heat by friction is as old as humans. The first scientific investigation of this phenomenon is sometimes credited to B. Thompson (1798) who studied the production of heat in the boring of cannons. He inferred from his observations that heat could not be a material substance.

9 Sometimes the notation ΔQ (or an analogous notation) is preferred to dQ, because the latter indicates in mathematical analysis an *exact differential*. We shall not make use of this distinction, and shall simply understand by dQ a sufficiently small amount of heat.

10 This represents the principle of conservation of energy, extended to include thermal phenomena. It is usually credited to R. J. Mayer, 1842, but a few other names should also be mentioned in this connection, as for instance, Clapeyron, 1834; Joule, 1845; Helmholtz, 1847.

11 As is generally known, $\ln x$ denotes the *natural logarithm* of x, i.e., the exponent that must be given to the number $e = 2.718 \ldots$, in order to obtain x. In other words, $y = \ln x$ means that $e^y = x$. We recall that e is a special number (with an infinite number of decimal digits), akin to π, which is very often encountered in mathematics and its applications.

12 Remember that $a \ln x = \ln x^a$, $-\ln x = \ln(1/x)$, and if $\ln x = \ln y$, then $x = y$.

13 With an admirable insight, S. Carnot derived his results in 1824, when he was still thinking in terms of the caloric and ignoring even the first law of thermodynamics!

14 From equations (3.28) and (3.7), it is readily derived that in an adiabatic expansion, $PV^\gamma = $ constant. Because $\gamma > 1$ and the isothermal transformation $1 \to 2$ is governed by Boyle's law $PV = $ constant, there results that the curve $2 \to 3$ is steeper than the curve $1 \to 2$, as indicated in Figure 3.4.

15 Such value could be fixed in an arbitrary way. But there exists the *Nernst theorem* or *third law of thermodynamics*, stating that the entropy at absolute zero is always zero. As a result, the ambiguity disappears if the initial point A is taken at absolute zero. We need not enter into the discussion here. We shall always put $S(A) = 0$, wherever the fixed point A is, and say that entropy is determined *up to an additive constant*.

16 Mathematical analysis shows that this integration is feasible because dS is an *exact differential*. This means essentially that the result of the integration does not depend on the path, but only on the initial and final states.

17 A forerunner of this attitude is plainly represented by illuministic rationalism or by d'Alembert and Lagrange's attempt to show that the whole of mechanics is amenable to mathematical analysis.

18 The value of k was first given by Planck in 1900.

19 This objection was raised by Buys-Ballot to Clausius in 1858.

20 This name was introduced by J. W. Gibbs, who in 1902 settled the subject with a fundamental treatise.
21 There is some uncertainty in the nomenclature. Laplace talked about *insufficient reason*, Poincaré about *sufficient reason*, and Keynes about *indifference*.
22 In mathematics, a famous problem, posed by J. Bertrand in 1889, is to find the probability for a chord drawn at random in a circle to be longer than the side of the inscribed equilateral triangle. That problem does not have a unique solution.
23 E. T. Jaynes, 1973, has recently defended the principle of indifference with interesting arguments. However, see the criticism advanced by K. S. Friedman, 1975.
24 For instance, F. Waismann observed:

It is well known by any mathematician that operating with infinite series means substantially nothing but operating with the laws that generate them. If, for instance, we talk about the convergence of a sequence, we refer uniquely to this law and not to the *actual* succession of the terms–because we can always encompass only a finite number of them. Briefly, a mathematical series is something essentially regular, of which we can manage to see thoroughly all properties. On the contrary, in a statistical series nothing is so clear as its irregularity, and this proves at once that a statistical series is not a mathematical concept (quoted by Weinberg, 1948). However, it will be recalled that A. Kolmogorov has recently given a reasonable mathematical characterization of the irregularity or randomness of a finite sequence (Kolmogorov, 1967).

25 A more detailed discussion shows that ϵ decreases as $k/\sqrt{n}$, where k is constant. This will be explained shortly.
26 As mentioned in §1.9, this is in close analogy with a general property of measurement. The result of a measurement of a physical quantity is expressed in the form $a \pm \epsilon$. If ϵ is properly chosen, the *probability* that a new measurement may fall outside the interval $a \pm \epsilon$, although not zero, is negligible.
27 Let us illustrate this point by an analogy. Consider the notion of derivative. A physicist, a geometer, and an economist might respectively maintain that a derivative is a velocity, the inclination of a tangent, the rate of increase of capital. An argument between them would be very awkward; they would all be right and wrong at the same time. In reality, the derivative is merely a mathematical concept. Regarding probability (denoted by P), Bunge says: "instead of asking 'What is the meaning of "P"'?, one should ask 'What meanings have, as a matter of fact, correctly been assigned to "P" up to now?'" (Bunge, 1967, p. 429).
28 Sometimes physicists talk about the probability of a single event, as when they say that the probability of a given process occurring in a given way has a given value. Obviously, they are then referring to an event that *can* belong to a set of repeated experiments, and of which we know the frequency of occurrence by virtue of previous experience. Of course, the probability that God may exist is not a physical problem.
29 It is of interest to notice that two of our most general sciences, that is, thermodynamics and information theory, have arisen mainly from the needs of technical applications.
30 When the emphasis is on the utilization of the information for the *control* of a machine, of an organism or of a group, rather than on the mere acquisition or processing of information, the theory is usually referred to as *cybernetics*, a science whose fundamental ideas were developed by N. Wiener (see Wiener, 1949; Ross Ashby, 1956).
31 We shall deal only with the *statistical* theory of information. Other kinds of theory are also possible. A theory of *semantic information* has been proposed by Y. Bar-Hillel and R. Carnap, 1953, and developed by J. Hintikka, 1970. A theory based on *complexity*, in place of probability, has been proposed by A. N. Kolmogorov, 1967.
32 Remember that $\log n^N = N \log n$.
33 To go from the logarithm to the base 2, to the natural logarithm, one must multiply by the constant $0.69314 \ldots = \ln 2$, whereas to go to the decimal logarithm, one must multiply by $0.30103 \ldots = \log_{10} 2$.

34 It would be erroneous to conclude from this that, for instance, the sound of a church bell does not carry information. The sexton can choose between ringing or not ringing the bell, therefore he chooses between two different messages.

35 Remember that whatever the base, $\log(1/a) = -\log a$. Furthermore, note that since $p < 1$, $\log p$ is negative, hence I turns out positive.

36 Of course, if the answer were *yes*, the amount of information gained would be enormous. But because the persons with *situs inversus* (tip of the heart on the right) are extremely rare, one can hardly expect that answer. The mathematical reason that the average information is virtually zero is that when p tends to zero, $-\log p$ tends to infinity *very slowly*, so that the product $-p \log p$ tends to zero.

37 One ordinarily makes the hypothesis that the average obtained from a very long sequence, emitted by a given source, should coincide with the average obtained by considering ensembles of different sources. In general, the hypothesis that the *time* average should coincide with the *ensemble* average is called the *ergodic* hypothesis. In our case it is not strictly valid, but can represent a reasonable approximation.

38 A further reduction is brought about by the fact that the probabilities of the various letters are not independent of one another. The probability of a letter depends obviously on the sequence of the letters that precede it. For instance, the letter u, when preceded by the letter q carries almost no information, for its occurrence is almost certain.

39 In fact, it would be like the message tapped by a monkey on a typewriter.

40 Because the probabilities of successive characters are not independent of one another, the expression of $\bar{I}$ would be a little more complicated than in equation (3.64).

41 It may become viable when the messages are comparatively few. For instance, in legal language, one can talk about the Fifth Amendment without writing down the entire amendment.

42 We are talking, of course, of all the microstates compatible with the macroscopic conditions requested (e.g., a given constant value for the energy, if the system is isolated).

43 As a matter of fact, it is usual in this context to consider the three components of *momentum*, rather than those of the velocity. However, this is immaterial for our discussion. Note that microstates belong, obviously, to microphysics. We should have to apply quantum mechanics. However, even the results of *classical statistical mechanics* are sufficient to clarify many points.

44 Boltzmann's (1871) own formulation was a little obscure. As was made clear by Maxwell in 1879, the assumption is: "that the system, if left to itself in its actual state of motion will, sooner or later, pass through every phase which is consistent with the equation of energy."

45 The fact that in equation (3.64) there appear logarithms to the base 2, whereas equation (3.77) contains logarithms to the base e, makes no difference.

46 This condition was automatically satisfied when we had assumed that all the molecules had the same constant velocity.

47 For the history of Boltzmann's program see, for example, Elkana, 1971.

48 As a matter of fact, Boltzmann defined H with the sign opposite to that adopted here for convenience. This is of very little importance.

49 It is because of this relation that Shannon has called *entropy* the equation (3.67) of the information theory. Shannon was uncertain how to call his H and asked J. von Neumann for advice. "You should call it entropy," said von Neumann "for two reasons. In the first place your uncertainty function has been used in statistical mechanics under that name, so it already has a name. In the second place, and more important, no one knows what entropy really is, so in a debate you will always have the advantage." This charming story is reported by Tribus and McIrvine (1971).

50 The Jesuit P. Teilhard de Chardin has resumed this conception in an attempt to merge it with modern science. There is a final point (point Ω) to which the world tends naturally, and not by an abrupt divine intervention (such as the final judgment). See Teilhard de Chardin, 1955.

51 As we already know, the *before–after* relation is relativistically invariant for two

events that can be connected by a signal. In particular, it is invariant for two events that both occur in the presence of the same observer (unless extremely brief times are involved).

52 Formulas such as equations (3.93) and (3.94) are commonly referred to as Bayes' formulas, and are said to furnish the probabilities of the *causes*. It will be observed that some of the probabilities involved represent rather *subjective* probabilities than relative frequencies. But, of course, one can surmise that subjective probabilities are assessed by the observer on the basis of how frequently certain things are observed to happen in the real world.

53 In technical jargon, physicists say that in solving an ordinary wave problem, one should use the *retarded potentials* and not the *advanced potentials*. Advanced potentials are used only in more sophisticated work.

54 The circular waves that have uniform amplitude along the circle represent only a particular case. There are more general kinds of circular waves.

55 Of course, the factor ln 2 appears because information is expressed with logarithms to the base 2, whereas H is expressed with natural logarithms.

Chapter 4. Microphysics

1 It is interesting to note that the Greek word is πράγμα; it is not neutral in respect to human activity, as *res* is.

2 I use this term (which may or may not be original) as quite distinct from *objectivation*. "Objectuation" is the activity that consists of dividing reality into objects, whereas objectivation consists of making objective, a concept, a course of reasoning, etc. As for the process of objectuation, the studies of Piaget on child psychogenesis are very interesting (see, e.g., Piaget, 1959, 1937, Chapter I).

3 As an example, let us consider Zinov'ev's treatise (Zinov'ev, 1973, p. 8). The author says: "We will use the word 'object' in its widest sense: an object is anything which can be perceived, represented, named, etc.; i.e., anything at all." One might be able to avoid circularity by introducing an object as a collection of points of space–time, as does Quine (to be published).

4 It is easy to note here that Russell completes a line of thought that has its roots in Berkeley, goes through Stuart Mill (according to whom things are *possibilities* of sensation), and to Mach (for whom things are *complexes* of sensations).

5 I prefer to use this term rather than that of *elementary objects*, because I also want to refer to atoms, molecules, and so on, and because contemporary physics has taught us that, far too often, objects that are at first considered elementary, turn out not to be so.

6 Perhaps it would be possible to observe that even in a Peano-like formal system, in which numbers are symbols or complexes of symbols, subject to certain axioms, it is necessary to be able to conceive every symbol or complex as an object distinct from the others.

7 Naturally, we know that in order to obey this definition, rational numbers, which do not have the power of the continuum, are sufficient. But here we are speaking of what may be called the *physical continuum*. The postulates of Dedekind and Cantor on the mathematical continuum represent a further abstraction.

8 Democritus believed that atoms had an infinite number of different shapes, because there was no reason that an atom should have one shape rather than another. Here is a good example *ante-litteram* of the fallacy of the principle of insufficient reason when applied a priori, without the support of experience.

9 Ostwald and his followers believed it necessary to give up building models (like that of the kinetic theory) and to limit oneself to studying the transformations of energy (*energetism*).

10 Ostwald honestly declared that these facts had convinced him of the existence of atoms. I have the impression (but I may be mistaken) that some modern historians tend to dramatize the history of atomism a bit, giving Ostwald an importance that in reality he did not have.

11 As generally known, in the periodic system, discovered in 1869 by D. Mendeleev, the elements fall into an ordered series in which the order number Z is called the *atomic number* and goes from 1 for hydrogen, to 2 for helium, up to 92 for uranium. The chemical properties of the elements show *more or less* periodic recurrences in respect to Z. The atomic mass, A, which indicates (except for a correction that we shall see in §4.23) how many times an atom is heavier than the hydrogen atom, grows from 1 for hydrogen, to 4 for helium, up to 238 for uranium.

12 J. J. Thomson is not to be confused with W. Thomson, *alias* Lord Kelvin, nor with B. Thompson, *alias* Count Rumford.

13 The name was coined by J. Stoney. The fact that electricity was a particle phenomenon had already been guessed by H. von Helmholtz in 1880.

14 We say as a first approximation, because electrons *repel* one another whereas the planets attract one another. Therefore the *perturbations* with respect to Kepler's laws are different.

15 More exactly, when it describes a circle or an ellipse, it is equivalent to *two* perpendicular and alternating currents, 90° out of phase with respect to one another.

16 One should not be surprised by this name. The blackbodies that we see are usually at low temperature, and practically do not emit visible radiation. But when heated beyond a few hundred degrees, they begin to give off light. The sun itself behaves almost exactly like a blackbody at 6000 K.

17 One can show that Kirchhoff's law is valid both for a true material surface and for an ideal surface, such as that of the hole. The device described was used for the first time by O. Lummer and W. Wien in 1895.

18 In reality, one cannot carry out the measurement on a *precisely monochromatic* radiation. Every spectroscope has a finite line width, therefore one measures the radiant energy conveyed by all frequencies within a small band. Conventionally, and in the interest of simplification, we can assume that for every frequency v, one measures the energy emitted in the band that goes from v to $v + 1$.

19 To be exact, if we express everything as a function of wavelength instead of frequency, the maximum wavelength is inversely proportional to the temperature, as was predicted by W. Wien in 1894.

20 Exactly, this is true when v is very large with respect to unity. In practice, this condition is always perfectly verified. In optics, e.g., v is on the order of 10^{14} or 10^{15} vibrations per second.

21 For the history and the conceptual development of quantum theory see Jammer, 1966, 1974. A suggestive collection of largely first hand evidence on this topic is contained in the articles of Rosenfeld, Heisenberg, van der Waerden, Jordan, Jauch, Wigner, Rohrlich, Peierls, Wentzel, Tomonaga, Schwinger, and Salam, edited by Mehra, 1973.

22 As a matter of fact, equation (4.17) is predicted by Maxwell's theory of electromagnetic waves, even without photons. We must accept, then, that the electromagnetic field has a momentum as well as an energy. The photon interpretation is perhaps more acceptable from an intuitive point of view.

23 For Boltzmann, two microstates obtained from one another by interchanging two molecules are generally considered to be two *different* microstates.

24 Because we could say the same thing for all the particles of modern physics, we notice that when two identical particles meet, the sentence – the intensions "incident particle" and "scattered particle" have the same extension – is undecidable, as anticipated in §1.6. Modern physics shows that the principle of *the identity of undiscernibles*, so dear to Leibniz and other philosophers, is not valid. Two equal particles found in the same system are not discernible, nor are they a single particle!

25 We shall now examine a particular case of Bose-Einstein statistics – that which applies to photons. In this case the total number of particles present in the system is not necessarily fixed. In fact, photons are continuously emitted and absorbed by the walls of the cavity.

26 Obviously, to put $n = \infty$ as an upper limit in the sum has formal value only. Because g_v is a finite number, we know that g_v^n will be zero for all values of n greater than a certain value.

27 We shall consider v as an index that assumes whole values from 0 to ∞. This is possible because the small values 0, 1, 2, . . . make an absolutely negligible contribution, compared with very large values, which are virtually continuous.

28 This term denotes the infinite sum $S = 1 + x + x^2 + x^3 + \cdots$, when $|x| < 1$. If one tries to multiply the expression S by $1 - x$, one finds that all the terms except the first cancel. One reaches the conclusion that it must be $S(1 - x) = 1$, or $S = 1/(1 - x)$. The proof can be rendered perfectly rigorous. If one makes $x = \exp(-hv/kT)$ in equation (4.24), one has a geometric series and so obtains for the sum the value $1/(1 - \exp(-hv/kT))$, which appears in equation (4.25).

29 That is, the series $R = 1x + 2x^2 + 3x^3 + \cdots$. If one multiplies this series by $(1 - x)^2 = 1 - 2x + x^2$, one finds that all the terms cancel except the first, which is equal to x. Thus one obtains $R(1 - x)^2 = x$, or $R = x/(1 - x)^2$. Making $x = \exp(-hv/kT)$ one obtains the result of the text.

30 The exponential law for the radiation of a blackbody had been proposed by W. Wien in 1896, but without very convincing arguments.

31 Bohr's idea was not altogether new. For example, A. E. Haas, who tried to apply quanta to Thomson's atom, is usually cited. But he did not have much success. Some eminent physicists of the time thought that he was trying to put together two unrelated things.

32 Potential energy is always defined up to an additive constant. This constant has no influence in the usual case, in which (like in our own) it is the *difference* between two values of E_p that has physical importance. When equation (4.41) is used, the constant is so chosen as to make E_p zero at infinity (i.e., for $r \to \infty$). Because of these considerations, one should not be surprised to see *negative* potential energy. It is a question of convention. A negative value can become positive if one adds a suitable constant.

33 In addition, the results obtained by N. Bohr agree with a hypothesis that we know today to be correct. According to this hypothesis, X rays are emitted in the quantum jumps of the electrons closest to the nucleus, as W. Kossel suggested in 1916. Formulas (4.46) and (4.47) (with a slight modification that will not be dealt with here) also account for a law found in 1913 by H. Moseley for the frequencies of X rays.

34 In regard to this singular scientist see the collection of articles, L. de Broglie (1973).

35 An attempt to determine the wavelength associated with a macroscopic body is hopeless. For example, a wavelength on the order of 10^{-27} cm is associated with a mass of 1 g, which moves at a speed of 1 cm/sec. This size is one hundred thousand billion times smaller than that of an atomic nucleus!

36 Naturally, one can use nonrelativistic mechanics only when the speed of the particle is much less than c. For electrons in the atom the speed is about 1/100 of that of light and the approximation is quite good.

37 Let us recall that, given a complex number $\psi = a + ib$ (where a and b are real and i represents the imaginary unit) the number $\psi^* = a - ib$ is called the *complex conjugate*. The product $\psi\psi^* = a^2 + b^2$ is indicated by $|\psi|^2$ and is called the square of the *modulus* of ψ.

38 Some philosophers and scientists, such as C. Renouvier, E. Boutroux, C. S. Pierce, F. Exner, and C. G. Darwin, had already guessed and maintained in various forms that strict physical determinism might not be valid and that room must be allowed for *chance*. For these precedents see Jammer, 1966. But only with Born, Bohr, and Heisenberg do these perceptions acquire a precise meaning.

39 This is the *superposition principle*.

40 Nevertheless, an attempt of A. Landé to derive the entire probabilistic system of quantum mechanics – including interference – from general a priori principles, is very interesting (A. Landé, 1970, p. 297). Whether or not the concept of probability used in microphysics is the same as that of classical physics, has been, and still is, a matter of controversy. See, for example; Margenau and Cohen, 1967, p. 71; Suppes, 1969, pp. 212, 227; Fine, 1971, p. 69.

41 We recall that angular momentum is represented by a vector in the direction of the axis of the revolving body, using the corkscrew rule.

42 Historically, the two hypotheses appeared in inverse order. First, Pauli formulated his principle of exclusion and only later, stimulated by Pauli's ideas, G. E. Goudsmith proposed the hypothesis of the revolving electron.

43 We choose the z axis arbitrarily. We could have referred to x or y or any other direction.

44 Naturally, the coordinate s_z can only assume the values $+\frac{1}{2}$ or $-\frac{1}{2}$. But this is immaterial for what we want to say.

45 Schrödinger's equation is a differential equation of the type that is called *linear* and has the property that a linear combination of two solutions is still a solution.

46 If ψ_m and ψ_n are normalized, $A = 1/\sqrt{2}$.

47 The fact that photons have a spin should not be surprising. Classically, this corresponds to the fact that an electromagnetic wave, besides having energy and momentum, can also have an angular momentum around its direction of propagation (circular polarization).

48 We refer to the sum of the rest energy and the kinetic energy, that is, to equation (2.72); we ignore the possible potential energy that, by definition, can also be negative.

49 As a matter of fact, this can easily be found when using complex notation. When we think of a real sinusoid, we must consider that ψ vibrates in a way that does not favor one point more than another.

50 To be exact, we use a *Fourier integral*. Fourier's theorem says that practically any function in physics can be represented as a superimposition of an infinite number of sinusoids, with suitable amplitudes, frequencies, and phases.

51 Heisenberg, as he explicitly affirmed, was inspired by the example of Einstein, who showed how wrong it is to discuss simultaneous events without giving the term a precise operational significance. In the same way, Heisenberg felt it necessary to analyze how x and p_x can be measured precisely. Ironically, Einstein, who had been one of the most strenuous opponents of the Copenhagen interpretation, inspired Born and Heisenberg to set down two fundamental elements of this interpretation: the probabilistic conception and the uncertainty principle!

52 A rigorous calculation would lead us to the same result as this approximate calculation.

53 We neglect the small correction due to Compton's effect.

54 What is important in the expression of the uncertainty principle is the *order of magnitude* of the product $\Delta x \, \Delta p_x$, not the precise value, which depends on the conventions we adopt (e.g., it is conventional to define the resolving power with the precise expression, equation (4.55)).

55 J. A. Wheeler says that the *observer* becomes a *participator*.

56 On this subject see also Caldirola (1974, p. 41).

57 Heisenberg later preferred to give uncertainty a vaguely Aristotelian interpretation, affirming that measurement actualizes one of the potential properties of the object. This concept will become clearer later on; see Heisenberg, 1958.

58 Naturally, the smaller the holes are, the lower is the probability that the photon will pass through both of them. But when a photon happens to pass through both, we know how to arrive at precise conclusions concerning its past.

59 These are the *linear vector spaces*, whose theory, introduced by H. Grassmann in the last century, has given rise to modern *functional analysis*.

60 The use of operators in quantum mechanics was introduced by M. Born and N. Wiener as early as 1926.

61 There are precisely two equations for the two unknowns u_1 and u_2. Be careful not to confuse the two *components* u_1, u_2 with the two *vectors* which for future convenience we shall call $\mathbf{u}_1$ and $\mathbf{u}_2$.

62 For simplicity, we shall always refer to the nondegenerate case. This is sufficient to understand general concepts.

63 Apparently G. Peano (1888) was the first to suggest that a linear space can have an infinite number of dimensions.

64 If, for example, we have $v_1 = 1$, $v_2 = 1/2$, $v_3 = 1/3$ and so on, the vector $\mathbf{v}$ is finite

in length. In fact, a generalization of Pythagoras' theorem gives this length as the square root of the infinite sum $1 + 1/4 + 1/9 + \cdots$ which is finite in value.

65 A state vector does not always exist. However, it exists in almost all the cases that the physicist meets in practical application.

66 Equation (4.84) is equivalent to Schrödinger's equation.

67 In quantum mechanics we often use the same symbol for the observable physical quantity and the hermitian operator that corresponds to it. It is even said, inappropriately, that the classical quantity becomes a hermitian operator. The operator A is constructed starting from the expression of the classical quantity, and applying rules of correspondence formulated by Dirac. This is a modern expression of an old *correspondence principle* formulated by Bohr in the first years of the theory.

68 Actually, this happens only in those processes of measurement that are called *ideal*. We shall always assume that we are working with ideal measurements.

69 For this reason, we often talk indifferently about wave function or state vector.

70 We must define the angular momentum as its maximum projection on any axis, in order for the result to be correct. But we shall not dwell on this fine point.

71 Some writers definitely reject the contents of this axiom. See, for example, Margenau, 1950, Chapter XIII.

72 This measurement can be carried out in several ways. The most obvious one is naturally to close hole F_2.

73 We use the term *algorithm* (which derives from the name of the Arab mathematician Al Chwarizmi) in a very loose sense, as a complex of rules that allow us to arrive in a finite number of steps from a set of initial data to the data that make up the result. For a more precise, technical meaning of the term see Hermes, 1969.

74 For Bohr, as already stated, it was a question of *complementary* aspects, both real, but which manifest themselves in a mutually exclusive way, according to the type of experiment we perform. It is doubtful that this conception is very useful today. I agree with Bunge, who declares that: "complementarity is not a part and parcel of the quantum theory." (Bunge, 1973, p. 116). Margenau even says that it is an absurd concept (Margenau, 1966, p. 330).

75 L. de Broglie adhered for some time to the school of Copenhagen. He then left it, going to openly critical positions (see de Broglie, 1967, p. 706).

76 An excellent introduction to all these problems is found in d'Espagnat, 1973. Volume 29 of the journal *Synthèse* (Reidel, 1974), entirely dedicated to quantum mechanics, is also important.

77 It is contrary to relativity.

78 To be precise, Einstein, Podolski, and Rosen said that in this case an "element of physical reality" must correspond to the measured quantity. We cannot say that this locution is a model of clarity.

79 D'Espagnat affirms: "we know for sure, contrary to Einstein's conception, that, in some respects at least, the world is non-separable." (d'Espagnat, 1973, p. 734). We can simply accept nonseparability by saying, with Bohm and Hiley: "that inseparable quantum interconnectedness of the whole universe is the fundamental reality and that relatively independently behaving parts are merely particular and contingent forms within this whole" (Bohm and Hiley, 1975, p. 102).

80 It is evident that the measuring devices must necessarily be macroscopic. This is a very important point, and has been emphasized many times by Bohr.

81 The significance of the product $|v_0\rangle |u_k\rangle$, of two vectors, each of which is in a different Hilbert space, deserves a precise description. In the interest of brevity we shall not go into this description, but shall instead consider an intuitive idea, that is, that this product is analogous to the product of the wave functions of two different particles as appears in equation (4.52).

82 I use "prejudice" in an etymological sense rather than a derogative one.

83 Besides, this is strictly related to what we have already noticed in regard to the uncertainty principle, which concerns the future and not the past.

84 Of course, the same observation can be made for Schrödinger's cat.

85 An excellent discussion is to be found in van Fraassen, 1974. See also, Suppes, 1969

he articles of J. Bub, B. C. van Fraassen, R. H. Greechey and S. P. Gudder,
C. A. Hooker, edited by Hooker, 1973.
do not want to ignore the fact that some authors have maintained that positivism
based on an idealist or idealist-tending world view." (Putnam, 1973, p. 211). But
is a matter of opinion.
oncerning this problem, one may find it useful to read Jauch (1973).
Needless to say, the most massive and often bitter criticism is directed at the school
of Copenhagen. Sometimes an unprepared reader can derive the impression from
such criticism that Bohr, Heisenberg, Born, and Dirac were mediocre and naïve
scholars who mainly accumulated a series of errors. Naturally, in physics there are
no workers inspired by God, immune to criticism. Any theory or interpretation can,
and must, undergo criticism. We can find that even the great have committed mistakes. But if we maintain some sense of proportion, we can say along with Popper
about these mistakes: "These, I hope, will soon be forgotten while the great physicists who happened to commit them will be for ever remembered by their marvelous
contributions to physics: contributions of significance and depth to which no philosopher can aspire" (Popper, 1967, p. 42).

89 In the place of "those who have not studied" we can substitute, in some cases,
"children" and in others, "the man on the street" or even, "eighteenth-century
scientists." It does not matter. In any event, the reasoning is false.

90 Sometimes instead of *real* facts, one talks about *concrete* facts, and complains that
quantum mechanics has replaced a concrete world with a completely abstract one.
But as Ne'eman rightly says: "This generation's abstract becomes the next generation's concrete, except that at the present rate of development a "generation"
may last as little as ten years!" (Ne'eman, 1974, p. 2).

91 Furthermore, if the world were necessarily made in that way, we could ask: "What
is the aim of study and scientific research? Is it not exactly that of investigating the
make-up of this reality, which we do not know a priori? Someone even gives a social
content to (1) reasoning more or less like this: "The great majority of humanity has
not had the opportunity to study. To study is a privilege. Therefore to think of the
world in a way that is different from that of those who have not studied is antisocial."
Here the premises are very correct. But the conclusion is absurd! In my opinion the
only reasonable conclusion that can be derived is that we must work to give everyone
the opportunity to study.

92 Maybe we could say that in this case the distinction between monadic predicates
and relations is a modern form of Locke's distinction between *primary* and *secondary*
qualities.

93 Many writers pose strong ontological demands. J. S. Bell says amusingly that quantum mechanics should not deal with "observables," but with "beables" (Bell, 1973,
p. 687).

94 See P. T. Matthews, 1971, p. 7. As Blokhintsev rightly observes: "We must accept
that the world is not built in the tidy or simple fashion envisaged by believers in the
various metaphysical systems" (Blokhintsev, 1968, p. 2).

95 Naturally, such a reasonable point of view can also be maintained by those who feel
the need of excluding the *observer* from quantum theory (see Bunge, 1973, p. 104).

96 For that matter, more than a few think that strict determinism is not even maintainable in classical physics (see Blokhintsev, 1968, Chapter I).

97 In regard to the search for T_{n+1}, see, for example, the collection of writings edited
by T. Bastin, 1971.

98 For a discussion and bibliography of these questions see, for example, Bunge, 1959;
Bergman, 1974. It will be good to recall W. Stegmüller's remark (Stegmüller, 1960)
that when we mention *the* causality problem, we can be induced by the definite
article to believe that it is one problem only. In reality, we are dealing with several
different problems. As Zinov'ev says: "When we use the word 'cause' we often
have in mind situations which are completely different from the logical point of view,
and this results in fruitless discussions" (Zinov'ev, 1973, p. 256). For that matter,
this was already known to Aristotle.

99 Even the study of psychogenesis can lead to this conclusion. J. Piaget says:
the age of three or four (the "why" years) to six or seven, causality is ess
assimilation to one's own action, whence an initially indissociable mixture of ti
and efficient causality, animism and phenomenism, etc." (Piaget, 1967, p. 6

100 In regard to the concept of cause Carnap says:

"It apparently arose as a kind of projection of human experience into the world
nature. When a table is pushed, tension is felt in the muscles. When somethin
similar is observed in nature, such as one billiard ball striking another, it is easy to
imagine that one ball is having an experience analogous to our experience of pushing
the table. The striking ball is the agent. It *does* something to the other ball that
makes it move. It is easy to see how people of primitive cultures could suppose that
elements in nature were animated, as they themselves were, by souls that willed
certain things to happen. This is especially understandable with respect to natural
phenomena that cause great harm. A mountain would be blamed for causing a land-
slide. A tornado would be blamed for damaging a village (Carnap, 1966, p. 189).

101 Let us think: What could a horse do, even if it had an intelligence equal to that of
a human? All animals transfer negentropy to the environment, but with much less
effective means than humans.

102 O. Costa de Beauregard says: "Physics accepts causal explanations in which improbability is given at the beginning and refuses final explanations in which
improbability is found at the end" (Costa de Beauregard, 1967, p. 74).

103 We already know that this relationship is relativistically invariant if A and B can be
joined by a signal (§2.23).

104 If I could influence the past, I could prevent an event that has happened, which is
already impressed on my memory, that is, irreversible, and that has left other irreversible traces.

105 It is clear that the problem of induction is found in all natural sciences. But here
we shall limit ourselves to the application in physics.

106 M. Black expressed the opinion that: "the problem of induction will eventually be
classified with such famous 'insoluble' problems as that of squaring the circle or that
of inventing perpetual-motion machines" (M. Black, 1967, p. 200).

107 Without finding *negative* instances, of course! A single negative instance *falsifies*
the general law.

108 To be precise, we refer to the *isotope* of atomic weight 238.

109 See, for example: C. G. Hempel (1966, p. 120). See also the articles of M. Black,
P. Suppes and G. H. von Wright, edited by J. Hintikka and P. Suppes (1966, pp.
175, 198, and 208, respectively).

110 In another paradox discovered by N. Goodmann (Goodmann, 1955) the predicate
B expresses a property that is not fixed, but changes, say, tomorrow or in 1990. If
all the objects of a certain class, which we have observed until today have property
B, shall we be able to make the induction from *many to all*? There is no paradox,
not only because we do not recognize the validity of the *many-to-all rule*, but also
because we shall rely on the space–time invariance of physics, as a necessary condition in order to make an inductive inference. Consequently, the predicate B is not
admissible in physics.

111 It has been surmised several times that physical constants may vary with time. One
notable investigator of this hypothesis is P. A. M. Dirac. But it was never possible
to reveal such variation. If it exists, it can only be extremely slow, measurable over
times of cosmic order (billions of years). For further information, see Yahil (1974,
p. 27). In any case, if a variation of this kind were detected one day, it would be
another example of the well-known circularities of physics. We would use the invariance of laws, in order to construct instruments, with which to reveal the
variation!

112 The distinction made here between macroscopic and microscopic, is a bit conventional, because as we have already pointed out, it does not coincide with the distinction between classical and quantum physics (see lasers, superconductivity, etc.).

can be seen in a quick and simple way. Let $p = a/n^k$ represent the probability negative instance, a and k being two positive constants. The probability of n itive instances will be given by $(1 - a/n^k)^n$. Now (as can be proved with calculus) s expression for $n \to \infty$ tends toward 1 if $k > 1$, toward e^{-a} if $k = 1$, and toward if $k < 1$. Consequently, if on the basis of a reasoning analogous to that of equation 3.94), we want to discard the hypothesis of having encountered an extremely rare case, we have to assume that $k \geq 1$, that is, that p is $O(1/n)$ or less.

4 It should be unnecessary to recall that O does not represent the *fact*, but the proposition that describes it.

15 Thomson might very probably have believed this, had he performed this experiment before the ones he actually did.

116 Nor does it seem that Popper's affirmation that "it is part of our *definition* of natural laws if we postulate that they are to be invariant with respect to space and time" (Popper, 1959, p. 278), can help him very much.

117 Often one picks the *equispaced* values of a_1. Normally, this is amply sufficient because *chance* tends to distribute the points uniformly along the a_1 axis.

118 I do not deny that the saying *natura non facit saltus* has a certain validity. But unfortunately, it is an a posteriori verification of known laws. It is not very useful a priori, if we do not know the steepness of the jumps that nature is willing to tolerate.

119 Among the various criticisms that can be made of the application of inductive logic to physics – in my opinion that of L. J. Cohen, according to which inductive logic is incapable of dealing with the "anomalies" that often occur in physics – is very interesting (Cohen, 1973).

120 This expression is probably not the best. It reflects the historical origin of the method and that certain amount of ambiguity that underlies the concept of the wave-particle *duality*. On this subject see P. Caldirola, 1974, p. 72.

121 In a sense, this happens already in classical electrodynamics. See Caldirola, 1974, p. 36.

122 Given the frequency with which we have to consider the creation and annihilation of particles in modern microphysics, we often express mass in units of energy, that is, we give directly the energy that the mass can generate with its annihilation. In this case it is useful to express the energies not in *ergs* but in *electronvolts* eV. An eV is the energy that an electron acquires by passing through the potential difference of one volt, and is precisely equal to 1.6×10^{-12} ergs. The multiples KeV (10^3 eV); MeV (10^6 eV) and GeV (10^9 eV) are often used. The mass of an electron is equal to about 0.5 MeV. Therefore to generate an electron–positron pair requires at least the energy of an MeV.

123 The lines represent symbolically some initial or final *states* of the particles, and not necessarily true *trajectories* or *world lines*. But conventionally, we shall speak of particles that go toward the right or left, or up or down.

124 Procedures of this type, which may concern both the charge and mass of the particle, are called procedures of *renormalization*. These have permitted the elimination from quantum electrodynamics of some uncomfortable *infinities* that appeared in the application of the perturbation method carried to higher orders. These methods work very well, but their mathematical justification is not entirely satisfactory.

125 As a matter of fact, the attitude of some "critics" of quantum mechanics reminds one of the person who affirmed that to enter the lion's cage is dangerous, because the door is low and one can bump one's head.

126 The word "real" is used here in an ontological sense rather than a technical one (*real* as opposed to *virtual*).

127 Comparison to the binding energy of a peripheral electron in the atom is useful. The latter is on the order of one or ten eV. Therefore nuclear energy can be a million times greater than chemical energy!

128 Naturally, we use an expression that is effective, although somewhat inexact.

129 For this reason, and because the proton and the neutron do not have equal mass, the weight of a given nucleus is not exactly equal to A times a given elementary

quantity. By international convention we have established 1/12 of the mas
atom of ^{12}C as the unit of atomic weight. This is *smaller* (1.6604×10^{-23}
the mass of hydrogen atom (1.6734×10^{-23} g).

130 From a logical point of view this faith is a little bit naïve. For also, the *nonoccu*
of that process represents in its own right a phenomenon that does not violat
conservation law. Nevertheless, the tenet can have great heuristic power.

131 Sometimes, these invariances are given the unfortunate term of space–time *i*
mogeneity or *isotropy*. In reality, the object with which we are concerned in th
context is not space–time, but the laws of physics.

132 We must point out a theory advanced by S. Weinberg and A. Salam (1967), whose
aim is to *unify* weak and electromagnetic interactions. This theory is more promising
today than ever because of recent experiments that have shown the probable existence of *neutral currents*, which it specifically requires (e.g., in the diffusion
$v + p \rightarrow p + v$). Efforts are currently being made to unify *all* interactions.

133 In other words, we deal only with what is relevant to strong interaction.

134 Think, for example, that if in Figure 4.25(b) the nucleus is changed into its anti-
nucleus, the latter will have negative charge. Therefore if it rotates in the direction
opposite to that of Figure 4.25(a), it will have a magnetic moment oriented in the
same direction.

135 If we were sure that T is not valid for some processes, we would have really discovered a *nonentropic irreversible process*, a concept we have already considered
in thermodynamics (§3.16).

136 Quine says: "We can improve our conceptual scheme, our philosophy, bit by bit
while continuing to depend on it for support; but we cannot detach ourselves from
it and compare it objectively with an unconceptualized reality. Hence it is meaningless, I suggest, to inquire into the absolute correctness of a conceptual scheme
as a mirror of reality." (Quine, 1953, p. 79).

137 For instance, one could think of the rise of those nonlinear phenomena that, according to Wigner, reduce the wave function, at the level of individual consciousness.

138 In reality, the opinability of these questions is very great, but not complete. For
example, it can be shown that if we limit ourselves to the application of classical
physics, a strict reductionism can lead to logical paradoxes. See Dalla Chiara and
Toraldo de Francia, 1974.

Chapter 5. The universe

1 As Wigner says: "It is, in fact, impossible to adduce reasons against the assumption
that the laws of nature would be different even in small domains if the universe had
a radically different structure." (Wigner, 1967, p. 3).

2 Since these considerations were first written, a system of five rings has been discovered about Uranus. Therefore the suspicion that Saturn is only a member of a
wider class, was quite reasonable!

3 Actually, the validity of the distinction between *analytic* and *synthetic* has been very
doubtful after Quine subjected it to a subtle and effective criticism (Quine, 1953, p.
20). There is a certain analogy and connection between these impossibilities of strict
distinction: law–fact, analytic–synthetic, nomological–nonnomological.

4 In the interest of thoroughness, we should also deal with *contradictions*. By definition, P cannot contain logically contradictory propositions, for example, the proposition $p \wedge \neg p$. But at least until we know perfectly a given physical law L, the
set P can contain propositions incompatible with that law. Let us consider, for
example, law (5.1), and a proposition $q \in Q_L$ which it indicates as a consequence
of p. It is clear that the proposition $p' = p \wedge \neg q$ is contradicted by the law. But
in many cases it can belong to P. In this case, in agreement with the Duns Scotus
law: *ex absurdo sequitur quodlibet*, we must assume that $f_L(p') = P - Q_0$, that
is, that any physical consequence can be derived from p'. This is the situation we
encounter when a hypothetical law is falsified by an experiment.

t, naturally, the case in which p contradicts p_1. As already remarked, we then
$f_L(p) = P - Q_0$.

he history of modern cosmology see, for example, North, 1965.

a comprehensive review of the primitive cosmologies and cosmogonies of var-
s populations see Blacker and Loewe (1975).

his doctrine is usually attributed to Pythagoras, whereas the Ionic philosophers
maintained that the earth was flat.

This example schematically illustrates the procedure of the sciences that describe
the environment. From here on we shall almost always avoid detailed illustrations
of the observations and reasoning that lead to the different conclusions. Thus our
discussions may sometimes appear *dogmatic*. But why bore the reader with so many
technical details which are not important for our purpose?

10 Some scholars believe that the physical laws that we know today may be changing
at the cosmological time scale. If this is true, there must be some unchanging superlaws, which prescribe how the laws change with time. One must then use those
superlaws in order to solve the cosmogonic problem.

11 This correspondence had already been considered surprising in 1620 by F. Bacon,
who, however, did not derive the right results from it. The first to suggest that the
continents had actually moved across the surface of the earth was Antonio Snider-
Pellegrini in 1858; but he thought that the cause had been the Great Flood!

12 How can we not smile when some poets (or even philosophers), speaking of human
things, use the adjective *eternal*? *Exegi monumentum aere perennius*. . . . Nevertheless, even without indulging in facile rhetoric, we cannot help being struck by
the fact that humans, so ephemeral and contingent, can conceive eternity and reconstruct, with a good probability of success, the life and evolution of the universe.

13 For obvious reasons we shall limit our remarks to those we consider essential.

14 The percentage of ozone in the atmosphere is valued on the order of 5 parts in 10
million.

15 Etymologically, ecology is the science of the *home*. It studies the relationships between living organisms and the environment in which they live.

16 Nitrogen monoxide can combine with ozone to produce nitrogen dioxide and oxygen:
$NO + O_3 \rightarrow NO_2 + O_2$.

17 Note that I have absolutely no intention of becoming an improvised biologist. I wish
only to point out some current concepts that, in my opinion, strictly pertain to our
theme.

18 Each of the four characters obviously can carry two bits of information. To identify
an amino acid, to select it from among 20 possible choices, requires more than 4
and less than 5 bits. Therefore a succession of 3 characters, that is, of 3 DNA bases,
is both necessary and more than sufficient (redundant) to identify an amino acid.
Nature achieves the aim by using ordered triplets of characters. This code has already been deciphered.

19 Recent exploration of Mars's atmosphere has furnished additional clues for a reconstruction of the primordial atmosphere of the earth.

20 As a matter of fact, a single normal type does not exist. Each species presents a
certain *polymorphism*, that is, a number of genetic varieties that coexist, each with
greater advantages or disadvantages, according to the environment.

21 Abused quotations are certainly annoying. But I must repeat Hamlet's phrase, "there
are more things in heaven and earth, Horatio, than are dreamt of in your
philosophy."

22 There are no doubts about this fact. But not everything is clear. In the chain of
reactions that we believe occur inside the sun, many high energy neutrinos should
be produced. A great number of these should reach the earth, and we should be able
to detect them. But the results of the difficult experiments performed with this aim
in mind, have fallen short of our expectations. This is one of the enigmas of modern
astrophysics.

23 However, the Chinese had noticed sunspots much earlier.

24 Curiously enough, the eleven-year cycle is not a constant feature of the sun. It has
been ascertained that sunspots virtually disappeared from 1645 to 1715.

25 The fact that we are not insisting on the enormous philosophical significance
   passage from the geocentric to the heliocentric system might seem surprisin
   has been done so many times, that it seems superfluous to discuss the them
   in this book. In addition to the works already cited, see, for example, Kuhn,
   Byard, 1977.
26 In reality the mass of Pluto is too small to cause these perturbations. The discov
   was in great part fortuitous.
27 As a matter of fact, we have reason to believe that the fall of meteorites and th
   formation of craters on the planets and the moon were much more intense a fev
   billion years ago, and that today it has almost stopped.
28 A light year, the distance that light travels in a year, is equal to about 10^{13} km.
29 In addition, as we know, the astronauts during these six or seven years would age
   much less than their counterparts on the earth.
30 Clearly, with our present technology we are still far from knowing how to accomplish
   this.
31 Someone has rightly said that we are like children brought to visit an art gallery;
   one may have to teach us, though, that in order to appreciate the paintings, it is not
   necessary to touch them.
32 It is interesting to recall that the existence of massive bodies for which the escape
   velocity is greater than c had already been suggested by Laplace in 1796. Even
   without relativity, it seemed inevitable to conclude that such bodies were invisible.
33 It is clear that a hypothesis of this kind can arise very naturally as a counterpart
   to that of the big bang. It is sufficient to think of Kant's first antinomy, whose
   thesis requires a world that has a beginning in time and is limited, and whose
   antithesis requires a world that is infinite in space and time.
34 As of today, the mean density of matter seems to be too small to account for a
   closed universe. An *open* and ever expanding universe seems more likely. However,
   we have been led to believe for various reasons that there may be a considerable
   quantity of matter besides what we are able to observe (perhaps found in blackholes,
   in invisible neutron stars, in intergalactic atoms). Therefore the problem of the *missing mass* is still open. A quite recent development is represented by the rise of a
   well-founded suspicion that neutrinos may not have an exactly zero mass. Due to
   the enormous amount of neutrinos that must be present in the universe, even an
   extremely small mass could make a tremendous difference.
35 For example, Alpha Centauri, the star closest to the earth, is a triple system and
   so is to be excluded.
36 Naturally, I wish to emphasize that I am discussing serious and sensible things,
   which have nothing to do with the countless absurdities concerning the appearance
   on the earth of extraterrestrial beings, with which we have been bombarded in the
   last few years, and which sometimes have found credence even among those who
   should through cultural background be more logical. On this subject see, for example,
   Klass, 1974.

apter 1. The method of physics

Ackermann, R. (1970). *The Philosophy of Science*. Pegasus, New York.
Agazzi, E. (1969). *Temi e problemi di filosofia della fisica*. Manfredi, Milano.
Agassi, J. (1975). *Science in Flux*. Reidel, Dordrecht.
Ageno, M. (1970). *La costruzione operativa della fisica*. Boringhieri, Torino.
Bachelard, G. (1965). *L'activité rationaliste de la physique contemporaine*. Presses Université, Paris.
Barber, B., and Hirsch, W., eds. (1962). *The Sociology of Science*. Free Press, New York.
Bernardini, G., and Fermi, L. (1965). *Galileo and the Scientific Revolution*. Fawcett, Greenwich, Conn.
Boas, M. (1962). *The Scientific Renaissance*. Collins, London.
Bohm, D. (1957). *Causality and Chance in Modern Physics*. Routledge, London.
Bohr, N. (1963). *Atomic Physics and Human Knowledge*. Interscience, New York.
Bondi, H. (1967). *Assumption and Myth in Physical Theory*. Cambridge University Press, Cambridge.
Born, M. (1943). *Experiment and Theory in Physics*. Cambridge University Press, Cambridge.
Braithwaite, R. B. (1953). *Scientific Explanation*. Cambridge University Press, Cambridge.
Bridgman, P. W. (1927). *The Logic of Modern Physics*. Macmillan, New York.
(1936). *The Nature of Physical Theory*. Princeton University Press, Princeton, N.J.
(1955). *Reflections of a Physicist*. Philosophical Library, New York.
Buchdal, G. (1969). *Metaphysics and the Philosophy of Science*. Blackwell, Oxford.
Bunge, M. (1963). *The Myth of Simplicity*. Prentice-Hall, Englewood Cliffs, N.J.
(1967). *Scientific Research, I and II*. Springer, Berlin.
(1973). *Philosophy of Physics*. Reidel, Dordrecht.
Butterfield, H. (1949). *The Origins of Modern Science*. Macmillan, New York.
Carnap, R. (1961). *Der Logische Aufbau der Welt*. Meiner, Hamburg.
(1966). *Philosophical Foundations of Physics*. Basic Books, New York.
Casimir, H. B. G. (1973). "*Physics and Society*." In: *Physics 50 Years Later*, S. C. Brown, ed. National Academy of Sciences, Washington, D.C.
Chiswell, B., and Grigg, E. C. M. (1971). *SI Units*. Wiley, New York.
Cole, E. A. B. (1973). "*Perception and Operation in the Definition of the Observable*." *Nuovo Cimento* 8, 155.
Crombie, A. C. (1959). *Medieval and Early Modern Science*. Doubleday, New York.
(1961). "Quantification in Medieval Science." *Isis* 52, 143.
Dalla Chiara, M. L., and Toraldo di Francia, G. (1973). "A Logical Analysis of Physical Theories." *Rivista del Nuovo Cimento* 3, 1.
Dijksterhuis, E. J. (1961). *The Mechanization of the World Picture*. Oxford University Press, London.
Dreyer, J. L. E. (1959). *A History of Astronomy from Thales to Kepler*. Dover, New York.
Einstein, A. (1950). *Out of My Later Years*. Thames and Hudson, London.
Einstein, A., and Infeld, L. (1942). *The Evolution of Physics*. Simon & Schuster, New York.
Feynman, R. (1965). *The Character of Physical Law*. MIT Press, Cambridge, Mass.
Gamow, G. (1970). "The Three Kings of Physics." In: *Physics, Logic and History*, W. Yourgrau and A. D. Breck, eds. Plenum, London.

449

Geymonat, L. (1960). *Filosofia e filosofia della scienza*. Feltrinelli, Milano.
(1965). *Galileo Galilei*. McGraw-Hill, London.
(1972). *Storia del pensiero filosofico e scientifico*, I-VII. Garzanti, Milano.
Grünbaum, A. (1973). *Philosophical Problems of Space and Time*. Reidel, Dord
Guitel, G. (1975). *Histoire Comparée des Numerations écrites*. Flammarion, Par
Haberer, J. (1969). *Politics and the Community of Science*. Van Nostrand, New
Hall, A. R. (1963). *From Galileo to Newton*. Collins, London.
Hanson, N. R. (1967). "Observation and Interpretation." In: *Philosophy of Scie
    Today, S. Morgenbesser, ed. Basic Books, New York.
(1969). *Patterns of Discovery*. Cambridge University Press, Cambridge.
Heinkin, L., Suppes, P., Tarski, A., eds. (1959). *The Axiomatic Method*. North Holland,
    Amsterdam.
Heisenberg, W. (1958). *Physics and Philosophy*. Allen & Unwin, London.
(1971). *Physics and Beyond*. Harper & Row, London.
Heitler, W. (1962). *Der Mensch und die naturwissenschaftliche Erkenntniss*. Vieweg,
    Braunschweig.
(1970). *Naturphilosophische Streifzüge*. Vieweg, Braunschweig.
Hempel, C. G. (1965). *Aspects of Scientific Explanation*. Free Press, New York.
Hermes, H. (1969). *Enumerability, Decidability, Computability*. Springer, New York.
Holton, G. (1973). "Johannes Kepler's Universe, Its Physics and Metaphysics." In:
    Thematic Origins of Scientific Thought. Harvard University Press, Cambridge,
    Mass.
Hutten, H. (1956). *The Language of Modern Physics*. Allen & Unwin, London.
Kleene, S. C. (1967). "Computability." In: *Philosophy of Science Today*, S. Morgen-
    besser, ed. Basic Books, New York.
Koyré, A. (1939). *Etudes Galiléennes*. Hermann, Paris.
(1968). *From the Closed World to the Infinite Universe*. Johns Hopkins University
    Press, Baltimore.
(1961). *Du monde de l' "à-peu-près" à l'univers de la précision*. Colin, Paris.
(1966). *Etudes d'histoire de la pensée scientifique*. Hermann, Paris.
Kuhn, T. (1957). *The Copernican Revolution*. Harvard University Press, Cambridge,
    Mass.
Lakatos, I. (1971). "History of Science and its Rational Reconstruction. In: *PSA 1970
    in Memory of Rudolf Carnap*. Reidel, Dordrecht.
Margenau, H. (1950). *The Nature of Physical Reality*. McGraw-Hill, New York.
McMullin, E., ed. (1963). *The Concept of Matter in Greek and Medieval Philosophy*.
    University of Notre Dame Press, South Bend, Ind.
Nagel, E. (1961). *The Structure of Science*. Routledge, London.
(1967). "The Nature and Aim of Science." In: *Philosophy of Science Today*. S. Mor-
    genbesser, ed., Basic Books, New York.
Neugebauer, O. (1952). *The Exact Sciences in Antiquity*. Princeton University Press,
    Princeton, N.J.
Neurath, O., Carnap, R., and Morris, C. (1971), *Foundations of the Unity of Science*.
    University of Chicago Press, Chicago.
Pap, A. (1962). *An Introduction to the Philosophy of Science*. Free Press, New York.
Pauli, W. (1958). *Aufsätze und Vorträge über Physik und Erkenntnistheorie*
Pearce, G., and Maynard, P., eds. (1973). *Conceptual Change*. Reidel, Dordrecht.
Piaget, J. (1970). *L'épistémologie génétique*. Presses Université, Paris.
Popper, K. (1959). *The Logic of Scientific Discovery*. Hutchinson, London.
(1968). *Conjectures and Refutations*. Harper & Row, New York.
(1972). *Objective Knowledge*. Oxford University Press, London.
Post, H. R. (1975). *The Misuse of Models*, Proceedings of Fifth International Congress
    of Logic Methodology and Philosophy of Science, VII-47.
Putnam, H. (1962). "What Theories Are Not." In: *Logic, Methodology and Philosophy
    of Science*, Nagel, E., Suppes, P., and Tarski, A., eds. Stanford University Press,
    Stanford, Calif.

"Is Logic Empirical?" In: *Boston Studies in the Philosophy of Science*. Reidel, [D]recht.

[Reichen]bach, H. (1951). *The Rise of Scientific Philosophy*. University of California Press, [Be]rkeley.

[(]). *Experience and Prediction*. University of Chicago Press, Chicago.

[,] P. (1962), *I filosofi e le macchine*. Feltrinelli, Milano.

[19]71). *Aspetti della rivoluzione scientifica*. Morano, Napoli.

[Sal]mon, J. J. (1970). *Science et politique*. du Seuil, Paris.

[San]tillana, G. de (1955). *The Crime of Galileo*. University of Chicago Press, Chicago.

[S]chrödinger, E. (1951). *Science and Humanism*. Cambridge University Press, London.

(1957). *Science, Theory and Man*. Cambridge University Press, Cambridge.

Shea, W. R. (1972). *Galileo's Intellectual Revolution*. Macmillan, London.

Singer, C. (1959). *A Short History of Scientific Ideas to 1900*. Oxford University Press, London.

Sneed, J. D. (1971). *The Logical Structure of Mathematical Physics*. Reidel, Dordrecht.

Stegmüller, W. (1974). *Probleme und Resultate der Wissenschaftstheorie und Analytischen Philosophie*. Springer, Berlin.

(1975). *Hauptströmungen der Gegengwartsphilosophie*. Kröner, Stuttgart.

Suppes, P. (1967). "What Is a Scientific Theory?" In: *Philosophy of Science Today*, S. Morgenbesser, ed. Basic Books, New York.

(1969). *Studies on the Methodology and Foundations of Science*. Reidel, Dordrecht.

Tarski, A. (1935). *Der Warheitsbegriff in den formalisierten Sprachen*. Studia Philosophica, I.

(1944). *The Semantic Conception of Truth and the Foundations of Semantics*. Philos. and Phenomenol. Research, IV.

Temple, G. (1959). "From the Relative to the Absolute." In: *Turning Points in Physics*. North Holland, Amsterdam.

Tondl, L. (1973). *Scientific Procedures*. Reidel, Dordrecht.

Walk, K. (1966). "Simplicity, Entropy and Inductive Logic." In: *Aspects of Inductive Logic*, I. Hintikka and P. Suppes, eds. North Holland, Amsterdam.

Weizsäcker, C. F. von. (1971). *Die Einheit der Natur*. Hauser, München.

Ziman, J. (1968). *Public Knowledge*. Cambridge University Press, Cambridge.

(1976). *The Force of Knowledge*. Cambridge University Press, London.

Zinov'ev, A. A. (1973). *Foundations of the Logical Theory of Knowledge*. Reidel, Dordrecht.

Chapter 2. The physics of the reversible

Agassi, J. (1971). *Faraday as a Natural Philosopher*. University of Chicago Press, Chicago.

(1973). *Continuity and Discontinuity in the History of Science*." *Journ. Hist. Ideas* 34, 609.

Andrade, E. N. da G. (1965). *Sir Isaac Newton*. Doubleday, New York.

Berkeley, G. (1948). "An essay toward a new theory of vision." In: *The Works of George Berkeley, Bishop of Cloyne*, A. A. Luie and T. E. Jessop, eds. Nelson, Nashville, Tenn., p. 203.

Bernal, J. D. (1953). *Science and Industry in the Nineteenth Century*. London.

Bridgman, P. W. (1927). *The Logic of Modern Physics*. Macmillan, New York.

(1936). *The Nature of Physical Theory*. Princeton University Press, Princeton, N.J.

Brillouin, L. (1970). *Relativity Reexamined*. Academic Press, New York.

Bunge, M. (1967). *Scientific Research*, I and II. Springer, Berlin.

Carnap, R. (1966). *Philosophical Foundations of Physics*. Basic Books, New York.

Cohen, J. B. (1974). "Newton's Theory vs. Kepler's Theory and Galileo's Theory." In: *The Interaction between Science and Philosophy*, Y. Elkana, ed. Humanities Press, Atlantic Highlands, N.J.

452   References

Cowperthwaite Graves, J. (1971). *The Conceptual Foundations of Contempor ativity Theories*. MIT Press, Cambridge, Mass.
Dijksterhuis, E. J. (1961). *The Mechanization of the World Picture*. Oxford Un Press, London.
Einstein, A. (1950). *The Meaning of Relativity*. Methuen, London.
Feyerabend, P. K. (1965). "Problems of Empiricism." In: *Beyond the Edge of Cert R. Colodny, ed. Prentice-Hall, Englewood Cliffs, N.J., p. 45.
    (1972). "Explanation, Reduction and Empiricism." In: *Scientific Explanation, Sp and Time*, H. Feige and G. Maxwell, eds. University of Minnesota Pre Minneapolis.
Galileo. *Dialogue Concerning the Two World Systems*, S. Drake, transl. (1962). Uni versity of California Press, Berkeley.
Gamow, G. (1970). "The Three Kings of Physics." In: *Physics, Logic and History*. W. Yourgrau and A. D. Breck, eds. Plenum, London.
Geymonat, L. (1972). *Storia del pensiero filosofico e scientifico*, I-VII. Garzanti, Milano.
Hall, A. R. (1963). *From Galileo to Newton*. Collins, London.
Hesse, M. B. (1961). *Forces and Fields*. Nelson, Edinburgh.
Hooker, C. A. (1974). "Defense of a Non-Conventionalist Interpretation of Classical Mechanics." *Boston Studies in the Philosophy of Science*, XIII, 123.
Jammer, M. (1954). *Concepts of Space*. Harvard University Press, Cambridge, Mass.
    (1957). *Concepts of Force*. Harvard University Press, Cambridge, Mass., p. 110.
Jánossy, L. (1970–72). "A New Approach to the Theory of Relativity." *Found. of Phys.* 1, 111, 251; 3, 9.
Koertge, N. (1973). "Theory Change in Science." In: *Conceptual Change*, G. Pearce and P. Maynard, eds. Reidel, Dordrecht, p. 167.
Kordig, C. R. (1971). *The Justification of Scientific Change*. Reidel, Dordrecht.
Koyré, A. (1965). *Newtonian Studies*. Chapman & Hall, London.
Kraft, V. (1968). *Der Wiener Kreis*. Springer, Wien.
Kuhn, T. (1962), "Energy Conservation as an Example of Simultaneous Discovery." In: *The Sociology of Science*. Free Press, New York.
    (1970). *The Structure of Scientific Revolutions*. University of Chicago Press, Chicago.
Lakatos, I. (1970). "Falsification and the Methodology of Scientific Research Pro- grammes." In: *Criticism and the Growth of Knowledge*. I. Lakatos and A. Mus- grawe, eds. Cambridge University Press, Cambridge.
    (1971). "History of Science and Its Rational Reconstruction." In: *PSA 1970 in Memory of Rudolf Carnap*. Reidel, Dordrecht.
Lanczos, C. (1974). *The Einstein Decade*. Academic Press, New York.
Pala, A. (1969). *Isaac Newton*. Einaudi, Torino.
Piaget, J. (1970). *L'épistémologie génétique*. Presses Université, Paris.
Popper, K. (1959). *The Logic of Scientific Discovery*. Hutchinson, London.
    (1968). *Conjectures and Refutations*. Harper & Row, New York.
    (1972). *Objective Knowledge*. Oxford University Press, London.
Post, H. R. (1971). "Correspondence, Invariance and Heuristics." *Stud. Hist. Phil. Sci.* 2, 213.
Quine, W. V. (1953). *From a Logical Point of View*. Harvard University Press, Cam- bridge, Mass.
    (1970). "Existence." In: *Physics, Logic and History*. W. Yourgrau and A. D. Breck, eds. Plenum, London.
Ramsey, F. P. (1954). *The Foundations of Mathematics*. Routledge, London.
Reichenbach, H. (1965). *The Theory of Relativity and a priori Knowledge*. University of California Press, Berkeley.
Russell, B. (1938). *The Principles of Mathematics*. Norton, New York, p. 473.
Schilpp, ed. (1949). *Albert Einstein: Philosopher Scientist*. Tudor, New York.
Sneed, J. D. (1971). *The Logical Structure of Mathematical Physics*. Reidel, Dordrecht.
Synge, J. L. (1970). *Talking About Relativity*. North Holland, Amsterdam.
Toulmin, S. (1961). *Foresight and Understanding*. Harper & Row, New York.
Weinberg, J. R. (1948). *An Examination of Logical Positivism*. Kegan Paul, London.

G. J., ed. (1967). *Einstein, the Man and His Achievements.* Dover, New York.

r, E. (1960). *A History of the Theory of Aether and Electricity.* Harper & Row, York.

, J. O. (1971). "Four Contemporary Interpretations of the Nature of Science." *und. of Phys.* 1, 296.

ev, A. A. (1973). *Foundations of the Logical Theory of Knowledge,* Reidel, Dordrecht.

hapter 3. The physics of the irreversible

Bar-Hillel, Y., and Carnap, R. (1953). "Semantic Information." *Brit. Journ. Phil. of Sci.* 4, 147.

Bellone, E. (1973). *I modelli e la concezione del mondo.* Feltrinelli, Milano.

Bergmann, H. (1929). *Der Kampf um das Kausalgesetz in der jüngsten Physik.* Vieweg, Braunschweig.

Brillouin, L. (1956). *Science and Information Theory.* Academic Press, New York.

Caldirola, P. (1974). *Dalla microfisica alla macrofisica.* Mondadori, Milano.

Costa de Beauregard, O. (1971). "Information and Irreversibility Problems." In: *Time in Science and Philosophy.* Elsevier, Amsterdam.

Costantini, D. (1970). *Fondamenti del calcolo delle probabilità.* Feltrinelli, Milano.

Davies, P. C. W. (1974). *The Physics of Time Asymmetry.* University of California Press, Berkeley.

De Finetti, B. (1970). *Calcolo delle probabilità.* Einaudi, Torino.

Elkana, Y. (1971). "Boltzmann's Scientific Research Program and its Alternatives." In *The Interaction between Science and Philosophy.* Humanities Press, Atlantic Highlands, N.J.

Friedman, K. S. (1975). "A Problem Posed." *Found. of Phys.* 5, 89.

Grünbaum, A. (1973). *Philosophical Problems of Space and Time.* Reidel, Dordrecht.

Herivel, J. (1975). *Joseph Fourier, the Man and the Physicist.* Clarendon Press, Oxford.

Hintikka, J. (1970). "On Semantic Information." In: *Information and Inference.* Reidel, Dordrecht.

Jaynes, E. T. (1973). "The Well-Posed Problem." *Found. of Phys.* 3, 477.

Jeffrey, H. (1961). *Theory of Probability.* Oxford University Press, London.

Kolmogorov, A. N. (1967). "Logical Basis for Information Theory and Probability Theory." *IEEE Trans.* I-T, 14, 662.

Kuhn, T. (1962). "Energy Conservation as an Example of Simultaneous Discovery." In: *The Sociology of Science.* Free Press, New York.

Linsay, P. H., and Norman, D. A. (1972). *Human Information Processing.* Academic Press, New York.

Mehlberg, H. (1961). "Physical Laws and Time's Arrow." In: *Current Issues in the Philosophy of Science.* Holt, Rinehart and Winston, New York.

Park, D. (1972). *The Myth of the Passage of Time.* In: *The Study of Time.* Springer, Berlin.

Piaget, J. (1947). *Le langage et la pensée chez l'enfant.*

Popper, K. (1959). *The Logic of Scientific Discovery.* Hutchinson, London.

Reichenbach, H. (1956). *The Direction of Time.* University of California Press, Berkeley.

Ross Ashby, W. (1956). *An Introduction to Cybernetics.* Chapman & Hall, London.

Schlick, M. (1948). *Grundzüge der Naturphilosophie.* Gerold, Wien.

Smart, J. J. (1954). "The Temporal Asymmetry of the World," *Analysis* 14, 81.

Teilhard de Chardin, P. (1955). *Le Phénomène humaine.* du Seuil, Paris.

Truesdell, C. (1971). *The Tragicomedy of Classical Thermodynamics.* Springer, Wien.

Walk, K. (1966) "Simplicity, Entropy and Inductive Logic." In: *Aspects of Inductive Logic.* I. Hintikka and P. Suppes, eds. North Holland, Amsterdam.

Weinberg, J. R. (1948). *An Examination of Logical Positivism.* Kegan Paul, London.

Weyl, H. (1949). *Philosophy of Mathematics and Natural Sciences.* Princeton University Press, Princeton, N.J.

Whitrow, G. J. (1972). "Reflections on the History of the Concept of Time.' *Study of Time*. Springer, Berlin.
(1973). *The Nature of Time*. Holt, Rinehart and Winston, New York.
Wiener, N. (1949). *The Human Use of Human Beings*. Houghton Mifflin, Bostc

Chapter 4. Microphysics

Bastin, T., ed. (1971). *Quantum Theory and Beyond*. Cambridge University Pres Cambridge.
Bellone, E. (1973). *I modelli e la concezione del mondo*. Feltrinelli, Milano.
Bergman, H. (1974). "The Controversy Concerning the Law of Causality in Contemporary Physics." In: *Boston Studies in the Philosophy of Science*, Vol. XIII. Reidel, Dordrecht.
Black, M. (1967). "The Justification of Induction." In: *Philosophy of Science Today*. Basic Books, London.
Blokhintsev, D. I. (1968). *The Philosophy of Quantum Mechanics*. Reidel, Dordrecht.
Bohm, D. J., and Hiley, B. J. (1975). "Intuitive Understanding of Nonlocality in Quantum Theory." *Found. of Phys.*, 5, 93.
Born, M. (1956). *Physics in my Generation*. Pergamon, London.
Broglie, L. de (1941). *Continu et discontinu en physique moderne*. Michel, Paris.
(1967). "Les représentations concrètes en microphysique." In: *Logique et connaissance scientifique*. Gallimard, Paris.
(1973). *Louis de Broglie physicien et penseur*. Michel, Paris.
Bunge, M. (1959). *Causality*. Harvard University Press, Cambridge, Mass.
(1973). *Philosophy of Physics*. Reidel, Dordrecht.
Caldirola, P. (1974). *Dalla microfisica alla macrofisica*. Mondadori, Milano.
Čapek, M. (1971). *The Significance of Piaget's Researches on the Psychogenesis of Atomism*, R. C. Book and R. S. Cohen, eds. Reidel, Dordrecht.
Carnap, R. (1950). *Logical Foundations of Probability*. University of Chicago Press, Chicago.
(1952). *The Continuum of Inductive Methods*. University of Chicago Press, Chicago.
(1966). *Philosophical Foundations of Physics*. Basic Books, New York.
and Jeffrey, R. (1971). *Studies in Inductive Logic and Probability*. University of California Press, Berkeley.
and Stegmüller, W. (1959). *Induktive Logik und Wahrscheinlichkeit*. Springer, Wien.
Chiswell, B., and Grigg, E. M. (1971). *SI Units*. Wiley, New York.
Costa de Beauregard, O. (1967). "La grandeur physique 'temps'." In: *Logique et connaissance scientifique*. Gallimard, Paris.
Dalla Chiara, M. L. (1974). *Logica*. Isedi, Milano.
and Toraldo di Francia, G. (1974). "Is Self-Reduction Paradoxical?" *Studia Logica*. 33, 345.
d'Espagnat, B. (1973). "Quantum Logic and Non-Separability." In: *The Physicist's Conception of Nature*, J. Mehra, ed. Reidel, Dordrecht.
(1976). *Conceptual Foundations of Quantum Mechanics*. Benjamin, Reading, Mass.
Dirac, P. A. M. (1973). "Development of the Physicist's Conception of Nature." In: *The Physicist's Conception of Nature*, J. Mehra, ed. Reidel, Dordrecht.
Fine, A. (1971). "*Probability in Quantum Mechanics and in Other Statistical Theories*." In: *Problems in the Foundation of Physics*. Springer, Berlin.
Goodman, N. (1955). *Fact, Fiction and Forecast*. Harvard University Press, Cambridge, Mass.
Grünbaum, A. (1971). "*Can We Ascertain the Falsity of a Scientific Hypothesis?*" In: *Observation and Theory in Science*." Mandelbaum, ed. Johns Hopkins University Press, Baltimore.
(1973), *Philosophical Problems of Space and Time*. Reidel, Dordrecht.
Heelan, P. A. (1974). "Quantum Logic and Classical Logic: Their Respective Roles." In: *Boston Studies in the Philosophy of Science*. Reidel, Dordrecht.

rg, W. (1958). *Physics and Philosophy.* Allen & Unwin, London.

Physics and Beyond. Harper & Row, London.

, C. G. (1966). "Recent Problems of Induction." In: *Mind and Cosmos,* R. G.
lodny, ed. Pittsburg University Press, Pittsburgh.

s, H. (1969), *Enumerability, Decidability, Computability.* Springer, New York.

, M. B. (1974). *The Structure of Scientific Inference.* University of California Press,
Berkeley.

ikka, J., and Suppes, P., eds. (1966). *Aspects of Inductive Logic.* North Holland,
Amsterdam.

ooker, C. A. (1973). *Contemporary Research in the Foundations and Philosophy of
Quantum Theory.* Reidel, Dordrecht.

Jammer, M. (1966). *The Conceptual Development of Quantum Mechanics.* McGraw-
Hill, New York.

(1974). *The Philosophy of Quantum Mechanics.* Wiley, New York.

Jauch, J. M. (1973). "The Problem of Measurement in Quantum Mechanics." In: *The
Physicist's Conception of Nature,* J. Mehra, ed. Reidel, Dordrecht.

(1973). *Are Quanta Real?* Indiana University Press, Bloomington.

Landé, A. (1970). "Non Quantal Foundations of Quantum Mechanics." In: *Physics,
Logic and History,* W. Yourgrau and A. D. Breck, eds. Plenum Press, London.

Margenau, H. (1950). *The Nature of Physical Reality.* McGraw-Hill, New York.

(1966). "The Philosophical Legacy of Contemporary Quantum Theory." In: *Mind and
Cosmos,* R. G. Colodny, ed. Pittsburgh University Press, Pittsburgh.

and Cohen, L. (1967). "Probabilities in Quantum Mechanics." In: *Quantum Theory
and Reality.* M. Bunge, ed. Springer, Berlin.

Matthews, P. T. (1971). *The Nuclear Apple,* St. Martin's Press, New York.

Mehlberg, H. (1967). "The Problem of Physical Reality in Contemporary Science." In:
Quantum Theory and Reality, M. Bunge, ed. Springer, Berlin.

Mehra, J., ed. (1973). *The Physicist's Conception of Nature.* Reidel, Dordrecht.

Ne'eman, Y. (1974). "Concrete vs. Abstract Theoretical Models." In: *The Interaction
Between Science and Philosophy,* Y. Elkana, ed. Humanities Press, Atlantic Highlands, N.J.

Niiniluoto, I., and Tuomela, R. (1973). *Theoretical Concepts and Hypothetico-Inductive
Inference.* Reidel, Dordrecht.

Piaget, J. (1937). *La construction de réel chez l'enfant.* Neuchatel.

(1943). *La Psychologie de l'intelligence.* Colin, Paris.

(1959). *La naissance de l'intelligence chez l'enfant.* Delachaux et Niestlé, Paris.

(1967). "Les données génétiques de l'épistémologie physique." In: *Logique et connaissance scientifique.* Gallimard, Paris.

and Szeminska (1941). *La genèse du nombre chez l'enfant.* Delachaux et Niestlé, Paris.

Popper, K. (1959). *The Logic of Scientific Discovery.* Hutchinson, London.

(1967). "Quantum Mechanics Without 'The Observer'." In: *Quantum Theory and
Reality,* M. Bunge, ed. Springer, Berlin.

(1968). *Conjectures and Refutations.* Harper & Row, New York.

(1972). *Objective Knowledge.* Oxford University Press, London.

Putnam, H. (1969). "Is Logic Empirical?" In: *Boston Studies in the Philosophy of Science.* Reidel, Dordrecht.

(1973). "Explanation and Reference." In: *Conceptual Change,* G. Pearce and P. Maynard, eds. Reidel, Dordrecht.

Quine, W. V. (1953). *From a Logical Point of View.* Harvard University Press, Cambridge, Mass.

Rosenfeld, L. (1974). "Statistical Causality in Atomic Theory." In: *The Interaction Between Science and Philosophy,* Y. Elkana, ed. Humanities Press, Atlantic Highlands, N.J.

Russell, B. (1926). *Our Knowledge of the External World.* Allen & Unwin, London.

(1948). *The Principles of Mathematics.* Allen & Unwin, London.

(1963). *Mysticism and Logic.* Allen & Unwin, London.

456   References

Salam, A. (1972). "Symmetry Concepts and the Fundamental Theory of Matt
    Scientific Thought. UNESCO, New York.
Salmon, W. C. (1966). "The Foundations of Scientific Inference." In: Mind and C
    R. G. Colodny, ed. Pittsburgh University Press, Pittsburgh.
Stegmüller, W. (1960). "Das Problem der Kausalität." In: Probleme der Wissensc
    theorie. Springer, Wien.
Suppes, P. (1966). "Concept Formation and Bayesian Decisions." In: Aspects oj
    ductive Logic, J. Hintikka and P. Suppes, eds. North Holland, Amsterdam.
    (1967). "What Is a Scientific Theory?" In: Philosophy of Science Today, S. Morge
    besser, ed. Basic Books, New York.
Van Fraassen, B. C. (1974). The Labyrinth of Quantum Logic. Reidel, Dordrecht.
Whitehead, A. N. (1925). An Enquiry Concerning the Principles of Natural Knowledge.
    Cambridge University Press, Cambridge.
Wigner, E. P. (1967). Symmetries and Reflections. Indiana University Press, Bloomington.
Yahil, A. (1974). "How Constant are the Constants of Physics?" In: The Interaction
    Between Science and Philosophy. Y. Elkana, ed. Humanities Press, Atlantic High-
    lands, N.J.
Zinov'ev, A. A. (1973). Foundations of the Logical Theory of Knowledge. Reidel,
    Dordrecht.

Chapter 5. The universe

Blacker, C., and Loewe, M. (1975). Ancient Cosmologies. Allen & Unwin, London.
Bracewell, R. L. (1974). The Galactic Club, Intelligent Life in Outer Space. Freeman,
    San Francisco, Calif.
Byard, M. M. (1977). "Poetic Responses to the Copernican Revolution." Scient. Am.
    236, 6, 121.
Klass, P. J. (1974). UFO's Explained. Random House, New York.
Kuhn, T. (1957). The Copernican Revolution. Harvard University Press, Cambridge,
    Mass.
North, J. D. (1965). The Measure of the Universe. Clarendon Press, Oxford.
Quine, W. V. (1953). From a Logical Point of View. Harvard University Press, Cam-
    bridge, Mass.
    (1966). "Necessary Truth." In: The Ways of Paradox. Random House, New York.
Sagan, C., and Drake, F. (1975). "The Search for Extraterrestrial Intelligence." Scient.
    Am. 22, 5, 80.
Weizsäcker, C. F., von (1971). Die Einheit der Natur. Hauser, München.
Wigner, E. P. (1967). Symmetries and Reflections. Indiana University Press, Bloomington.

e index

Subject index